"Sprites, Elves and Intense Lightning Discharges"

NATO Science Series

A Series presenting the results of scientific meetings supported under the NATO Science Programme.

The Series is published by IOS Press, Amsterdam, and Springer in conjunction with the NATO Public Diplomacy Division

Sub-Series

I. Life and Behavioural Sciences	IOS Press
II. Mathematics, Physics and Chemistry	Springer
III. Computer and Systems Science	IOS Press
IV. Earth and Environmental Sciences	Springer

The NATO Science Series continues the series of books published formerly as the NATO ASI Series.

The NATO Science Programme offers support for collaboration in civil science between scientists of countries of the Euro-Atlantic Partnership Council. The types of scientific meeting generally supported are "Advanced Study Institutes" and "Advanced Research Workshops", and the NATO Science Series collects together the results of these meetings. The meetings are co-organized bij scientists from NATO countries and scientists from NATO's Partner countries – countries of the CIS and Central and Eastern Europe.

Advanced Study Institutes are high-level tutorial courses offering in-depth study of latest advances in a field.
Advanced Research Workshops are expert meetings aimed at critical assessment of a field, and identification of directions for future action.

As a consequence of the restructuring of the NATO Science Programme in 1999, the NATO Science Series was re-organised to the four sub-series noted above. Please consult the following web sites for information on previous volumes published in the Series.

http://www.nato.int/science
http://www.springer.com
http://www.iospress.nl
http://www.wtv-books.de/nato-pco.htm

Series II: Mathematics, Physics and Chemistry – Vol. 225

"Sprites, Elves and Intense Lightning Discharges"

edited by

Martin Füllekrug
Centre for Space Atmospheric and Oceanic Science,
University of Bath, United Kingdom

Eugene A. Mareev
Institute of Applied Physics,
Russian Academy of Sciences,
Nizhny Novgorod, Russian Federation

and

Michael J. Rycroft
CAESAR Consultancy,
Cambridge, United Kingdom

Published in cooperation with NATO Public Diplomacy Division

Proceedings of the NATO Advanced Study Institute on
"Sprites, Elves and Intense Lightning Discharges"
Corte, Corsica, France
24–31 July 2004

A C.I.P. Catalogue record for this book is available from the Library of Congress.

ISBN-10 1-4020-4628-6 (PB)
ISBN-13 978-1-4020-4628-5 (PB)
ISBN-10 1-4020-4627-8 (HB)
ISBN-13 978-1-4020-4627-8 (HB)
ISBN-10 1-4020-4629-4 (e-book)
ISBN-13 978-1-4020-4629-2 (e-book)

Published by Springer,
P.O. Box 17, 3300 AA Dordrecht, The Netherlands.

www.springer.com

Printed on acid-free paper

All Rights Reserved
© 2006 Springer
No part of this work may be reproduced, stored in a retrieval system, or transmitted in any form or by any means, electronic, mechanical, photocopying, microfilming, recording or otherwise, without written permission from the Publisher, with the exception of any material supplied specifically for the purpose of being entered and executed on a computer system, for exclusive use by the purchaser of the work.

Printed in the Netherlands.

Contents

Contributing Authors   xi
Preface   xiii

INTRODUCTION TO THE PHYSICS OF SPRITES, ELVES AND INTENSE LIGHTNING DISCHARGES   1
Michael J. Rycroft
1.1 Basic Properties of the Atmosphere   1
    1.1.1 Global Scale Variations (Horizontal Scale greater than 10^4 km)   1
    1.1.2 Regional Variations (Horizontal Scale between 30 and 300 km)   3
1.2 Basic Theory of Electrical Phenomena Occurring in the Atmosphere   3
    1.2.1 Introduction   3
    1.2.2 The DC Global Atmospheric Electric Circuit   4
    1.2.3 The AC Circuit   5
    1.2.4 Thundercloud Charges and their Screening   6
    1.2.5 Spatial and Temporal Variations of the Global Circuit   7
1.3 The Properties of Sprites, Elves and Intense Lightning Discharges   7
    1.3.1 Observations and their Interpretation – An Overview   8
    1.3.2 ELF Radiation by Sprites   10
    1.3.3 Summary of Observations   11
1.4 Introduction to Theories and Numerical Modelling of Sprites   12
    1.4.1 Basic Physical Concepts   12
    1.4.2 Computer Modelling Results   12
1.5 Conclusions   13
Acknowledgments   13

THE METEOROLOGY OF TRANSIENT LUMINOUS EVENTS- AN INTRODUCTION AND OVERVIEW   19
Walter A. Lyons
2.1 Introduction   19
    2.1.1 Scales of Atmospheric Motion   19
    2.1.2 Basic Concepts of Atmospheric Vertical Stability   21
    2.1.3 Convective Cloud Nomenclature   22
2.2 Observations of Convective Phenomema   24
    2.2.1 Conventional Convective Storm Monitoring   25
    2.2.2 Lightning Observation Techniques and Terminology   26
2.3 A Brief History of TLE Observations   30
2.4 Characteristics of TLE-Parent Lightning and Storms   34
    2.4.1 The Phenomenology of TLEs   34
    2.4.2 Convective Storm Types and TLEs   39
2.5 Research Frontiers   41

2.5.1	Importance	42
2.5.2	Outstanding Research Questions	43

Acknowledgments 44

THE MICROPHYSICAL AND ELECTRICAL PROPERTIES OF SPRITE-PRODUCING THUNDERSTORMS 57
Earle Williams and Y. Yair

3.1	Introduction	57
3.2	The Non-Inductive Charging Process in Thunderclouds	58
3.3	Cloud Scale Charge Structure Possible with the Non-Inductive Mechanism	61
3.4	The Electrical Structure Inside Sprite-Producing Storms in Summertime	64
3.5	The Electrical Structure inside Sprite-Producing Storms in Wintertime	68
3.6	Gaps in Knowledge and Future Needs	73

Acknowledgments 74

GLOBAL THUNDERSTORM ACTIVITY 85
Colin Price

4.1	The Earth's Energy Balance	85
4.2	The General Circulation of the Atmosphere	87
4.3	Frontal Thunderstorms in Mid-Latitude Regions	90
4.4	Global Observations of Lightning	92
4.5	The Global Atmospheric Electric Circuit	93
4.6	Future Directions	96

IMAGING SYSTEMS IN TLE RESEARCH 101
Thomas H. Allin, T. Neubert and S. Laursen

5.1	Introduction to Low Light Imaging	101
	5.1.1 Optics	102
	5.1.2 Electronic Imaging Sensors	102
	5.1.3 Noise and Dynamic range	104
	5.1.4 Intensified versus Non-intensified Imaging	105
	5.1.5 Summary	106
5.2	The Spritewatch Systems	107
	5.2.1 The 2003 Spritewatch System	107
	5.2.2 Hardware in the 2003 system	108
	5.2.3 Software in the 2003 system	112
	5.2.4 Results from 2003	114
	5.2.5 The 2004 Spritewatch System	115
	5.2.6 Hardware in the 2004 system	115
	5.2.7 The 2004 Spritewatch Software	117
	Event Detection	117
5.3	Conclusions	118
	5.3.1 Summary	118
	5.3.2 Future work	120

SPACECRAFT BASED STUDIES OF TRANSIENT LUMINOUS EVENTS 123
Stephen B. Mende, Y. S. Chang, A. B. Chen, H. U. Frey, H. Fukunishi, S. P. Geller, S. Harris, H. Heetderks, R. R. Hsu, L. C. Lee, H. T. Su and Y. Takahashi

6.1	Introduction	124

6.2	FORMOSAT-2 Satellite and the ISUAL Instrument		124
	6.2.1	The FORMOSAT-2 Satellite	125
	6.2.2	The ISUAL Imager	127
	6.2.3	The Spectrophotometer	131
	6.2.4	Data Interpretation of the Spectrophotometer	136
	6.2.5	The Array Photometers	138
6.3	Initial Observations with ISUAL		140
6.4	Summary		144
Acknowledgments			146

OBSERVATIONS OF SPRITES FROM SPACE AT THE NADIR: THE LSO (LIGHTNING AND SPRITE OBSERVATIONS) EXPERIMENT ON BOARD OF THE INTERNATIONAL SPACE STATION 151

Elisabeth Blanc, T. Farges, D. Brebion, A. Labarthe and V. Melnikov

7.1	Introduction	152
7.2	Spectral Differentiation of Sprite and Lightning Emissions	153
7.3	Experiment	153
7.4	Observations	156
7.5	Perspectives	160

REMOTE SENSING OF THE UPPER ATMOSPHERE BY VLF 167

Craig J. Rodger and R. J. McCormick

8.1	Ionospheric Conductivity		167
8.2	Sources of VLF Electromagnetic (EM) Waves		168
	8.2.1	Thunderstorms and Lightning	168
	8.2.2	Man-Made VLF Radiation	169
8.3	VLF Propagation in the Earth-Ionosphere Waveguide		169
	8.3.1	Variations in Subionospheric Propagation	170
	8.3.2	TLE Associated Perturbations on VLF Transmissions	172
8.4	Relaxation of High-Altitude Ionospheric Modifications		181
	8.4.1	Temperature Relaxation	181
	8.4.2	Ionization Relaxation	181
8.5	Summary		183

MEASUREMENTS OF LIGHTNING PARAMETERS FROM REMOTE ELECTROMAGNETIC FIELDS 191

Steven A. Cummer

9.1	Background and Motivation		191
9.2	Remote Lightning Parameter Measurements		195
9.3	Data Analysis Techniques		198
	9.3.1	Electromagnetic Field Modeling	200
	9.3.2	Data Inversion	202
9.4	Summary		205

LOCATION AND ELECTRICAL PROPERTIES OF SPRITE-PRODUCING LIGHTNING FROM A SINGLE ELF SITE 211

Yasuhide Hobara, M. Hayakawa, E. Williams, R. Boldi and E. Downes

10.1	Introduction	212
	10.1.1 Lightning Activity	212
	10.1.2 Terrestrial ELF Electromagnetic Signals	212
	10.1.3 Sprites and Elves and Causative Lightning Properties	214
	10.1.4 Contents of this Chapter	214

10.2	Locating Distant ELF Sources and Quantifying their Electrical Properties	215
	10.2.1 Theory	215
	10.2.2 Global Map of Lightning by Rhode Island Station	218
10.3	Winter TLEs and Associated Electromagnetic Phenomena in Japan	220
	10.3.1 Winter Thunderstorm Activity and TLEs in the Hokuriku Region	220
	10.3.2 ELF Radiation Associated with the Hokuriku TLEs and the Generation Condition of Winter Sprites	224
	10.3.3 Atmosphere-Mesosphere-Ionosphere Coupling	227
10.4	Conclusion	228
Acknowledgments		229

CALIBRATED RADIANCE MEASUREMENTS WITH AN AIR-FILLED GLOW DISCHARGE TUBE: APPLICATION TO SPRITES IN THE MESOSPHERE   237
Earle Williams, M. Valente, E. Gerken and R. Golka

11.1	Introduction	237
11.2	Methodology	238
11.3	Optical Spectrum	240
11.4	Radiance Response to Power and Current Density	242
11.5	Estimates of Bulk Plasma Conductivity	243
11.6	Application to Current Flow in Sprites	244
11.7	Conclusion	247
Acknowledgments		247

THEORETICAL MODELING OF SPRITES AND JETS   253
Victor P. Pasko

12.1	Introduction	253
	12.1.1 Phenomenology of Sprites	254
	12.1.2 Phenomenology of Blue Jets, Blue Starters and Gigantic Jets	255
12.2	Classification of Breakdown Mechanisms in Air	256
	12.2.1 Concept of Electrical Breakdown	256
	12.2.2 Classification of Breakdown Mechanisms in Terms of pd Values	257
	12.2.3 Classification of Breakdown Mechanisms in Terms of Applied Electric Field	261
	12.2.4 Similarity Relations	265
12.3	Physical Mechanism and Numerical Modeling of Sprites	267
	12.3.1 Large Scale Electrodynamics	267
	12.3.2 Altitude Structuring of Optical Emissions	271
	12.3.3 Large Scale Fractal Models of Sprites	272
	12.3.4 Modeling of Small-scale Sprite Streamer Processes and Photoionization Effects	274
	12.3.5 Optical Emissions Associated with Sprite Streamers	277
12.4	Physical Mechanism and Numerical Modeling of Blue Jets, Blue Starters and Gigantic Jets	279
	12.4.1 Blue Jets as Streamer Coronas	280
	12.4.2 Thundercloud Charge and Current Systems Supporting Blue Jets, Blue Starters and Gigantic Jets	283
	12.4.3 Numerical Simulation of Blue Jets and Blue Starters	285
	12.4.4 Modeling of Optical Emissions from Blue Jets and Blue Starters	286

12.5	Unsolved Problems	287
	12.5.1 Relationship of Sprites and Jets to High Air Pressure Leader Processes	287
	12.5.2 Initiation of Sprite Streamers in Low Applied Fields	288
	12.5.3 Propagation of Sprite Streamers	289
	12.5.4 Branching of Sprite Streamers	291
	12.5.5 Thermal Runaway Electrons in Streamer Tips in Sprites	292
Acknowledgments		293

ON THE MODELING OF SPRITES AND SPRITE-PRODUCING CLOUDS IN THE GLOBAL ELECTRIC CIRCUIT   313
Eugene A. Mareev, A. A. Evtushenko and S. A. Yashunin

13.1	Introduction	314
13.2	Time-Dependent Electric Field in the Conducting Atmosphere	315
13.3	Modeling of the Lower Positive Charge Layer in the Stratified Region	325
13.4	Global Electric Circuit Implications	330
13.5	Conclusion and Outlook for Promising Future Work	335
Acknowledgments		335

ACTUAL PROBLEMS OF THUNDERCLOUD ELECTRODYNAMICS   341
Victor Y. Trakhtengerts and Dmitry I. Iudin

14.1	Introduction	341
14.2	Electric Field Generation in an Atmospheric Convective Cell	343
	14.2.1 The Case of a Cylindrical Convective Cell	343
	14.2.2 Some Generalizations	346
	14.2.3 Dynamics of the Large-Scale Electric Field in a Convective Cloud	348
14.3	Fine Structure of Electric Fields in a Thundercloud	351
	14.3.1 Multi-Flow Electrical Instability in a TC	351
	14.3.2 A Mechanism of Electric Field Fast Growth during the TC Mature Stage	353
	14.3.3 Fractal Dynamics of Micro-Discharges in a TC	356
14.4	Acceleration of Relativistic Electrons during a Thunderstorm	360
	14.4.1 Runaway Breakdown in a Constant Electric Field	360
	14.4.2 Acceleration by Stochastic Electric Fields	365
14.5	Conclusions	369
Acknowledgments		370

POSTER ABSTRACTS   377
T. Farges

15.1	Introduction	377
15.2	Observations from the Ground	377
	15.2.1 Automated, Remote-Controlled Optical Observation Systems in TLE Research	377
	15.2.2 Observation of Schumann Resonance Transients at Nagycenk, Hungary	378
	15.2.3 Post Filtering of Unwanted Powerline and Lightning Effects in VLF	378
	15.2.4 Infrasonic Signatures of Thunder	379
	15.2.5 On the Absorption of ELF Signals in the Earth-Ionosphere Waveguide	379

		15.2.6	A Global Lightning Location Algorithm Based on Electromagnetic Signatures in the Schumann Resonance Band	381
		15.2.7	Neutral and Charged Particles at Low Latitudes. Is their Connection with Thunderstorms Possible?	381
		15.2.8	Sprites Observed over France on 23 July 2003 in Relation to their Parent Thunderstorm System	382
		15.2.9	Sprite Observations from Langmuir Laboratory, New Mexico	383
		15.2.10	VLF Signatures Associated with Sprites	383
		15.2.11	Stratospheric Electric Field, Magnetic Field and Conductivity Measurements Above Thunder- storms: Implications for Sprite Models	384
		15.2.12	Triggering of Positive Lightning and High-Altitude Atmospheric Discharges	384
	15.3	Observations from Space		385
		15.3.1	Transient Luminous Events Explored by the ROCSAT-2/ISUAL Instrument: Observation with the Array Photometer	385
		15.3.2	Fractal Analysis Method Applicability to Terrestrial Gamma-Ray Flashes	386
		15.3.3	Searching for Lightning-Induced Terrestrial Gamma Ray Bursts on CORONASF Satellite	387
		15.3.4	Seismo-electromagnetic Emissions	388
		15.3.5	First Results of Transient Luminous Event Observations by ISUAL	389
		15.3.6	Detection of Terrestrial Gamma-ray Flashes with the RHESSI Spacecraft	389
		15.3.7	ENVISAT Capabilities of Observing High Altitude Optical Phenomena	390
	15.4	Theoretical Modelling		391
		15.4.1	Do sprites Impact Climate? An atmospheric Coupling Approach.	391
		15.4.2	The Sodankylä Ion Chemistry model: Application of Coupled Ion-neutral Chemistry Modelling	391
		15.4.3	Simulation of Streamer Propagation Using a PIC-MCC Code: Application to Sprite Ignition	392
		15.4.4	Characteristics of Transient Luminous Event Streamers in Weak Electric Fields	392
		15.4.5	Three-Dimensional Subionospheric VLF Electro- magnetic Field Scattering by a Highly Conducting Cylinder and Its Application to the Trimpi Effect Problem	393
		15.4.6	Changes of the Lower Ionospheric Electron Concentration due to Solar Cosmic Rays	394
Index				395

Contributing Authors

Thomas H. Allin. *Technical University of Denmark, Copenhagen, Denmark.*

Elisabeth Blanc. *Commissariat à l'Energie Atomique, Bruyères le Châtel, France.*

Steven A. Cummer. *Duke University, Durham, USA.*

Thomas Farges. *Commissariat à l'Energie Atomique, Bruyères le Châtel, France.*

Yasuhide Hobara. *Swedish Institute of Space Physics, Kiruna, Sweden.*

Dmitry I. Iudin. *Russian Academy of Sciences, Nizhny Novgorod, Russian Federation.*

Walt A. Lyons. *FMA RESEARCH, Fort Collins, USA.*

Eugene A. Mareev. *Russian Academy of Sciences, Nizhny Novgorod, Russian Federation.*

Stephen B. Mende. *University of California, Berkeley, USA.*

Victor P. Pasko. *The Pennsylvania State University, University Park, USA.*

Colin Price. *Tel Aviv University, Ramat Aviv, Israel.*

Craig J. Rodger. *University of Otago, Dunedin, New Zealand.*

Michael J. Rycroft. *CAESAR Consultancy, Cambridge, UK.*

Earle Williams. *Massachusetts Institute of Technology, Cambridge, USA.*

Yoav Yair. *The Open University of Israel, Ra'anana, Israel.*

Preface

Particularly intense lightning discharges can produce transient luminous events above thunderclouds, termed sprites, elves and jets. These short lived optical emissions in the mesosphere can reach from the tops of thunderclouds up to the ionosphere; they provide direct evidence of coupling from the lower atmosphere to the upper atmosphere. Sprites are arguably the most dramatic recent discovery in solar-terrestrial physics. Shortly after the first ground based video recordings of sprites, observations on board the Space Shuttle detected sprites and elves occurring all around the world. These reports led to detailed sprite observations in North America, South America, Australia, Japan, and Europe. Subsequently, sprites were detected from other space platforms such as the International Space Station and the ROCSAT satellite.

During the past 15 years, more than 200 contributions on sprites have been published in the scientific literature to document this rapidly evolving new research area. The need for international information exchange was quickly recognized, and sprite sessions became a permanent feature with a constantly growing number of contributions in the scientific communities of the American Geophysical Union (AGU), the International Union of Radio Science (URSI), the International Association of Geomagnetism and Aeronomy (IAGA) and the European Geosciences Union (EGU).

The idea for a summer school on sprites was born during an informal meeting at Tohoku University in 2001, where a number of young scientists gave extended lectures on the details of their scientific work to interested colleagues following an invitation from Yukihiro Takahashi. This experience was felt to be so rewarding that, during the third SPECIAL Scientific Network meeting of the European Science Foundation held in Cambridge, 2002, an organising committee was formed to attract support for a NATO Advanced Study Institute on "Sprites, Elves and Intense Lightning Discharges". The proposal was accepted by NATO shortly thereafter and the Advanced Study Institute took place at the University of Corsica, in Corte on Corsica, from 24-31 July, 2004. The activity was co-sponsored by the European Science Foundation (ESF), the Research Training Network (RTN) "Connecting Atmospheric Layers" (CAL) of the European Commission (EC) and the International Union of Radio Science

(URSI). About 80 scientists attended this first summer school on sprites, which was a scientifically excellent, exciting, interactive, and memorable event.

The lectures delivered are presented in this book, along with the abstracts of the poster presentations by the young scientists. The book begins with an introduction to the electrical properties of the atmosphere, which sets the scene. Subsequently, the book is organised into four general themes of sprite research. The first theme describes the meteorological conditions for intense lightning discharges which are needed to produce sprites. The second theme describes the experiments for making optical observations of sprites from the ground and space. The third theme deals with radio waves in the ULF/ELF/VLF frequency range from 3 Hz to 30 kHz, which occur in association with intense lightning discharges and sprites. The book finally wraps up the results within theme four, on theory and modelling in all three experimental areas of sprite research.

The Editors thank the lecturers for their strong committment to deliver their scientific knowledge in oral and written from, and also the young scientists for their poster presentations. The Editors also thank Bettina Zapfe for her careful copy-editing of the book and, in particular, for her organisation of the index and the references. The organisers of the summer school express their gratitude to all the sponsors, who helped with substantial travel support. The organisers also thank the University of Corsica for hosting the meeting and Dominique Grandjean for her local support on site. The sprite summer school would not have been so successful without the helping hands of Norma Crosby, Thomas Farges, Ingrid Hörnchen, Tai-Yin Huang, Thomas Ulich, and many others.

MARTIN FÜLLEKRUG, EUGENE A. MAREEV AND MICHAEL J. RYCROFT

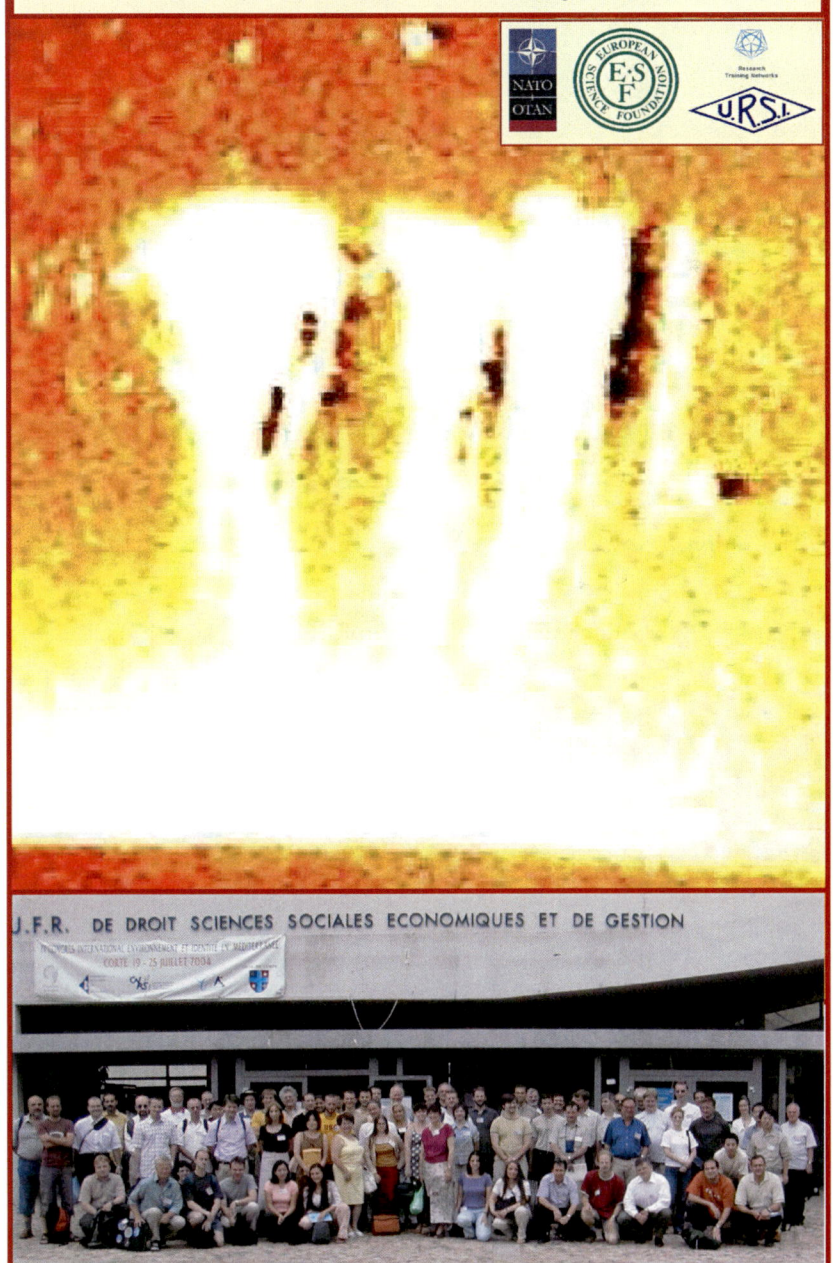

INTRODUCTION TO THE PHYSICS OF SPRITES, ELVES AND INTENSE LIGHTNING DISCHARGES

Michael J. Rycroft
CAESAR Consultancy, 35 Millington Road, Cambridge CB3 9HW, U.K.

Abstract   This chapter introduces the fundamentals of the subject, the laboratory for which is the Earth's atmosphere. All of its important physical properties vary significantly with altitude; its electrical conductivity varies greatly with altitude and also with latitude. The *modus operandi* of the global atmospheric electric circuit, in part powered by thunderstorms, is outlined. Some observations of transient luminous events, which occur above certain energetic thunderstorm systems just after a strong lightning discharge (the effect of which is usually to take positive charge to the ground), are introduced and their characteristics mentioned. Different theoretical ideas that have been put forward to explain crucial aspects of these phenomena are introduced, and some relevant numerical simulations are briefly discussed.

1.1   Basic Properties of the Atmosphere

1.1.1   Global Scale Variations (Horizontal Scale greater than 10^4 km)

The Earth's atmosphere is bound gravitationally to our home planet. The pressure and density of the air decrease exponentially upwards from the surface, the scale height being $\sim$7 km (Rycroft, 2003). This scale height varies slightly through the atmosphere because it is not isothermal. Relatively, the hottest regions of the Earth's neutral atmosphere are at its surface (global average temperature T=288 K, pressure p=1013 hPa, density ϱ=1.2 kg/m^3), at the stratopause ($\sim$250 K, near 50 km altitude) and in the thermosphere (above $\sim$110 km, from where the temperature rises rapidly up to $\sim$2000 K at $\sim$300 km altitude), depending on the phase of the cycle of solar activity, with successive maxima $\sim$11 years apart. About half of the atmosphere resides at and below an altitude of 5 km; 90% of the atmosphere lies below an altitude of $\sim$15 km. The pressure of the air at 32 km altitude is $\sim$1% of its surface value and, at 100 km altitude, it is about one millionth of its surface value. The coldest regions are at the top of the troposphere, termed the tropopause (at 18 km altitude over the tropics, 14 km at middle latitudes, and only 9 km at high latitudes), and at the top of the mesosphere, termed the mesopause (∼140 K near 85 km altitude in summer, and ∼210 K in winter). Above the rotating Earth, winds blow away from the hotter regions, essentially horizontally – zonally (at constant latitude) or meridionally (at constant longitude). The winds carry heat through the atmosphere, in an attempt to bring the atmosphere into thermal equilibrium.

The atmosphere near the Earth's land surface is an electrically charged gas; $\sim 2 \cdot 10^6 / m^3 s^{-1}$ molecular ion pairs are formed by radon emanating from the land but, since recombination is rapid, the electron density is only $\sim 1\ m^{-3}$. Bursts of gamma-radiation and galactic cosmic rays, which are extremely energetic (GeV) charged particles (electrons, protons and helium nuclei) from outside our galaxy, come down into the atmosphere, ionising atmospheric molecules by collision. The maximum ionisation production rate is reached near 20 km altitude. Thus, from the Earth's surface up to ∼20 km, the electrical conductivity (the reciprocal of the electrical resistivity of the atmosphere) increases from $\sim 10^{-14}\ Sm^{-1}$ with a scale height ∼5 km; above 20 km, it increases with an ever increasing scale height. The concentration of molecular ion pairs has a maximum value $\sim 10^9 m^{-3}$ at 15-20 km altitude (Harrison and Carslaw, 2003). The electrical conductivity of the stratosphere varies somewhat, and that of the mesosphere varies enormously, according to the prevailing geophysical conditions.

Near solar minimum, the solar wind flowing outwards from the Sun is "quiet", with few irregularities whereas, at solar maximum, it is disturbed, with many large amplitude fluctuations. These fluctuations scatter galactic cosmic rays coming into the solar system so that their flux is smaller at solar maximum than at solar minimum. Because of the dipolar character of the Earth's magnetic field, cosmic rays preferentially enter the Earth's environment at polar and high latitudes. These two considerations explain the interesting results of Ney (1959), that there is a significant variation of atmospheric ionisation through the 11 year cycle, and that this is especially marked at higher latitudes (∼50°-60° geomagnetic latitude).

A complicating effect is that the Sun is more prone to generate very energetic charged particles (termed solar particle events, SPEs) at, and after, solar maximum than at solar minimum. These enter the Earth's environs preferentially in the polar regions, polewards of the auroral oval, at ∼67° magnetic latitude.

The ionosphere is formed in the thermosphere by the ionising action of the Sun's extreme ultraviolet radiation and X-rays; both of these are ∼100 times stronger at solar maximum than at solar minimum. The lowest region of the ionosphere, the D-region, is formed at and above 80 km, primarily by the action of solar Lyman-alpha radiation on nitric oxide. At night, the ionospheric plasma density decreases most rapidly at the lowest altitudes (decreasing from 10^{11} to $10^8 \mathrm{m}^{-3}$ at 90 km altitude), and less rapidly higher up where the effective recombination process takes much longer (Schunk and Nagy, 2000).

The ionosphere and the magnetosphere above it respond rapidly to solar wind disturbances, nowadays termed space weather events. Often associated with solar flares, coronal mass ejections (CMEs) or shocks in interplanetary space, these can manifest themselves as geomagnetic storms or magnetospheric substorms (Baker, 2000).

1.1.2 Regional Variations (Horizontal Scale between 30 and 300 km)

For our purposes, the most important phenomena in this category are thunderstorms (Rakov and Uman, 2003). Thunderstorms form under strongly convective meteorological conditions, where cumulus clouds become cumulonimbus clouds, and reach up to the tropopause. In the simplest case, the bottom of a thundercloud is negatively charged, and the top (which may evolve into an "anvil") is positively charged. More complex arrangements of charge generally occur; there can be a region of positive charge below the negatively charged region. Several different electrification mechanisms have been proposed.

Within a thundercloud, the vertical electric field ranges within $\pm 100 \, \mathrm{kVm}^{-1}$, and varies considerably over a fraction of 1 km due to the presence of narrow charged layers. Directly above a thundercloud, this field deviates from the mean field by a few kVm^{-1}; it changes between such values when a lightning discharge occurs. Just after the stroke, even larger fields can be reached.

Individual convective cells are typically >1 km in radius, and may constitute individual thunderstorms. Alternatively, they may develop into mesoscale convective systems (MCSs), large scale thunderstorms; their radius is typically up to 100 km, or even more.

There exist regional scale variations of the troposphere and the stratosphere, due to meteorological processes, and of the mesosphere, the thermosphere and the ionosphere, some of which are due to various different effects of spatially structured charged particle precipitation from the magnetosphere.

1.2 Basic Theory of Electrical Phenomena Occurring in the Atmosphere

1.2.1 Introduction

Throughout the atmosphere, the electric (E) and magnetic (B) fields, charge density $q(N_i - N_e)$ and total electric current density J are related by the four equations of electromagnetism known as Maxwell's equations. (Here the magnitude of q is the charge of an electron, the number density of positive ions is

N_i, and the number density of electrons and/or negative ions is N_e.) Together with laws expressing the continuity of mass, momentum and energy, an appropriate equation of state p(N,T), an appropriate constitutive relation between J and E, and appropriate parameterisation schemes, all these equations can, in principle, be solved to determine the overall behaviour of the atmosphere. However, this goal may never be fully achieved because many of the physical processes occurring are nonlinear, and because of the extremely broad range of scales involved. Spatially, these range from turbulence to the global scale and, temporally, from microseconds to centuries.

1.2.2 The DC Global Atmospheric Electric Circuit

One important generator for the global circuit (for reviews see Rycroft et al., 2000; Williams, 2002, pp. 9; Harrison, 2004; Rycroft, 2005) is linked to thunderstorms and electrified shower clouds. Wilson (1920) suggested that a current should flow up from the top of a thunderstorm to the upper atmosphere. Upward (or Wilson) current passes through a charging resistor (R_1 in Figure 1) to the ionosphere which is almost an equipotential. Electric currents flow to different latitudes, mostly through the highly conducting ionosphere as indicated in Figure 1, but partly through the mesosphere. Currents flow along geomagnetic flux tubes into and out of the magnetosphere, especially at high latitudes. Currents flow downward in the fair weather regions of the Earth, remote from thunderstorms; in these regions, most of the electrical resistance (r in Figure 1) is within a few km of the surface. This current closes through the land and sea portions of the Earth's surface and through the atmosphere below the thunderclouds. To a first approximation, the region between the good conducting Earth and ionosphere acts as a spherical capacitor, the atmosphere being an insulator. This capacitor is charged up by the upward currents associated with thunderstorms, and by other generators of potential difference. These other sources of upward currents include processes linked to thermal convection currents and to point discharge (or corona) currents (Wormell, 1953).

Cloud-to-ground discharges transferring negative charge-to- ground (termed -CG) are effectively another source of upward currents. Also to be considered is the displacement current (Roble, 1991), but this cannot charge up the capacitor, neither can this discharge it. Chauzy (personal communication, 2003) has measured upward currents due to precipitation (rain) below thunderclouds and shower clouds, as postulated by Wilson (1920) – see also Williams and Satori (2004). However, the electric current from rain clouds can be downwards rather than upwards, as mentioned by Williams and Heckman (1993). Boundary layer aerosol particles due to the burning of fossil fuels (Adlerman and Williams, 1996) complicate the situation over land areas, especially in winter.

INTRODUCTION TO SPRITES

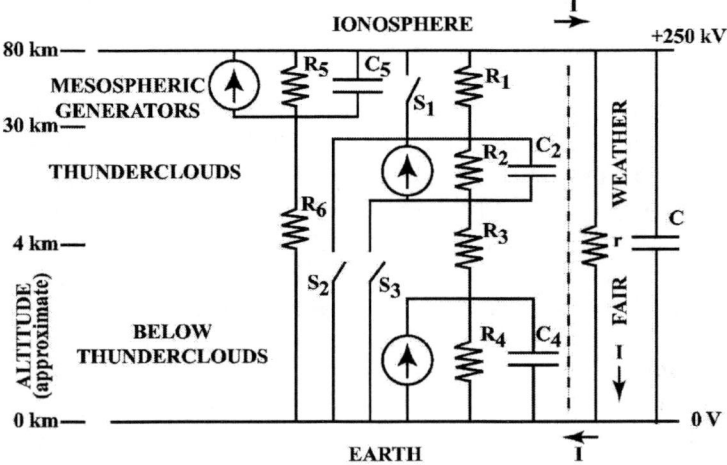

Figure 1. In this schematic equivalent circuit for the global atmospheric electric circuit, the current generators shown to the left of the vertical dashed line act over <1 % of the Earth's surface. The load on this circuit is the fair weather region, shown to the right of the dashed line, ~99 % of the Earth's surface. The total current flowing is ~1 kA. The time constant of the circuit is rC, a few minutes. When a +CG discharge occurs, the S_2 for that storm closes for ~1 ms and then reopens; when a sprite occurs, its S_1 closes for a few ms, and then reopens.

1.2.3   The AC Circuit

The AC circuit operates on time scales smaller than the relaxation time, which is defined as the permittivity of free space divided by the conductivity, σ, at the point under consideration. It is a strong function of altitude. For example, from 60 to 80 km on a typical night, the relaxation time is 10 ms (Hale, 1984). Atmospherics (or "sferics") are radio signals of ~1 ms duration and which contain frequencies from a few kHz to 30 kHz (in the Very Low Frequency (VLF) band) radiated as an ElectroMagnetic Pulse (EMP) by a lightning discharge; these propagate up to 10^4 km or further in the Earth-ionosphere waveguide. Also excited by lightning are Schumann resonances (at 8, 14, 20, 26, ... Hz) of the insulating spherical shell between the Earth and the ionosphere. These are in the ELF (Extremely Low Frequency) part of the spectrum, at <3 kHz, down to 3 Hz. The properties and uses of these ELF/VLF radio waves are discussed in considerable detail in Chapters 8 and 10 by Rodger and Hobara, respectively. Temporal changes (occurring on time scales longer than minutes) to the current sources for, and the properties of, the DC circuit are not considered as contributing to the AC circuit discussed here.

1.2.4 Thundercloud Charges and their Screening

As is evident from Figure 2, there should be a large electrostatic field above the large positive electric charge ($\sim$100 C) at the top of a thundercloud. However, such large fields do not always exist in practice due to screening. Screening is a well known concept in plasma physics; in a highly ionised gas, electrons screen the Coulomb field produced by a single positive charge at distances > the Debye length. In the atmosphere, which is a weakly ionised gas, where the electron-neutral collision frequency is very large, the conductivity is finite. In the real atmosphere, screening from the positive charge at the top of a thundercloud takes place first at $\sim$60 km altitude, where the relaxation time is $\sim$10 ms; at 40 km, the relaxation time is $\sim$100 ms, and it is a few s above the cloud top. Only on these time scales does there exist a downwards quasistatic electric field, from the ionosphere to the cloud top. Screening reduces the electric field above a thundercloud to a value which is almost comparable with the fair weather electric field, i.e. the field shown in Figure 2, with horizontal equipotentials.

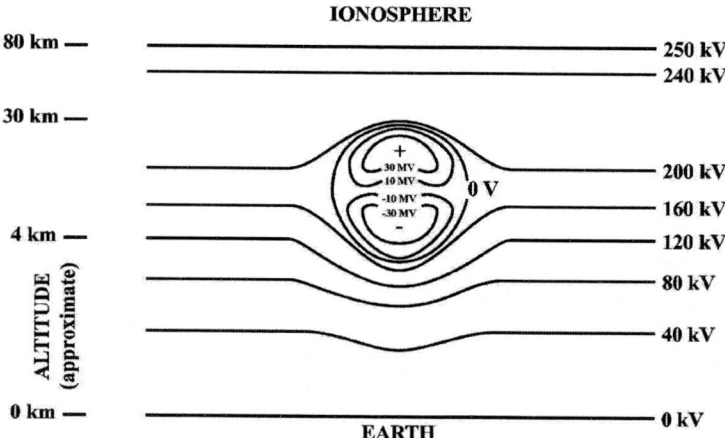

Figure 2. Schematic diagram showing the distribution of equipotential surfaces between the Earth and the ionosphere near an idealised thundercloud, having +100 C at 15 km altitude and -100 C at 5 km, at +100 MV and -100 MV, respectively, before a lightning discharge occurs (with acknowledgements for valuable discussions to L.Sorokin).

1.2.5 Spatial and Temporal Variations of the Global Circuit

Both in the charging (where J·E is negative) and the load (positive J·E) parts of the circuit, Ohm's law can be applied; the conduction current J=σ·E. Four different types of variability may occur:

i) if σ is constant, J and E are linearly related

ii) if J is constant and σ is increased, E decreases

iii) if E is constant and σ is increased, J increases in proportion to σ

iv) J, σ and E may vary independently.

Thus, the interpretation of the apparently simple Ohm's law is not straightforward.

Regarding spatial variability, different processes can dominate depending on geographic location (e.g., latitude, land or ocean, type of weather, the presence or absence of pollution). Temporal variability occurs over a huge period range, from $\sim$ μs (lightning discharge) to $\sim$ ms (VLF/ELF radio phenomena), $\sim$0.1 s (Schumann resonance phenomena), and minutes to hours (evolution of thunderstorm cells and thunderstorms) to diurnal (both with local time and with Universal Time) and 27 day (solar rotation). Changes of tropospheric current sources and solar/geomagnetic activity effects exhibit seasonal, semi-annual and annual variations. Quasi-biennial oscillations (of the stratosphere), solar cycle and longer term climatic variations also exist.

Consider, for example, possible changes as solar activity increases from solar minimum to solar maximum:

i) the height of the lower ionosphere decreases slightly

ii) the electrical conductivity of the stratosphere is reduced appreciably.

If the current generator does not vary with the solar cycle, the charging resistor R_1 in the generator part of the circuit will increase, raising the ionospheric potential. In the fair weather part of the circuit the current will be unchanged, and the electric field will be increased on all three counts. These qualitative statements need to be made more quantitative. Markson (1981) has reported some interesting observations in this regard.

1.3 The Properties of Sprites, Elves and Intense Lightning Discharges

Having outlined some of the relevant properties of the laboratory in which the phenomena of our interest are observed, and also some basic theoretical concepts, attention is now focused on observations, i.e., on what theory has to explain. An overview of what is observed, and how, is presented; this includes a discussion of where and when sprites, elves and the intense discharges, which produce them, occur. Their signatures across the electromagnetic spectrum, from radio waves of the lowest to highest frequencies and into the visible part of the spectrum are mentioned, including typical intensities. Attention is paid to both spatial structure and temporal structure, i.e. their evolution. Many more details are given in subsequent chapters of this book.

1.3.1 Observations and their Interpretation – An Overview

Intense Lightning Towards the end of an intense Mesoscale Convective Complex ahead of a cold front observed near local midnight over the High Plains of the mid West of the U.S.A., strong lightning discharges that produce sprites are often observed to carry positive charge from the stratiform region of a thundercloud at ∼5 km height to ground (termed a +CG discharge). The large charge moment formed by the charge in the cloud (∼100 C) and its image in the ground, ∼1000 C·km, is destroyed (Lyons et al., 2003). That is equivalent to a current of 100 kA flowing over a horizontal distance of, 15 km at a height of 5 km, and then flowing to ground (a current moment of 1000 kA·km), in a discharge lasting ∼1 ms. These values correspond closely to the model values used by Cho and Rycroft (2001), to be discussed later. Huang et al. (1999) measured the charge moment change for the entire lightning discharge which initiates a sprite, and found it to exceed 300 C·km. Hu et al. (2002) reported that sprites are triggered by intense CG discharges, with charge moments >100 C·km.

Elves A +CG discharge radiates a strong electromagnetic pulse which heats the atmosphere at ∼90 km altitude. The heated electrons excite nitrogen molecules which then radiate in the N_2 first positive band (red). Geometry determines that this light appears as a ring (at 90 km altitude) which expands at the speed of light (e.g., Inan et al., 1997; Cho and Rycroft, 1998; Rycroft and Cho, 1998; Nagano et al., 2003). This is termed an "ELVE", which stands for Emission of Light and VLF perturbation from an EMP source (Fukunishi et al., 1996). It occurs ∼0.3 ms after the onset of the discharge.

Sprites From about 1 ms up to 10 ms after the lightning discharge, a bright spot of red light can typically appear at ∼70 km altitude. Within the next 1 ms, discharges moving both upwards and downwards are generated, with lengths of several (up to 10) km and widths ∼0.1 km, or even less; these evolve for up to 100 ms. Many similar discharges can appear simultaneously over a horizontal distance of several km, up to 30 km. Such spectacular "transient luminous

INTRODUCTION TO SPRITES

events" (TLEs), which last for a few ms, and which may be reignited, are termed sprites. On a frame from a video camera observation, sprites look like "celestial fireworks".

Sprites may develop other interesting and notable features, such as halos at their tops and tendrils at their bottoms, down to $\sim$50 km typically. Different sprites may exhibit different features (termed, e.g., "carrot" sprites, or "columniform" sprites), and may evolve differently on time scales of $\gtrsim$1 ms. Some sprites may last for $\gtrsim$ 100 ms.

These properties are known from video images (generally taken at a repetition period of 33 ms) or from photometer traces (which can have 1 ms resolution, or even better). For example, Gerken et al. (2000) have presented a telescopic image of a sprite observed on 13 July 1998 over the North of Mexico which exhibits considerable structure on the 0.1 km scale, and a brightness up to 650 kR (kiloRayleighs), caused by a 128 kA peak current +CG discharge. Neubert et al. (2001) observed sprites over Europe which evolved markedly during 33 ms, and which were caused by +CG discharges between 7 and 124 kA. Hardman et al. (2000) observed "dancing" sprites on a time scale of 20 ms over active summer thunderstorms, having unusually high cloud tops, in Australia; their duration was up to 1 s. And Stenbaek-Nielsen et al. (2000) reported a considerable evolution of a large sprite over Nebraska, U.S.A., on a time scale of only 1 ms.

Sprites observed over the oceans around Taiwan in the summer of 2001 were brighter than those seen over land; "the brightness of some of the oceanic sprites was estimated to exceed 5 MegaRayleighs" (Hsu et al., 2003). During three winter seasons, sprites were recorded photometrically (with $\sim$1 ms resolution) off the coast of Japan (Takahashi et al., 2003). On 27 January 1999 the causative discharge was identified as being in the cloud band of a cold front, the top of which was at a height of 5 km.

Besides being observed from the ground, sprites have also been studied using various instrumental techniques from aircraft, from balloons and from space, e.g., from the International Space Station (Blanc et al., 2004) and from the ill-fated Columbia Space Shuttle mission (Yair et al., 2003; Price et al., 2004).

Sprite spectra were studied by Hampton et al. (1996) and Wescott et al. (2001); theoretical spectra were computed by Milikh et al. (1998). They are strong in the first positive band of nitrogen (red) and also in the N_2^+ first negative band and in the second positive band of nitrogen (blue).

1.3.2 ELF Radiation by Sprites

It has been reported (Cummer et al., 1998; Cummer, 2003) that sprites can radiate signals at up to 1 kHz (ELF). The sprite discharge enhances the local conductivity appreciably, allowing currents of up to 5 kA to flow for $\sim$2 ms over a height interval of, say, 25 km. With its image in the ionosphere, this current moment amplitude of 250 kA·km is equivalent to a charge moment change of 500 C·km, which is about half of the charge moment of the causative discharge (in the ELF band). Bering et al. (2002) discuss comparable numerical values of these parameters in terms of their sprite observations made from a balloon at 32 km altitude. Destroying this charge moment of 500 C·km extracts $\sim$10 C from the reservoir of $2 \cdot 10^5$ C stored in the spherical capacitor which represents the global atmospheric electric circuit (Rycroft et al., 2000). Thus, one sprite removes up to $0.5 \cdot 10^{-4}$ of the charge stored in the global circuit.

Price et al. (2004) reported 7 sprites and 7 elves observed from the Space Shuttle; they detected "ELF transients, with accurate geolocation, for 5 out of 7 elves, but for no sprite events." This result may be "contrary to the present theories of TLE formation, and may require some new thinking into the mechanisms that produce sprites and elves." "Is it possible that, if CG discharges did produce the sprites, they were too weak to produce ELF transients?" "If this is the case, the discharges had charge moments less than 100 C·km." "Are the lightning discharges that produce sprites in the tropics different from those studied in mid-latitude storms? Is the conductivity of the mesosphere in the tropics significantly different from that in mid-latitudes, allowing for weaker discharges to initiate sprites?" Further research is clearly needed.

The electric field between the Earth and ionosphere is positive in the downward direction in the fair weather region. For a downward flowing current, in either a +CG discharge or a sprite, J·E is positive. Thus, both represent dissipation in the global electric circuit. Davydenko et al. (2004) have found that, for a MCS, the quasi- stationary vertical current is $\sim$25 A, downwards. If there is one such positively charged cloud created somewhere in the world each 15 s, this is equivalent to a charge transfer to ground of +100 C per minute. (This is equivalent to a current of <2 A, which is small compared with the DC current of $\sim$1 kA flowing in the global circuit.) Füllekrug (2004) has recently shown that intense positive discharges removing $\sim$100 C on the $\sim$1 minute time scale should decrease the geoelectric field near the Earth's surface by $\sim 10^{-4}$ on the 1 minute time scale, which could be measurable. Two sprites occurring in one minute would remove this same amount of charge.

It is not yet evident whether sprites radiate elsewhere in the electromagnetic spectrum. In particular, it is not known whether photons more energetic than those in the ultraviolet part of the spectrum (wavelength $\sim$100 nm) that are

INTRODUCTION TO SPRITES

required to photoionise molecules at the head of a sprite streamer discharge (Liu and Pasko, 2004) could be generated and, if so, by what process.

1.3.3 Summary of Observations

The main features of the observations discussed in this section may be summarised as follows.

- Elves may occur at 90 km altitude 0.3 ms after an intense discharge in the tropics or at middle latitudes; after 1 ms, their radius is $\sim$300 km.

- Sprites may be generated at 70 ($\pm$15) km altitude, above intense thunderstorms which have discharges bringing positive charge to ground.

- The spatial structure of sprites in the mesosphere ranges from $\lesssim 0.1$ to 30 km.

- The temporal evolution of sprites occurs on time scales ranging from $\lesssim 1$ ms to $\sim$100 ms.

- Sprites radiate up to $\sim$1 MR (MegaRayleigh) of light, primarily in the first positive band of molecular nitrogen (red) and in the second positive band (blue), in the first ms of their life.

- Considering their electrostatic image in the ground, +CG discharges which generate sprites have a charge moment $\gtrsim$100 C·km, and a current moment $\gtrsim$ 100 kA·km.

- A 1000 C·km charge moment change is equivalent to a lightning discharge current of 100 kA flowing horizontally for $\sim$15 km at a height of 5 km, and then flowing to ground, for $\sim$1 ms.

- Sprites can radiate a burst of ELF radio energy lasting $\sim$2 ms; considering their image in the ionosphere, a charge moment change of 500 C·km is equivalent to a current of $\sim$5 kA flowing over a distance of 25 km, which corresponds to a current moment change of 250 kA·km.

- Four intense mesoscale convective systems (MCSs), having positive charges, remove $\sim 10^{-4}$ of the potential difference stored in the global circuit on a time scale of one minute.

- Sprites also represent dissipation in the global circuit; two sprites occurring in one minute also remove $\sim 10^{-4}$ of the potential difference across the global electric circuit.

1.4 Introduction to Theories and Numerical Modelling of Sprites

1.4.1 Basic Physical Concepts

The energy source for sprites and elves is electric field energy associated with lightning. This can be in the form of the quasi-static field due to the distribution of charge in a thunderstorm, or the EMP pulse from a lightning discharge. There are two basic theories for the formation of sprites above energetic thunderstorms; these are:

i) conventional (thermal) discharge physics, and

ii) runaway (relativistic) electron discharge physics.

These theories have been outlined by Rowland (1998) and Rycroft (2005), amongst others. The latter theory is more complex than the former (although, when considering details, both are complex). By Occam's razor, the former is preferred until it has been demonstrated that it fails to explain a significant observation.

1.4.2 Computer Modelling Results

Computer codes, both electrostatic and electromagnetic, have been developed by a number of groups to model the response of the upper atmosphere to thunderstorm fields and to lightning discharge currents (the electrostatic code results are included within the predictions of an electromagnetic model). Maxwell's equations are solved self consistently through the atmosphere which has a modelled conductivity profile. When the transient electric field becomes large, the accelerated electrons heat atmospheric molecules by collision; when it becomes even larger, the greatly accelerated electrons create additional ionisation via collisions with neutral molecules.

In the conventional discharge picture, the crucial parameter is the transient (either quasi-static or EMP pulse) electric field at a certain altitude in the mesosphere divided by the neutral gas density there; this parameter is termed the "reduced" electric field. When this parameter exceeds a certain value, a discharge spontaneously occurs. Examining the spatial and temporal development of this "reduced" electric field parameter, Figures 7 and 9 of Cho and Rycroft (1998) show clearly how an elve is launched and travels horizontally at $\sim$90 km altitude. By having a +CG discharge transferring 100, 200 or 300 C, Cho and Rycroft (1998) also show that the process by which the optical emissions are generated is very nonlinear. Rycroft and Cho (1998) modelled the ELF/VLF spectrum radiated by a +CG discharge, and found it to be rich near $\sim$10 Hz (about the fundamental Schumann resonance frequency) and also from 0.1 to 0.3 kHz.

Currents ∼100 kA flowing in a strong horizontal lightning discharge (which create large oscillatory vertical electric fields above it) were considered in an electromagnetic code model by Cho and Rycroft (2001). They emphasised the importance of the interference at VLF/LF between electromagnetic waves radiated directly by the lightning, waves reflected from the ground, and waves reflected by the ionosphere as a means of creating a spatially structured region of enhanced "reduced" electric field, the position of which changes appreciably on a time scale of $\lesssim 0.03$ ms. These could generate propagating streamers, see Figure 26 of Cho and Rycroft (2001), and so account for the multiplicity of sprites in a region of ∼30 km horizontal extent above an active MCS thundercloud (Lyons et al., 2003).

Liu and Pasko (2004) have modelled in detail the photoionisation phenomena occurring at the head of the sprite, where the electric field is very large, and the spatial and temporal structure so caused; these topics are considered in Chapter 12 by Pasko.

1.5   Conclusions

The purpose of this chapter has been to introduce the reader to the atmosphere, and its spatially and temporally varying properties over a wide range of scales, to thunderstorms, to lightning discharges and their ELF/VLF radiation, to optical emissions termed elves lasting <1 ms at ∼90 km altitude, and to sprites with a time scale from ∼1 ms up to 0.1 s at altitudes ∼70 (±15) km. Further, the chapter has presented in outline some important published results; it gives some key references.

Acknowledgments

The author appreciates the comments expressed by both the formal and the informal referees, which have led to significant improvements in the content of this chapter.

Bibliography

Adlerman, E. J. and Williams, E. R. (1996). Seasonal variation of the global electrical circuit. *J. Geophys. Res.*, 101(D23):29679–29688.

Baker, D. N. (2000). Effects of the Sun on the Earth's environment. *J. Atmos. Sol.-Terr. Phys.*, 62:1669–1681.

Bering, E. A., Benbrook, J. R., Garrett, J. A., Paredes, A. M., Wescott, E. M., Moudry, D. R., Sentman, D. D., Stenbaek-Nielsen, H. C., and Lyons, W. A. (2002). The electrodynamics of sprites. *Geophys. Res. Lett.*, 29:doi:10.1029/2001GL013267.

Blanc, E., Farges, T., Roche, R., Brebion, D., Hua, T., Labarthe, A., and Melnikov, V. (2004). Nadir observations from the International Space Station. *J. Geophys. Res.*, 109(A2):doi:10.1029/2003JA009972.

Cho, M. and Rycroft, M. J. (1998). Computer simulation of the electric field structure and optical emission from cloud-top to the ionosphere. *J. Atmos. Sol.-Terr. Phys.*, 60:871–888.

Cho, M. and Rycroft, M. J. (2001). Non-uniform ionisation of the upper atmosphere due to the electromagnetic pulse from a horizontal lightning discharge. *J. Atmos. Sol.-Terr. Phys.*, 63:559–580.

Cummer, S. A. (2003). Current moment in sprite-producing lightning. *J. Atmos. Sol.-Terr. Phys.*, 65:499–508.

Cummer, S. A., Inan, U. S., Bell, T. F., and Barrington-Leigh, C. P. (1998). ELF radiation produced by electrical currents in sprites. *Geophys. Res. Lett.*, 25:1281–1284.

Davydenko, S. S., Mareev, E. A., Marshall, T. C., and Stolzenberg, M. (2004). On the calculation of electric fields and currents of mesoscale convective systems. *J. Geophys. Res.*, 109(D11):doi:10.1029/2003JD003832.

Fukunishi, H., Takahashi, Y., Kubota, M., Sakanoi, K., Inan, U. S., and Lyons, W. A. (1996). Elves, lightning-induced transient luminous events in the lower ionosphere. *Geophys. Res. Lett.*, 23:2157–2160.

Füllekrug, M. (2004). The contribution of intense lightning discharges to the global electric circuit during April 1998. *J. Atmos. Sol.-Terr. Phys.*, 66:1115–1119.

Gerken, E. A., Inan, U. S., and Barrington-Leigh, C. P. (2000). Telescopic imaging of sprites. *Geophys. Res. Lett.*, 27:2637–2640.

Hale, L. C. (1984). Middle atmosphere structure, dynamics, and coupling. *Adv. Sp. Res.*, 4:175–186.

Hampton, D. L., Heavner, M. J., Wescott, E. M., and Sentman, D. D. (1996). Optical spectral characteristics of sprites. *Geophys. Res. Lett.*, 23:89–92.

Hardman, S. F., Dowden, R. L., Brundell, J. B., Bahr, J. L., Kawasaki, Z., and Rodger, C. J. (2000). Sprite observations in the Northern Territory of Australia. *J. Geophys. Res.*, 105(D4):4689–4697.

Harrison, R. G. (2004). The global atmospheric electrical circuit and climate. *Surv. Geophys.*, 25:441–484.

Harrison, R. G. and Carslaw, K. S. (2003). Ion-aerosol-cloud processes in the lower atmosphere. *Rev. Geophys.*, 41:doi:10.1029/2002RG000114.

Hsu, R. R., Su, H. T., Chen, A. B., Lee, L. C., Asfur, M., Price, C., and Yair, Y. (2003). Transient luminous events in the vicinity of Taiwan. *J. Atmos. Sol.-Terr. Phys.*, 65:561–566.

Hu, W., Cummer, S. A., Lyons, W. A., and Nelson, T. E. (2002). Lightning charge moment changes for the initiation of sprites. *Geophys. Res. Lett.*, 29:doi:10.1029/2001GL014593.

Huang, E., Williams, E., Boldi, R., Heckman, S., Lyons, W., Taylor, M., Nelson, T., and Wong, C. (1999). Criteria for sprites and elves based on Schumann resonance observations. *J. Geophys. Res.*, 104:16943–16964.

Inan, U. S., Barrington-Leigh, C., Hansen, S., Glukhov, V. S., Bell, T. F., and Rairden, R. (1997). Rapid lateral expansion of optical luminosity in lightning-induced ionospheric flashes referred to as 'elves'. *Geophys. Res. Lett.*, 24:583–586.

Liu, N. and Pasko, V. P. (2004). Effects of photoionization and branching of positive and negative streamers in sprites. *J. Geophys. Res.*, 109(A04):doi:10.1029/2003JA010064.

Lyons, W. A., Nelson, T. E., Williams, E. R., Cummer, S. A., and Stanley, M. A. (2003). Characteristics of sprite-producing positive cloud-to-ground lightning during the 19 July 2000 STEPS Mesoscale Convective Systems. *Mon. Wea. Rev.*, 131:2417–2427.

Markson, R. (1981). Modulation of the Earth's electric field by cosmic radiation. *Nature*, 291:304–308.

Milikh, G., Valdivia, J. A., and Papadopoulos, K. (1998). Spectrum of red sprites. *J. Atmos. Sol.-Terr. Phys.*, 60:907–915.

Nagano, I., Yagitani, S., Miyamura, K., and Makino, S. (2003). Full-wave analysis of elves created by lightning-generated electromagnetic pulses. *J. Atmos. Sol.-Terr. Phys.*, 65:615–625.

Neubert, T., Allin, T. H., Stenbaek-Nielsen, H., and Blanc, E. (2001). Sprites over Europe. *Geophys. Res. Lett.*, 28:3585–3588.

Ney, E. P. (1959). Cosmic radiation and the weather. *Nature*, 183:451–452.

Price, C., Greenberg, E., Yair, Y., Satori, G., Bor, J., Fukunishi, H., Sato, M., Israelivitch, P., Moalem, M., Devir, A., Levin, Z., Joseph, J. H., Mayo, I., Ziv, B., and Sternlieb, A. (2004). Ground-based detection of TLE-producing intense lightning during the MEIDEX mission on board the space shuttle Columbia. *Geophys. Res. Lett.*, 31:doi:10.1029/2004GL020711.

Rakov, V. A. and Uman, M. A. (2003). *Lightning: physics and effects*. Cambridge University Press. 687 pp.

Roble, R. G. (1991). On modeling component processes in the Earth's global electric circuit. *J. Atmos. Terr. Phys.*, 53:831–847.

Rowland, H. L. (1998). Theories and simulations of elves, sprites and blue jets. *J. Atmos. Sol.-Terr. Phys.*, 60:831–844.

Rycroft, M. J. (2003). The Earth and its atmosphere. In Davies, M., editor, *The standard handbook for aeronautical and astronautical engineers*, pages 16.1–16.12. McGraw-Hill.

Rycroft, M. J. (2005). Electrical processes coupling the atmosphere and ionosphere: An overview. *J. Atmos. Sol.-Terr. Phys.* (submitted).

Rycroft, M. J. and Cho, M. (1998). Modelling electric and magnetic fields due to thunderclouds and lightning from cloud-tops to the ionosphere. *J. Atmos. Sol.-Terr. Phys.*, 60:889–893.

Rycroft, M. J., Israelsson, S., and Price, C. (2000). The global atmospheric electric circuit, solar activity and climate change. *J. Atmos. Sol.-Terr. Phys.*, 62:1563–1576.

Schunk, R. W. and Nagy, A. F. (2000). *Ionospheres: physics, plasma physics and chemistry*. Cambridge University Press. 570 pp.

Stenbaek-Nielsen, H. C., Moudry, D. R., Wescott, E. M., Sentman, D. D., and Sao Sabbas, F. T. (2000). Sprites and possible mesospheric effects. *Geophys. Res. Lett.*, 27:3829–3832.

Takahashi, Y., Miyasato, R., Adachi, T., Adachi, K., Sera, M., Uchida, A., and Fukunishi, H. (2003). Activities of sprites and elves in the winter season, Japan. *J. Atmos. Sol.-Terr. Phys.*, 65:551–560.

Wescott, E. M., Stenbaek-Nielsen, H. C., Sentman, D. D., Heavner, M. J., and Sao Sabbas, F. T. (2001). Triangulation of sprites, associated halos and their possible relation to causative lightning and micro-meteors. *J. Geophys. Res.*, 106(A6):10467–10477.

Williams, E. R. (2002). Global electric circuit. In *Encyclopedia of Atmospheric Science*. Academic Press.

Williams, E. R. and Heckman, S. J. (1993). The local diurnal variation of cloud electrification and the global diurnal variation of negative on the Earth. *J. Geophys. Res.*, 98(D3):5221–5234.

Williams, E. R. and Satori, G. (2004). Lightning, thermodynamic and hydrological comparison of the two tropical continental chimneys. *J. Atmos. Sol.-Terr. Phys.*, 66:1213–1231.

Wilson, C. T. R. (1920). Investigations on lightning discharges and the electric field of thunderstorms. *Phil. Trans. Roy. Soc. Lond.*, 221(A):73–115.

Wormell, T. W. (1953). Atmospheric electricity: some recent trends and problems. *Quart. J. Roy. Met. Soc.*, 79:3–50.

Yair, Y., Price, C., Levin, Z., Joseph, J., Israelevitch, P., Devir, A., Moalem, M., Ziv, B., and Asfur, M. (2003). Sprite observations from the space shuttle during the Mediterranean Israeli dust experiment (MEIDEX). *J. Atmos. Sol.-Terr. Phys.*, 65:635–642.

THE METEOROLOGY OF TRANSIENT LUMINOUS EVENTS - AN INTRODUCTION AND OVERVIEW

Walter A. Lyons
FMA RESEARCH Inc., Yucca Ridge Field Station, Fort Collins, CO 85024, USA.

Abstract This contribution reviews the basics of atmospheric deep convection and electrification as it pertains to the generation of stratospheric and mesospheric transient luminous events (TLEs). Emphasis is placed on sprites and sprite-producing lightning, and the meteorological regimes in which they are found.

2.1 Introduction

This introduction provides a brief overview of key concepts in the atmospheric sciences relevant to investigations of transient luminous events (TLEs), with emphasis on convective clouds and thunderstorms. Our purpose is to familiarize those TLE researchers with little grounding in either the theoretical or operational aspects of meteorology with the terminology that may be encountered in their further explorations and readings of the literature (American Meteorological Society, 2000). Extensive reference resources will allow in depth pursuit of concepts introduced herein.

2.1.1 Scales of Atmospheric Motion

The Earth's atmosphere is a fluid whose energetics is almost entirely derived from the unequal distribution of solar energy upon the planet. The sun's input fluctuates over the time scales of climate change due to variations in the solar output plus eccentricities in the Earth's orbit. Seasonal changes arise from the 23.5° inclination of the axis of rotation to the ecliptic plane, along with a slightly elliptical orbit. The diurnal fluctuations of energy resulting from the planet's rotation are the result of short term imbalances of incoming short wave solar radiation with outgoing long wave radiation. The heterogeneity of solar energy gain at any point on the surface is further modulated by surface characteristics, principally the 75% of the surface covered by water, with its high specific heat, or highly reflective ice packs. Solar heating of land is greatly influenced by surface characteristics including land use, soil characteristics, moisture and snow cover.

Unequal temperatures, especially through deeper atmospheric layers, give rise to differences in hydrostatic pressure. Pressure gradient forces, modulated by the Earth's rotation (the Coriolis force), and retarded by friction near the surface, give rise to wind. The Earth's winds transport large amounts of heat and water vapor quasi-horizontally (advection). The patterns of atmospheric motion occur over a series of interlinked and somewhat arbitrarily differentiated scales. Global scale motions (the general circulation) are defined by well defined regions of quasi-steady flow (the polar easterlies, the mid-latitude westerlies, the sub-tropical easterlies and equatorial doldrums). Hemispheric flows, best visualized using polar coordinate charts, reveal the dominant Rossby wave troughs and ridges separating polar air masses from more temperate tropical air. Also found are the major jet streams (polar, mid-latitude, and sub-tropical) just below the tropopause, the demarcation between the 10-20 km deep weather-containing troposphere and the overlying stable, generally quiescent stratosphere. The synoptic, or macro-scale is defined by structures such as the familiar high and low pressure centers and warm and cold fronts of daily weather maps (scale: $\sim$1000 km). But most weather experienced on a day-to-day basis is associated with the mesoscale, systems with scale lengths of tens to hundreds of kilometers (Ray, 1986; Fujita, 1992, pp. 298). Individual thunderstorms are sometimes considered as sub-mesoscale or cloud scale events (Cotton and Anthes, 1989, pp. 883). The microscale refers to intense gradients of temperature and wind found within the planetary boundary layer (PBL) which experiences diurnal changes in depths ranging from tens to several thousands of meters.

Yet it is the molecular scale which in many ways controls key cloud formation processes. The maximum amount of moisture contained by a volume of atmosphere is a strong, non-linear function of temperature (the Clausius-Clapeyron equation). Moisture content can be specified in a variety of ways. Relative humidity is the percentage of moisture compared to its maximum carrying capacity at a given temperature (RH = 100% = saturation). Moisture is also defined by the dewpoint temperature, the temperature at which saturation is reached if air is cooled at constant pressure. The ratio of water vapor to dry air is the mixing ratio. Water is highly unusual in that it exists in all three phases, gaseous, liquid and solid, at common atmospheric temperatures and pressures. Any phase change results in either releasing or absorbing energy (sensible heat). Thus, water vapor is often said to possess latent heat, i.e., $2.501 \cdot 10^6$ J·kg^{-1} released upon condensation, plus an additional $3.337 \cdot 10^5$ J·kg^{-1} upon freezing. This process is reversible during melting and evaporation. Latent heat is the primary "fuel" of thunderstorms.

Upon reaching saturation, water vapor can not spontaneously condense into liquid without cloud condensation nuclei (CCN), specks of matter ranging greatly in size and origin including crustal materials, sea salt and industrial pollutants. Ice nuclei also play a similar role in allowing direct deposition of ice in the formation of ice crystals. Most cloud particles form in the liquid phase. Cloud droplets may remain supercooled to temperatures as cold as -40°C. The process of converting supercooled droplets to ice crystals is termed glaciation. Many thunderstorm clouds are comprised of liquid droplets near their base, fully glaciated ice crystals at the top, but with a deep mixed phase region between in which electrification processes occur (MacGorman and Rust, 1998, pp. 422).

2.1.2 Basic Concepts of Atmospheric Vertical Stability

The primarily vertical atmospheric transport of heat (sensible and latent) is called convection. Forced convection may result from mechanically driven flows over mountain (orographic) barriers or by horizontally converging air streams. Free (or gravitational or buoyant) convection, the dominant process in thunderstorm growth, arises when a parcel of air is locally less (more) dense than its environment and is thus accelerated upwards (downwards). If the parcel is unsaturated, it will cool (warm) at 1°C per 100 meters of upward (downward) displacement. Due to the effects of latent heat, if the parcel is saturated, it warms or cools at a rate <1°C per 100 meters (typically ~0.6°C/100 m in the middle atmosphere). Once initially displaced, the fate of this parcel, however, depends on the environmental lapse rate of temperature, which can be highly variable. If the local environmental lapse rate is greater than that of the displaced parcel, it will continue to accelerate away from its initial position. This is termed an unstable atmosphere, and conditionally unstable if the parcel must be saturated for it to continue to accelerate. If the environmental lapse rate is less than the parcel lapse, then any displacement results in the parcel becoming denser (if moving upward) or lighter (if moving downward) than its environment, with restoring forces returning it to its initial position, often accompanied by oscillatory motions about that point. Atmospheres in which the temperature change with height is constant (isothermal) or increasing (inversion) are always absolutely stable.

The potential for deep convective motions in the atmosphere is often analyzed by applications of parcel theory. Given an observed atmospheric sounding of temperature and moisture, one can easily compute the fate of a parcel (or layer) of air initially displaced from near the surface, or some other arbitrary level in the lower atmosphere. If the parcel is able to continue rising to the point of saturation, it has reached the lifted condensation level (LCL), which marks the cloud's base. Continued ascent may bring it to the level at which it is now more buoyant than its environment, the level of free convection (LFC). From there, upward acceleration continues until reaching a stable layer, most often the tropopause, the abrupt change in environmental lapse rate demarcating the troposphere from the stratosphere above. Most thunderstorm tops flatten out on the tropopause, accounting for the familiar anvil-shaped thunderstorm cloud. The more intense updrafts, however, penetrate into the lower stratosphere for up to several kilometers before becoming negatively buoyant and subsiding. The overshooting domes are often a sign of severe weather.

There are numerous ways to quantify the thermal instability of an air mass. The Lifted Index (LI) is the computed temperature difference between the ascending parcel and its environment at the 500 hPa ($\sim$5500 m) level. Unstable air will have negative LI values, with -10°C being the most extreme buoyancy which may be expected. It is also possible to compute the Convective Available Potential Energy (CAPE, $J \cdot kg^{-1}$) which represents the integrated energy in the free convection regime between the LFC and the tropopause penetration. Modestly unstable air may have values in the 500 to 1000 $J \cdot kg^{-1}$ range, with the extremes approaching 5000 $J \cdot kg^{-1}$ being associated with the most severe thunderstorms. In order for an atmosphere to become unstable enough to support deep convection, various combinations of warming of the lower layers, cooling of the upper layers, or moistening (primarily the lower portion) of the air mass are required.

It should be noted that to induce convection, some form of triggering mechanism is often needed. This is often supplied by a warm air "bubble" near the ground, a product of differential heating. But mechanical lifting induced by flow over terrain, fronts and often the outflows from an adjacent storm, can provide enough initial lift to reach first the LCL and then finally, the LFC. The amount of energy required to reach the LFC, or "breaking the cap" in storm chaser parlance, is the Convective Inhibition (CIN, $J \cdot kg^{-1}$). In most cases, CIN $<<$ CAPE, but without this initial energy input the atmosphere's convective potential can not be realized. This, in part, explains why, on warm, humid summer days, convection occurs only in localized regions with discernible patterns often strongly controlled by the triggering mechanisms.

2.1.3   Convective Cloud Nomenclature

With the advent of modern remote sensing systems such as radar during World War II and later, meteorological satellites, convective storm research entered an era of unprecedented activity. Much has been learned through organized field campaigns, the prototype for which is the U.S. Thunderstorm Project held in Florida and Ohio in the late 1940s (Byers and Braham, 1949). Recent programs such as the Severe Thunderstorm Electrification and Precipitation Study (STEPS) (Lang et al., 2004) and the Bow Echo and Mesoscale Convective Vortex Experiment, BAMEX (Davis et al., 2004) illustrate the increasingly specialized nature of such campaigns, not only in the U.S. but, increasingly, worldwide. Such efforts have highlighted the complex and varied structures of convective storm systems, which we can only begin to outline below (Cotton and Anthes, 1989, see for example).

Convective clouds were first formally classified by Luke Howard in 1803. He devised a Latin-based cloud nomenclature system. Howard's "clouds of vertical development" include the modest cumulus cloud, the cumulus congestus (towering cumulus or TCU) and the cumulonimbus (thundercloud or "Cb"). However, convective storms are often organized into systems vastly larger and more multifaceted than any individual cumulonimbus element.

Convective storms are sometimes characterized by their "triggering" or forcing function, such as cold frontal thunderstorms. Mountain, or orographic, thunderstorms commonly occur in the late afternoon during warm seasons over higher terrain. Large stratiform precipitation regions in extratropical cyclones sometimes contain "embedded" convective cells, often poleward of warm fronts. The inventory includes sea and lake breeze storms, urban effect storms, and even forest fire storms. Thunderstorms also occur when winter polar air masses stream over warmer bodies of water such as the Sea of Japan, the Gulf Stream and the "lake effect" storms of the Great Lakes.

Over the past decades it has become common practice to classify atmospheric deep convection according to its structural and morphological features. Among the first to be documented are short-lived (30 to 60 minutes) "air mass" storms, occurring quasi-randomly in air masses, often far from frontal boundaries. Their single cell updrafts quickly transform into downdrafts which then terminate the convective motions. Somewhat longer-lived multicellular clusters also occur. Under conditions of extreme thermodynamic instability (high CAPE) and low level wind shear (high helicity), supercellular storms develop. These storms, which persist for many hours, are prolific producers of large hail and tornadoes. They are accompanied by quasi-steady state, rotating updrafts. They are further sub-classified in low precipitation (LP), classic, and high precipitation (HP) supercells. Squall lines refer to narrow, linear storm systems extending from 100 km to >1000 km in length, comprised of nearly continuous or discrete cellular elements. While often associated with cold fronts, they can also propagate far ahead of any frontal zone. Larger clusters of convective storms (>2,000 km^2) with lifetimes exceeding several hours are collectively called mesoscale convective systems (MCSs). They typically have intense updrafts and high radar reflectivities in convective cores which, as the MCSs mature, become embedded within much larger stratiform precipitation regions which may trail, surround or lead the intense convection. Some MCSs develop "bow echoes" at their leading edge. These can be associated with intense, sustained straight line winds (derechos). The largest and best organized MCS is the Mesoscale Convective Complex (MCC) appearing in satellite images as a quasi-elliptical cloud mass often well in excess of 100,000 km^2 (Maddox, 1980). Often lasting over 12 hours, they are among the world's most prolific producers of rainfall and lightning. MCCs occur worldwide, but most frequently over mid-latitude land masses during night (Laing and Fritsch, 1997). Nocturnal MCSs and MCCs often evolve upscale from daytime supercellular or squall line convection. Most convective systems produce gust fronts or outflow boundaries which often propagate far from the storm and persist long after its demise but which are critical in triggering new convection, especially when colliding with other outflow boundaries.

Numerical models are increasing our understanding of deep convection processes (Pielke Sr., 2002, pp. 673). Operational weather forecasting models generally do not explicitly realize the form of the convection per se, but rather rely on parameterizations to deduce thunderstorm occurrence. As increasing computer power allows reducing the horizontal computational grids to <10 km and hundreds of meters in the vertical, models will simulate in considerable detail the interplay of dynamics and microphysics of the various classes convection. Experimental modeling runs with horizontal mesh sizes <100 m are beginning to provide meaningful insights into storm scale atmospheric processes. Charge generation, separation, and lightning discharge parameterizations are achieving ever greater sophistication.

2.2 Observations of Convective Phenomema

For the first half of the last century, thunderstorm observations were limited to government weather observing stations. In most nations, a thunderstorm was defined as any period in which thunder could be heard at an observing site, typically an airport weather station. Given that the audible range of thunder is usually limited to 10-25 km, a small fraction of the typical spacing between stations, many thunderstorms went unreported. Visual observations of cumulonimbus clouds were provided at some stations. Most global summaries of convective frequency were cast in terms of "thunder days," days in which one (or more) periods of thunder occurred. Given the scarcity of marine observations, convective climatologies were necessarily biased towards land observations. Given such limitations, it is rather remarkable that Brooks (1925), using estimates based on the global electrical circuit (Bering, 1997), estimated global lightning flash rates to be on the order of 100 s^{-1}. This was remarkably prescient given that today's satellite estimates of global lightning are only less than a factor of two smaller.

2.2.1 Conventional Convective Storm Monitoring

During the wartime development of anti-aircraft radar surveillance systems, anomalous returns were soon recognized as areas of precipitation. The meteorological applications became immediately evident. During the 1940s and much of the 1950s, various military systems were reconfigured to detect atmospheric phenomena, notably precipitation. Many early systems operated at 3 cm wavelengths, ideal for detecting snow and wet clouds, but suffering from attenuation during heavier rain events. A series of hurricanes along the U.S. east coast in the mid-1950s lead to the development of the WSR-57 radar, a minimal attenuation 10 cm radar ideal for probing thunderstorms. Similar systems became commonplace worldwide. Conventional radar returns were described in terms of decibels of reflectivity (dBZ), which are related to precipitation rate. Several hundred Z/R (reflectivity/rainfall) rate algorithms have emerged over the years, so one to one correlations between precipitation rates and reflectivity remain elusive. By the 1980s, computers allowed conversion of radar grey scale CRT displays into colorized images using various dBZ contouring schemes. Typically, very light precipitation is associated with 0-10 dBZ values. Echoes achieving 30 dBZ indicate moderate rain and the possibility of lightning. Intense rain, often accompanied by hail, is indicated for >50-55 dBZ. The most extreme convection peaks around ~ 70 dBZ. Certain reflectivity patterns such as hook echoes, bow echoes and "notches" were qualitatively associated with tornadoes, straight line winds and hail. But means to directly detect potentially damaging surface wind speeds became increasingly urgent. To this end, by the mid-1990s, the U.S. began deploying 10 cm Doppler radars (the NEXRAD WSR-88D) which have since been integrated into a nationwide network. Similar national and regional networks are emerging elsewhere. In addition to reflectivity, a single Doppler radar can measure winds, although only along the beam's radial direction. However, algorithms have been developed which allow detection of supercell storm-scale rotation (mesocyclones), and on occasion, the circulation of very large tornadoes (the TVS, tornado vortex signature). Multiple research Doppler radars can be run in tandem to provide 2-D and 3-D wind fields. Terminal Doppler Weather Radars at airports specialize in the detection of a key aviation hazard, the microburst. New, portable Doppler on Wheels (DoW) are becoming key components of tornado field research programs, providing high resolution views of storm dynamics. Dual-polarization radars can derive 3-D volumetric estimates of major precipitation types within storms, greatly facilitating microphysical investigations.

Based upon brief film records of the Earth's cloud cover obtained using surplus World War II rockets, great hope was placed in the capability of weather satellites to map synoptic scale cloud patterns, such as fronts. With the launch of TIROS I (the television and infrared observation satellite) on 1 April 1960, that hope was realized. Since that date, no tropical storm, typhoon or hurricane has gone unmapped anywhere in the world. But with each successive satellite launch, even more amazing was the degree to which the mesoscale structure of the atmosphere could be deduced. Polar orbiting satellites, which typically provide twice daily views of a given area, were joined in the mid-1960s by geostationary systems. Given hourly or more frequent updates, animations of the cloud patterns revealed structure and organization which greatly aided our understanding of convective processes. Presently, continuous global cloud monitoring is provided by a minimum of five geostationary satellites (two U.S., Europe, India and Japan). High resolution visible imagers provide useful detailed information on daytime convective clouds of all sizes. Somewhat lower resolution infrared scanners provide 24-hour coverage, including estimates of cloud top heights. Middle and upper tropospheric water vapor is another widely used product. Most geostationary satellite data are readily available on line to researchers. Polar orbiting research satellites provide multispectral views of the Earth's surface, clouds and aerosol patterns. NASA's advanced MODIS system provides extremely high resolution snapshots of cloud systems.

Meteorological satellite imagery graphically portrays major convective events such as tropical cyclones (including the convective outer bands), frontal and squall line convection, the intertropical convergence zone (ITCZ), polar lows, and organized MCS and MCC cloud canopies. Smaller scale convective organization along sea breeze fronts and mountains, during cold air advection over warmer waters, and random air mass convection are easily determined. Cumulus clouds demarcating outflow boundaries are routinely visible. Key details of intense convection such as overshooting domes, jet-stream sheared anvil clouds, and V-notch patterns in the top of supercells are among the many clues afforded by higher resolution meteorological satellite images.

Additional remote sensing technologies continue to evolve. Infrasound has shown promise in detecting tornadic circulations as well as signatures for sprites (Bedard Jr. et al., 1999) and Chapter 15. Additional research is required before such techniques become routine.

2.2.2 Lightning Observation Techniques and Terminology

This section will not review the physics of the lightning discharge per se. Such information can be found in a large number of textbooks (Rakov and Uman, 2003; MacGorman and Rust, 1998; Williams, 1988; Uman, 1987; National Research Council, 1986; Volland, 1982; Golde, 1977). Rather, we briefly summarize key terminology used in discussing lightning and provide a brief overview of techniques used to investigate and observe lightning.

The human eye was certainly the first lightning sensing system, and human observations have remained a mainstay of aviation weather observation reporting (though they are gradually being replaced by less descriptive, automated lightning detection systems). Terminology evolved to include cloud-to- ground (CG), cloud-to-cloud (CC), in-cloud (IC) and cloud-to-air (CA) lightning. Today it is more common to distinguish between CG and non-CG (collectively IC) events, which in combination are referred to as total lightning.

The earliest known photographs of lightning were taken by a Philadelphia amateur photographer, William Jennings, between 1883 and 1890. By the following decade, photographs of lightning spectra had been obtained. By the 1920s, the Boys camera was among the early streak photography techniques which began to unmask the complex temporal series of events and spectra comprising the CG flash (Uman, 1987; Salanave, 1980; Golde, 1977, pp. 496). Investigations of lightning striking the Empire State Building in the 1930s by K. B. McEachron, instrumented towers in Switzerland in the 1960s and 1970s (Berger et al., 1975), and rocket triggered lightning (Uman and Krider, 1989), are highlights of earlier studies of lightning. The CG discharge to this day remains a phenomenon of intense research interest. The initial in-cloud breakdown processes may include a stepped leader which, as it approaches Earth, is met by an upward propagating streamer. When the connection is made, a brilliant return stroke transports negative charge (negative polarity CG) or positive charge (+CG) to ground. Much of the current is transferred impulsively within less than a millisecond in peak currents ranging from 10 to >100 kA. In many cases, lesser amounts of current flow in a continuing current which can last for tens and, in some cases, hundreds of milliseconds. Negative CGs are usually composed of multiple strokes (3 to 5 being typical, over 40 have been reported) with inter-stroke intervals on the order of tens of milliseconds. These repetitive strokes make the lightning appear to flicker. The entire multi-stroke process is collectively termed a flash. Positive CGs tend to have stroke multiplicities of one, on average have higher peak currents and longer, more intense continuing currents. It is estimated that -CGs outnumber +CGs by nearly an order of magnitude. In multistroke CGs, most strokes are preceded by a dart leader that follows the initial channel to ground. However, recent research suggests that multi-attach point flashes, either branching strokes or strokes following different channels, are quite common (Figure 1). The lightning discharge and storm electrification have been investigated using a wide variety of techniques including networks of automated lightning flash counters, electric field mill networks, acoustic reconstruction of lightning channels, rocket triggered lightning, balloon and aircraft soundings of *in situ* electric fields in clouds and radio frequency (RF) signature analysis of "sferics" (Rakov and Uman, 2003). During the last quarter century, however, major advances have arisen from two primary technologies, terrestrial lightning detection networks and space-borne lightning sensors.

Though the majority of the world's lightning occurs in the tropics, the first practical CG lightning detection network (LDN) evolved in response to forest fire threats in Alaska (Krider et al., 1980). Broadband VLF magnetic direction finding (MDF) systems have been employed in ever growing numbers since the early 1980s for operational detection and location of CG events. Using CG wave form discrimination criteria, the CG return stroke can be distinguished from IC events for a large majority of discharges. The current U.S. National Lightning Detection Network (NLDN) arose from the gradual assembly of regional networks sponsored by various research and utility interests. During the late 1980s, a second technology, time of arrival (TOA) also was developed into a network (Lyons et al., 1989). By the mid-1990s, the benefits of exploiting both MDF and TOA technology resulted in the creation of the hybrid NLDN that exists today (Cummins et al., 1998). The network has subsequently expanded into a North American Lightning Detection Network (Orville et al., 2002), and similar regional networks are emerging worldwide. A typical LDN can provide the following information, either in real time or from archives: return stroke time (millisecond or better), estimated peak current (kA), polarity, location (latitude and longitude), and stroke wave form parameters. Software systems can simulate LDN performance. Estimates of typical locational

Figure 1. A cloud-to-ground lightning flash composed of three separate strokes, each with their own attach points separated by several kilometers. Image courtesy of Tom Nelson.

accuracy (LA) for modern systems are approximately 500 m. Stroke and flash detection efficiency (DE) vary, but typically average 50-70% and 80-90%, respectively (higher for +CGs). Especially for the new hybrid systems, IC rejection is less than perfect. In analyses of LDN data, ICs misidentified as low peak current +CGs (<10 kA) are sometimes excised from the data. Currently, LDNs are concentrated on land masses. New experimental networks (Dowden et al., 2002) are providing nearly global coverage of large peak current CG events.

While the peak currents of sprite parent +CGs (SP+CGs) average 25-50% larger than other +CGs in the same storm, peak current is a poor predictor of the TLE potential of any given lightning stroke (Lyons, 1996b). Based upon sprite modeling theory, the charge moment change, a parameter not measured by conventional LDNs, is perhaps the most important lightning metric for sprite researchers (Wilson, 1925, 1956; Huang et al., 1999; Hu et al., 2002). The charge moment change,

$$\Delta Mq\,(t) = Z_q \cdot Q(t) \text{ (units, C·km)}$$

is defined as the product of Z_q(km), the mean altitude above ground level (AGL) from which the charge is lowered to ground, and Q(t) (unit, C), the amount of charge lowered. Note this second term is most appropriately considered as a function of time, t.

Numerous investigators have explored the use of ELF transient analysis to detect those SP+CGs capable of initiating sprites. Boccippio et al. (1995), Huang et al. (1999) and Williams (2001) measured essentially the entire ΔMq, including that from continuing currents (tens of ms). Based upon analyses of high speed video (Stanley et al., 1999), it appears that most sprites initiate in the 70-75 km altitude. This allows computation of a ΔMq value on the order of 500-1000 C·km required to induce dielectric breakdown in the mesosphere (Williams, 2001). Cummer and Inan (2000) developed a related approach at ELF/VLF which allows routine measurements as a function of time. It is especially effective for determining the impulse charge moment change, $i\Delta Mq$, covering the first 2 milliseconds of the stroke from almost all CGs over long ranges (>1000 km) (Cummer and Lyons, 2004, 2005). In addition, ΔMq for longer periods can also be extracted for the more powerful events (Hu et al., 2002; Lyons et al., 2003a). Though it may vary somewhat from night to night, reflecting changes in ionospheric conditions, it appears that there exists a fairly firm threshold in $i\Delta Mq$ for short delay sprites, those occurring less than ~6 ms after the return stroke (Cummer and Lyons, 2005). This value appears to be in the 100 to 500 C·km range. Long delay sprites, in which the continuing current plays a more significant role in imitating breakdown, will require somewhat larger threshold ΔMq values (300-500 kC·km).

Another major development in lightning measurements is VHF 3-D lightning mapping systems. Research applications date to the pioneering work of Proctor in South Africa. The lightning hazard to Space Shuttle launches resulted in the development of an operational Lightning Detection and Ranging System (LDAR) at the Kennedy Space Center in the 1980s. More recently, New Mexico Tech's pioneering Lightning Mapping Array (LMA) (Krehbiel et al., 2000) has been deployed in several major convective storms field programs in the U.S., along with the French 3-D lightning mapping system (SAFIR). Commercial versions of the LDAR II are now gradually coming into use (Demetriades et al., 2003). These systems provide not only the IC flash rates, but the horizontal and vertical extent of electrified clouds as well as the volume from which charge is removed for each discharge. When combined with an LDN, both the CG and IC components can be measured to produce storm total lightning rates.

Satellite lightning measurements are becoming increasingly important. The U.S. military polar orbiting Defense Meteorological Satellite Program (DMSP) pioneered with an optical flash counter in the 1970s. Early analyses revealed a startling order of magnitude land/ocean asymmetry in total lightning counts (Orville and Henderson, 1986). Why lightning is far more prevalent over land masses, remains an area of active research today. DMSP detection of optical "superbolts" (Turman, 1977) also provided early hints that the tail of the statistical lightning distribution may contain extraordinary events far exceeding the "normal" discharge. During the past decade, NASA's Optical Transient Detection (OTD) polar orbiter has mapped global total lightning (Christian et al., 2003). The Tropical Rainfall Measurement Mission (TRMM) has likewise mapped total lightning, though orbital constraints limited the coverage to tropical regions. To date, no civilian geostationary satellite has been equipped with a lightning sensor, a matter of considerable dismay to atmospheric electricians. The U.S. Department of Energy's FORTE satellite (Smith et al., 2002) obtains specialized data on lightning. Spacecraft specifically designed to detect TLEs will be discussed in Chapters 7 and 6 by Blanc and Mende, respectively, notably the ISUAL experiment on the Taiwanese ROCSAT II satellite.

2.3   A Brief History of TLE Observations

For almost 120 years, unexplained luminous phenomena above thunderstorms have been reported in the literature, beginning with MacKenzie and Toynbee (1886), who described what today might be regarded as a giant jet. Davidson (1893), reporting from tropical Queensland, Australia, noted lightning visible on the horizon when suddenly "...a patch of ...rosy light...5° to

6° in diameter... rose up from above the thunderstorm and mounted upwards; disappearing at an elevation of from 40°-45° ...there were about... twenty-five of the patches in about an hour...." Malan (1937) in South Africa reported "...a long and weak streamer of reddish hue... some 50 km high." Corliss (1977, pp. 542), Vaughan Jr. and Vonnegut (1989) and Vonnegut (1980) are credited with compiling numerous eyewitness accounts from credible observers. Even Nobel laureate Wilson (1956) reported seeing "... diffuse, fan-shaped flashes... extending up into the clear sky...." Yet, without documented evidence, such anecdotal sightings received little attention from the atmospheric electricity community – until the night of 6 July 1989.

Prof. J. R. Winckler and his graduate students were testing a Xybion ISS-255 low-light television (LLTV) for an upcoming research rocket flight. Playback of the video tape revealed a star field, distant "heat lightning" – and two video fields revealing brilliant twin columns of light, extending tens of kilometers into the atmosphere (Franz et al., 1990). These first-ever images of a sprite were presumed to originate over a large MCS in northern Minnesota, several hundred kilometers distant. For the U.S. manned space flight program, the prospect of a new form of "upward lightning" was unsettling, given several unfortunate encounters of spacecraft with lightning (Uman and Rakov, 2003). Similar LLTV cameras were then being flown on the Space Shuttle as part of a mesoscale lightning mapping program. A careful re-inspection of the tapes revealed both a transient airglow enhancement (almost certainly an elve) and more than a dozen "upward lightnings" above thunderstorms (Boeck et al., 1992, 1995; Vaughan Jr. et al., 1992; Lyons and Williams, 1993). By 1993, both ground-based and airborne investigations had been funded by NASA. On the night of 7 July 1993, using Winckler's same Xybion LLTV installed at FMA's Yucca Ridge Field Station (YRFS) near Ft. Collins, CO, almost 250 events were detected during a several hour period above an MCC located 400 km to the east over Kansas (Lyons and Williams, 1993; Lyons, 1994a,b). Within 24 hours, a NASA DC-8 using an LLTV equipped with fish eye lenses detected similar phenomena over an MCC in Iowa (Sentman and Wescott, 1993). Within a year, the FMA and University of Alaska (U of A) teams had noted distinctive VLF audio signatures associated with these events. By this time, at the suggestion of D. D. Sentman, the phenomenon had been named a sprite (after the fleeting spirits in Shakespeare plays) as opposed to "cloud-to-stratosphere lightning," "cloud-to-space lightning" and various other misleading terms that had begun to be used. By 1994, the U of A mounted an airborne campaign which produced the first color images of sprites (Sentman et al., 1995) and the totally unexpected blue jets (Wescott et al., 1995).

During the 1994 and 1995 YRFS sprite campaigns, the correlation between sprites and +CGs became obvious (Lyons, 1996b). An estimated 10,000 optically confirmed sprite observations to date reveal only several confirmed sprites from -CGs (Barrington-Leigh et al., 1999). Also, during the 1994 campaign, YRFS LLTV video data coordinated with ELF Schumann resonance transient (Q-burst) measurements at the Massachusetts Institute of Technology facility in Rhode Island, revealed the vast majority of sprites, or more properly, sprite parent +CGs, had distinctive ELF signatures (Boccippio et al., 1995).

Sprites and blue jets were soon accompanied by other transient luminous events. The predictions of Taraneko et al. (1993) of intense, very brief (<1 ms) glows at the base of the ionosphere associated with lightning EMP (now called elves) were confirmed optically at YRFS in 1995 by Fukunishi et al. (1996). Optical spectra also identified the key $N_2 1P$ red emissions (Mende et al., 1995; Hampton et al., 1996). Evidence of ionization in some, but not all, sprites was obtained by photometric analyses (Armstrong et al., 1998). Ongoing optical programs at YRFS and New Mexico Tech's Langmuir Lab revealed that what many early observers thought to be elves were, in fact, the halo which precedes some, but not all sprites (Barrington-Leigh et al., 2001; Bering et al., 2004). High Speed Imagers (HSI) operating at 1000 fps at Socorro and YRFS documented the first millisecond-scale structure of sprites (Stanley et al., 1999; Armstrong and Lyons, 2000) and notably, that the initiation point of sprites appears to be in the 70-75 km altitude range. The finer temporal and spatial scale structure of sprites, elves and halos have continued to be explored using a variety of photometric sensors including the fly's eye (Inan et al., 1997), telescopic imagery (Gerken et al., 2000; Gerken and Inan, 2003), high speed, high resolution cameras (Stenbaek-Nielsen et al., 2000; Moudry et al., 2003) high speed telescopic imagery (Marshall and Inan, 2005). U of A flights during the mid-1990s confirmed sprites over South American MCSs. Global sprite watching rapidly expanded with confirmations from Australia (Hardman et al., 2000), above Sea of Japan winter snow squalls (Fukunishi et al., 1999), Europe (Neubert et al., 2001), the Caribbean (Pasko et al., 2002) and East Asia (Su et al., 2002; Hsu et al., 2003) to name but a few. Stratospheric TLE balloon missions (Bering et al., 2004; Holzworth et al., 2005) are being pursued to obtain TLE optical signatures with minimal atmospheric absorption as well as *in situ* electric field transient data directly above the parent lightning discharge. Infrasound measurements have also detected apparent signatures from TLEs of the SP+CG (Bedard Jr. et al., 1999) and Chapter 15.

Sprite observations from space have resumed in recent years, highlighted by the MEIDEX experiment on board STS-107 (Isrealevich et al., 2004; Yair et al., 2004). Nadir observations of sprites have been obtained from the International Space Station (Blanc et al., 2004). As mentioned, the ROCSAT-II ISUAL experiment began observations in mid-2004.

In addition to sprites, elves and halos, a variety of electrical discharges emanating from cloud-tops have been uncovered, further expanding the TLE family. Since 1989, more than 10,000 low-light television (LLTV) images of sprites have been obtained by various research teams (Lyons, 1996b; Sentman et al., 1995; Lyons et al., 2000, 2003a). Many anecdotal reports in the literature (Vonnegut, 1980; Vaughan Jr. and Vonnegut, 1989; Lyons and Williams, 1993; Heavner, 2000; Lyons et al., 2003a) described TLEs which simply can not be categorized as sprites:

"... vertical lightning bolts were extending from the tops of the clouds... to an altitude of approximately 120,000 feet... they were generally straight compared to most lightning bolts..."

"... at least ten bolts of lightning went up a vertical blue shaft of light that would form an instant before the lightning bolt emerged..."

"...a beam, purple in color... then a normal lightning flash extended upwards at this point... after which the discharge assumed a shape similar to roots in a tree in an inverted position..."

"... an ionized glow around an arrow-straight finger core..."

"... an American Airlines captain... near Costa Rica... saw from an anvil of a thunderstorm... several discharges vertically to very high altitudes... the event was white..."

"... the top of the storm was not flat... looked like a dome of a van de Graff generator...clearly saw several bolts of lightning going upwards... dissipating in the clear air above the storm... all in all 5 or 6 occurrences ..."

Upward extending white channels topped by blue, flame-like features were captured on film near Darwin, Australia (Lyons et al., 2003b) and over the Indian Ocean (Wescott et al., 2001). This latter event reached a height of $\sim$35 km. Welsh geographer Tudor Williams, who in 1968 was residing near Mt. Ida, Queensland, Australia, visually observed a series of lightning-like channels rising at least several kilometers above the top of a large nocturnal thunderstorm. He photographed several of the approximately 15 events (using 50 ASA 35 mm transparency film, long exposures) that occurred at fairly regular intervals over a 45 minute period. Upward-extending electrical discharges from a supercellular thunderstorm over Colorado were observed during the Severe Thunderstorm Electrification and Precipitation Study (STEPS) on 22 July 2000 (Lyons et al., 2003b).

Eyewitness recollections of lightning-like channels emanating from overshooting convective domes of very active storm cells often share common characteristics. They appear bright white to yellow in color, are relatively straight, do not flicker, extend above cloud tops to heights equal to or exceeding the depth of the cloud (10-15 km), are notably long lasting ($\sim$1 second) and can be observed during *daylight*. It is difficult to understand how these might represent the faint blue jet phenomenon reported by Wescott et al. (1995).

On 15 September 2001, a team of scientists familiar with sprites and blue jets were investigating the effects of lightning on the ionosphere at the Arecibo Observatory in Puerto Rico (Pasko et al., 2002). At 0325.00.872 UTC, above a relatively small ($\sim$2500 km^2) storm cell 200 km northwest of Arecibo, the LLTV video captured an amazing upward discharge, blue in color (see the full animation at http://pasko.ee.psu.edu/Nature). Seen as brilliant blue to the human eye, it appeared as a series of upward and outward expanding streamers which rose from the storm top (16 km). The event reached a terminal altitude of 70 km, the estimated lower ledge of the ionosphere. The event lasted almost 800 ms, including several re-brightenings. This case marked the first hard evidence of a direct electrical link between a tropospheric thunderstorm cell and the ionosphere. A series of similar giant upward jets have since been reported emanating from thunderstorm tops over the Pacific near the Philippines (Su et al., 2003). While sprites are believed to occur with a global frequency of several per minute, the frequency of upward jets and lightning- like cloud top discharges remains unknown. It is becoming clear, however, that they are less rare than once believed.

2.4 Characteristics of TLE-Parent Lightning and Storms

Our understanding of TLE characteristics is evolving at a rapid pace. Investigations into the nature of their unusual parent lightning and, in turn, the types of convective storms which give rise to such atypical discharges are likewise an area of accelerating research. We here briefly summarize the current consensus, but readers should note that new details are emerging with each passing month, potentially dating certain aspects of this summary relatively quickly.

2.4.1 The Phenomenology of TLEs

Red Sprites Based upon surface measurements, these are apparently the most frequently observed of the TLEs. Initiation appears to occur most frequently at 70-75 km altitude, with highly structured branching streamers often first propagating downward, followed by upward expansion in luminosity with the top portion of the sprite a more diffuse glow (Pasko et al., 2002). The tendrils sometimes extend below 40 km, and evidence suggests they can reach below 30 km. The lower portion of the sprites often has a distinct blue coloration. It is uncertain if any tendrils might actually make a physical connection with the parent storm top. Like snowflakes, no two sprites are visually identical, but there are several morphological shapes that are repeated. The columnar or c-sprites are very narrow (order 1 km), quasi-continuous, nearly vertical columns, often with downward and upward extending filaments. They can occur in clusters, sometimes of a dozen or more spread out over several tens of km (Wescott et al., 1998). The classic "carrot" sprite has groups of streamers tapering downwards with outward flaring elements above, causing it to resemble its namesake. Larger clusters of sprites often resemble "A-bombs" or "angels." A "typical" storm may produce a sprite every several minutes. In the U.S. High Plains, after initial onset, sprites usually continue for several hours. Several dozen would be a typical number though, on occasion, storms have produced 400 to 750 sprites within 4-5 hours. In certain storms with very frequent sprites, they appear in video as amorphous glows (Gerken and Inan, 2004). The reason for such hyper-active storms is unclear.

The overwhelming majority of sprites are induced by +CG flashes. On occasion, a massive horizontal cloud discharge, called spider lightning (Mazur et al., 1998), can propagate through the cloud for >100 km, sometimes with several +CG attach points. These often trigger successive sprites in a "dancer" sequence. Most information on sprite durations has been obtained from video with 16.7 ms resolution. Some sprites, often the brightest, occur within a single video field, but some may persist for ten or more fields, slowly dimming from their peak luminosity early in the event. High speed imagery (1000 fps or better) shows the brightest elements persist for only several milliseconds, although subsequent bright features sometimes emerge from those sprites which evolve structure over time. The delay time between the CG return stroke, which can be determined from HSIs or photometric measurements, is highly variable. Often the brightest events occur approximately 1 ms after the return stroke. However, delays of tens of milliseconds are common. Approximately 10% of sprites can not be associated with +CG from a lightning detection network. Given that the DE for +CGs is $\sim$90%, it is suspected that such sprites did have a parent +CG which remained undetected by the LDN. Some sprites can be seen with the dark-adapted, human eye, though they are sometimes perceived as green, white or yellow due to the vagaries of human vision at such low light levels. The inherent brightness of sprites is usually estimated at around 1.0 MR, but briefly can be several times brighter, with some rare cases thought to reach as high as 10-30 MR.

Peak current is a poor predictor of the sprite potential of a CG. A sprite has been detected with an SP+CG as small as 9 kA. Even in the most "sprite efficient" storm, rarely will more than one in five +CGs initiate a sprite. An ongoing question has been "What is different about those +CGs which initiate sprites?" Cummer (Chapter 9) describes in detail the procedures to extract ΔMq from ELF and VLF signals. During the 2000 STEPS program (Lang et al., 2004; Lyons et al., 2003b), detailed analyses of ΔMq (out to 10 ms after the return stroke) suggested that at 600 C·km, there was a 10% of sprite initiation, reaching to 90% by 1000 C·km (Hu et al., 2002). Using more impulsive (2 ms) $i\Delta Mq$ estimates, a threshold of 100 to 500 C·km for rapid onset sprites (within several ms after a +CG) was found in several storms, with the minimum varying somewhat from night to night (Cummer and Lyons, 2004, 2005).

Experience at YRFS has focused on nocturnal MCS and MCC convection. As illustrated in several papers (Lyons, 1996b; Lyons et al., 2000), SP+CGs tend not to occur until the storm has approached its mature stage and developed a considerable stratiform precipitation region. The SP+CGs tend to cluster in a portion of the stratiform region, sometimes towards the trailing edge where clear cloud electrification processes are very different from those experienced in the high reflectivity convective cores. The MCS stratiform area usually reaches a minimum of $10\text{-}20 \cdot 10^3$ km^2 before significant sprite activity can be expected. Detailed analyses of 3-D lightning patterns from STEPS storms (Lyons et al., 2003b; Lyons and Cummer, 2004) have revealed several possible signatures. The main centers of VHF emissions, representing IC discharges, remained high in the cloud (8-12 km) during its active growth stage. But as the stratiform precipitation region expanded, a low-level secondary center of VHF activity developed and the +CGs began initiating sprites (Figure 2). As suggested by Williams (1998), this low level positive charge pool is located around 4 km AGL, near the melting layer. Thus, for the cases studied to date, the average Z_q has resided at 3-5 km. This is in marked contrast to the 10 to 20 km many earlier theoretical sprite modeling studies invoked, in part to achieve sufficiently large ΔMq values. However, evidence is accumulating that some SP+CGs can lower 100 to 300 C of charge. Based on the work of Boccippio et al. (1995) and many subsequent papers, it appears that most SP+CGs produce a globally detectable ELF transient signature (Sato et al., 2003; Price et al., 2002b; Füllekrug and Constable, 2000). This allows a crude estimate of global TLE rates of several per minute, assuming most "Q-bursts" represent TLEs. However, it is not known how many ELF transients (Q-bursts) result from non-sprite producing events. The recent puzzling results from MEIDEX, in which optically confirmed sprites could not be matched with ELF signatures (Price et al., 2004) also requires further investigation.

Elves It is likely that the first elve (the singular is *not* elf, to avoid obvious confusion with ELF radio waves) was recorded by the Space Shuttle low light camera. Though theoretically predicted in the early 1990s, the phenomenon was not documented from ground sensors until 1995 (Fukunishi et al., 1996). Elves, though perhaps as bright as a typical sprite ($\sim$1000 kR), are very brief (hundreds of microseconds), and thus invisible to the human eye. They are also difficult to detect using video systems, and is thus better investigated using photometer arrays (Barrington-Leigh et al., 2001). Red in color, an elve is a rapidly expanding toroidal disk, the result of the EMP pulse from a CG discharge. Thus the lag between return stroke and the onset of luminosity is that of the propagating speed of light ($\sim$300 microseconds). In conventional video, an elve will persist only for a single field. The altitude is in the range of 80-100 km and the expanded disk can attain a diameter of 400 km or even larger. While the most accepted theoretical explanations for both sprites and elves are independent of the CG polarity, only elves seem to be associated with a significant percentage of -CG strokes. Typically these have higher peak currents than those for sprites, indicative of the more impulsive nature of the CG source. There has not yet been a systematic survey of rise times for elve parent CGs. Typical ΔMq values approach those for sprites, though more data is required to determine if any threshold may be systematically lower. Using ground cameras, sprites and elves occur intermingled in the same storms, though the ratio varies considerably from storm to storm. It is rare to have just elves or just sprites in a given storm. Given the recent findings from the STS-107 MEIDEX mission (Yair et al., 2004) and preliminary returns from ROCSAT II, elves may be more common than expected from analysis of ground-based video. This

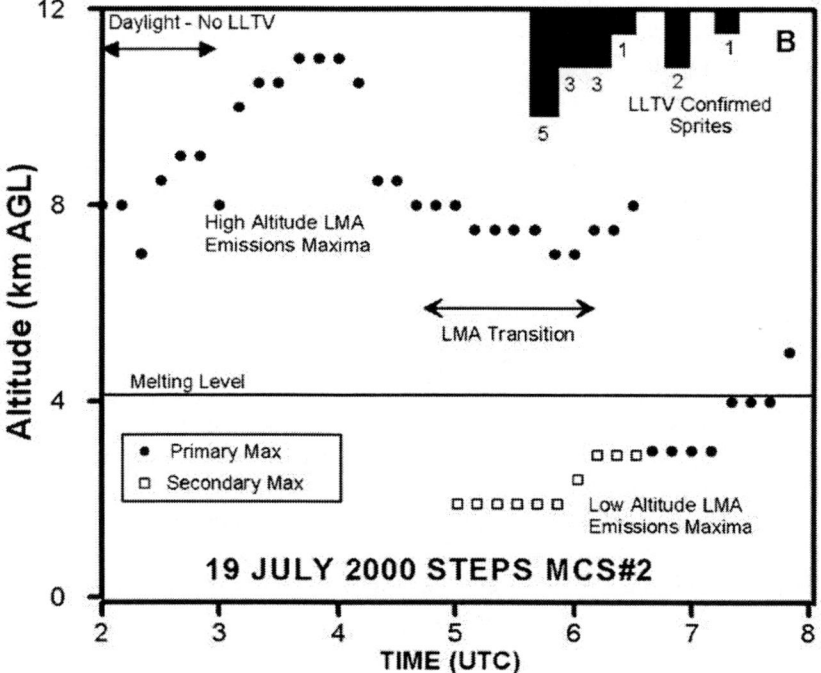

Figure 2. Sequence of events in a sprite-producing MCS When VHF lightning returns from a 3-D lightning mapper remain between 8-11 km AGL, +CGs do not initiate sprites. When a low-level maximum of VHF returns develops as the storm matures, +CGs begin producing sprites. (Lyons et al., 2003a, Courtesy of the American Meteorological Society).

may result from better viewing angles (limb views) but also from more systematic monitoring over the world's oceans. There is growing suspicion that large peak current (and perhaps more impulsive) -CGs are more prevalent over salt water than land (Lyons et al., 1998; Füllekrug et al., 2002).

Halos In the early days of LLTV monitoring, it was common to observe an apparent elve followed by a sprite, sometimes called a "sprelve." While elves do indeed precede some sprites, the "sprelve" in most cases was found to represent a "sprite halo" (Barrington-Leigh et al., 2001; Moudry et al., 2003; Miyasato et al., 2003). High speed video (Armstrong and Lyons, 2000; Stanley et al., 1999; Stenbaek-Nielsen et al., 2000) have shown the halo to be a downward descending, lens shaped amorphous glow that initiates typically one or 2 ms after the return stroke, and persists for 1 to 3 ms. Halos tend to be much smaller (maximum diameter 100 km) and lower (80 to 65 km) than elves. Their brightness is similar to an average sprite (500-1000 kR) and is red in color. Spites often initiate from the underside of the descending halo feature. Unlike sprites, halos have been observed in association with -CGs from the ground. The 1999 balloon campaign (Bering et al., 2004) suggested that numerous halos associated with -CGs were detected by the balloon's optical sensors. While many +CG halos followed by sprites have been noted, no -CG halo is known to have been followed by a sprite. Like elves, halos tend to be centered more or less directly over the parent CG, whereas the sprite centroid can often be offset by up to 50 km (Wescott et al., 2001; Lyons, 1996b).

Blue Jets Perhaps the most distinguishing feature of the blue jet is its rarity. Emerging at speeds of 200 km s^{-1} from thunderstorm anvils, they gradually decelerate as they extend to heights of 30-40 km. Their grainy appearance is distinctive, and their color is almost pure blue (Wescott et al., 2001). The inherent brightness is close to 1000 kR. Thus, they can be seen with the dark-adapted human eye under ideal conditions. First confirmed from the U of A aircraft in 1994 when a hail-producing supercell over Arkansas produced several dozen within an hour, they have been documented only rarely since. In nine years of ground LLTV monitoring at YRFS, none have been recorded. The reasons are several. First, many standard LLTV spectral responses are red biased. But more importantly, most TLEs are usually monitored above storms several hundred kilometers distant, with transmission of the weak blue signal severely attenuated. Evidence to date suggests the blue jet, and their shorter "blue starter" cousins (Wescott et al., 1996), do not appear associated with specific CG or IC discharges, though this later assertion is more difficult to confirm. Blue jets may more likely be associated with intense supercellular storms in which significant amounts of electrical charge are penetrating into the stratosphere by overshooting convective domes. This assertion, too, remains a conjecture to be proven. Space-borne optical sensors may prove more suitable for blue jet detection.

Upward Lightning from Cloud Tops   Less is known about this class of TLE than any other. A detailed synopsis of known reports appears in Lyons et al. (2003b). The mere handful of observations does not yet permit their association with specific lightning characteristics. Ranging in vertical extent from upward a few hundred meters to tens of kilometers, it is unknown whether there are a host of distinct classes of upward discharges or merely a broad range of appearances resulting from a basic underlying mechanism. Perhaps the most intriguing is the true "upward lightning" which may be a feature of especially intense deep convection. Visible in daylight, the lightning-like column extends upwards from the cloud top, does not flicker, is yellow to white in color, exhibits little tortuosity, and can persist for 1 to 2 seconds before the entire column fades. Some events may reach above 30 km. Not yet captured on video, only a handful of still photos and eyewitness accounts exist. Their relationship to giant jets (Pasko et al., 2002; Su et al., 2003) remains unknown at this time.

A variety of theoretical papers have detailed TLE mechanisms in great detail. Readers are referred to the following representative papers for in depth discussions of TLE theory (Wilson, 1925, 1956; Taraneko et al., 1993; Rowland, 1998; Rodger, 1999; Huang et al., 1999; Stanley, 2000; Williams, 2001; Pasko et al., 1995, 1996, 1997, 1998, 2000, 2001, 2002; Roussel-Dupré and Gurevich, 1996; Cho and Rycroft, 2001; Bering et al., 2002).

2.4.2   Convective Storm Types and TLEs

TLEs are global phenomena. They have been recorded over all oceans and every continent, save Antarctica. With the gradual assembly of a TLE census using the ROCSAT II satellite, we suspect that the TLE density will bear some resemblance to maps of global lightning (Price, Chapter 4), but there will not be a one to one correspondence. This results from the realization that only few CGs possess the unique characteristics required to initiate sprites (high ΔMq), elves (highly impulse with large ΔMq) and halos (similar to sprites?). Especially in regard to sprites, it is clear that only certain classes of convective storms, and then only during certain parts of their life cycle, generate CGs with the requisite large ΔMq. Much of our understanding of the meteorology of TLE-producing lightning has been gained during field programs in the central U.S., though more recent programs in Europe, East Asia, the Middle East, Japan and Australia have greatly expanded the geographic domain of our understanding. We will summarize our best estimates as to which meteorological regimes will, and will not be, prolific TLE producers, with an emphasis on sprites, the most common and best understood of the phenomena.

Stratiform Cloud Systems : All TLEs appear in some way related to tropospheric lightning. The vast majority of cloud systems, while possibly containing some space charge, do not meet the requirements for lightning. Convective clouds, with vertical motions of at least 5 to 10 ms^{-1} and a substantial depth of mixed phase precipitation (-10°C to -40°C) are required for most lightning. A notable exception is the stratiform precipitation region of MCSs and MCCs, where weaker vertical motions are found, along with unique *in situ* charge generating processes quite dissimilar to those in convective updrafts. Prognosis: No lightning, no TLEs.

Air Mass and Multicellular Thunderstorms: Perhaps the most common of all thunderstorm systems, these cumulonimbi are small ($\sim$10-100 km horizontal dimension) and short lived. While producing ample IC and CG lightning, initial studies suggest they rarely produce large ΔMq strokes. Prognosis: Sprites extremely rare; upward lightning events possible.

Supercells: These severe weather machines, often prolific producers of large peak current +CGs (especially the LP variety), do not typically generate sprites, except during the dying stages when significant stratiform rain areas develop, perhaps reaching a size of 5,000-10,000 km^2. On rare occasions, extremely powerful and impulsive +CGs occurring during the most intense phase of supercell growth have produced sprites due to the extremely large ΔMq they achieved. But these are exceptions that tend to prove the rule (Lyons and Cummer, 2004, 2005, pp. 6). Prognosis: Sprites are rare, except at the end of the storm; may be source of blue jets, giant jets and upward lightning discharges.

Squall Lines: Squall lines can often be relatively continuous zones of convection, or can be composed of discrete supercellular elements. When individual elements are more or less connected at middle and upper levels, it appears sprites and elves can and do occur, as was the case over the U.S. High Plains on 12 October 1997 (Lyons and Nelson, 1998). If contiguous cloudy areas in squall line do attain a size of $\sim$20,000 km^2, sprites may be fairly common. Prognosis: Sprites likely in varying numbers, possibly blue jets, giant jets and upward lightning.

Midlatitude MCSs and MCCs: Multicellular, long-lived clusters of deep convection generally produce large areas of stratiform precipitation which generate TLEs upon reaching 20,000 km^2 area. The more intense the system, such as in the largest MCCs, the higher the sprite and elve rates. Usually reaching their maturity during the night, large MCCs may be the world's most prolific sprite producers. Prognosis: A nocturnal MCC is the best producer of sprites and elves, but not of blue jet or cloud top discharges.

Tropical MCSs and MCCs: Based upon Space Shuttle observations and other limited sampling, it appears that tropical storm clusters should also be notable sprite producers, though whether the frequency would match mid-latitude systems remains unknown. Some evidence suggests that -CGs over salt water may qualitatively differ from their overland counterparts, producing more frequent elves. There have been suggestions that giant jets may be more common above intense maritime convective cores. This remains to be verified. Prognosis: Sprites; possibly giant jets and elves, especially over salt water.

Tropical Cyclones: Mature hurricanes and typhoons do not produce much lightning, and therefore are not likely to be major sources of TLEs. There are several key exceptions, however (Lyons and Keen, 1994). The outer spiral bands of many storms are electrically active. Also, on occasion, explosive supercells will develop within the eye wall circulation. One such event during Hurricane Georges produced a series of blue jets observed by an over-flying ER-2 pilot. Prognosis: Few TLEs, except for the special cases mentioned.

Winter Monsoon Clouds: Cold continental air flowing over adjacent warm waters often results in shallow but intense convection, often with +CGs of extraordinary large peak currents and apparently large ΔMq as well. Sprites and elves have been observed in winter over the Sea of Japan (Fukunishi et al., 1999), have been predicted over the Gulf Stream (Price et al., 2002a) and are likely to occur over the Great Lakes and other open water bodies experiencing extreme cold advection episodes. Embedded convection within intense mid-latitude extratropical cyclones and oceanic polar lows may also be a source of infrequent but large ΔMq flashes and TLEs. Prognosis: While far less frequent than during summer convection, TLEs, especially sprites and elves, can occur during "winter weather" regimes.

2.5 Research Frontiers

Why conduct research on transient luminous events? Little practical justification is required for curiosity-driven research. When John Winckler viewed his very first sprite images, there were no immediately evident "uses" for such knowledge. The initial interest of the scientific community was purely scientific, not pragmatic. Yet, as is so often the case, there are indeed significant implications for this new discipline that was initiated in the night sky over Minnesota on 6 July 1989.

2.5.1 Importance

As discussed, an early driving force for TLE research was the concern for the Space Shuttle's safety during the launch and recovery phase (Lyons, 1996a). A preliminary estimate was that there was on the order of a 1 in 100 chance that a Shuttle recovery trajectory over the central U.S. on a summer night could encounter a sprite. The apparent comparatively low energy density suggested that the potential hazard might be minimal. Yet the ill-informed public speculation that STS-107 was knocked out of the sky by a "sprite" (NASA, 2003, pp. 38) suggests many are not comfortable with our level of understanding of the energetics of TLE phenomena. The recent discovery of giant jets and upward discharges further clouds the issue.

The potential impacts of TLEs on atmospheric chemistry, in particular NO_x production, has been investigated (Lyons and Armstrong, 1997). Initial results suggested that sprites might be locally important in mesospheric chemistry. The role of TLEs in generating terrestrial gamma-rays (Fishman et al., 1994; Inan et al., 1996), and the implications of high energy processes, remains largely unresolved.

TLEs may provide a source of undocumented "optical clutter" for sensors on space-borne military assets (Armstrong and Lyons, 2000). As scientific and defense platforms expand their domain into the stratosphere, it is imperative that the dynamic electrical nature of the region be considered. Sprites, jets and related TLEs are also a potential source of "optical clutter" for space-borne monitoring and missile detection systems. To the extent that their optical signatures are not well characterized, the potential remains for natural phenomena to be misidentified. The potential for intense electrical fields and direct interactions with blue jets, giant jets and upward lightning are an emerging concern for those designing and operating stratospheric station-keeping platforms (unpiloted aerial vehicles [UAVs], high altitude airships [HAAs]) (Lyons and Armstrong, 2004). TLE-related processes appear to be generating infrasound signatures at frequencies similar to those of small nuclear detonations (Bedard Jr. et al., 1999) and Chapter 15. This has potential implications for Comprehensive Test Ban Treaty monitoring activities.

If, as some suggest (Williams, 1992), increasing global temperatures will result in increased global lightning frequency, might not long-term monitoring TLEs be of potential value for global change studies?

And in the broadest sense, do we not need to understand TLEs simply to complete our understanding of the global electrical circuit (Rycroft et al., 2000)? While the optical effects of TLEs may terminate around 100 km, are there significant interactions of TLEs with the radiation belts? Simply stated, TLEs are just one more piece of the giant puzzle of how our world actually works. Today the long term implications of this new knowledge are unknown – but if the history of science is a guide, they may well be important.

2.5.2 Outstanding Research Questions

Every TLE researcher can prepare an expansive listing of unanswered questions. Some of those on the mind of this author would include:

- Are large ΔMq values a necessary AND sufficient cause of sprites?
- Is there a minimum ΔMq sprite threshold, and does it vary over space and time?
- What are the transient electric fields above clouds from SP+CGs?
- What are the meteorological environments which create large ΔMq CGs?
- Do sprites occur during the daytime (Stanley et al., 2000)?
- Why do some 10% of sprites appear to contain significant current flows (Cummer and Stanley, 1999)?
- Do sprite tendrils ever physically connect with cloud tops?
- Do theoretically predicted (Lehtinen et al., 2001) conjugate sprites occur?
- What criteria can be used to discriminate elve-producing CG lightning?
- What accounts for the notable lack of -CG sprites? Is it simply a matter of negative CGs so rarely having large ΔMq values? In U.S. storms examined to date this appears to be the case, yet global ELF monitoring suggests large negative ΔMq are fairly common, especially over the oceans (Füllekrug et al., 2002).
- As indicated by balloon observations, are negative halos are far more common than ground observations suggest (Bering et al., 2004)?
- Why do sprites never follow negative halo events?
- What is the global rate and distribution of TLEs?
- Do all sprites and elves generate ELF Q-bursts? Are all Q-bursts sprites or elves?
- Are giant jets related mostly to maritime storms as has been initially suggested?
- Is there a RF signature to discriminate blue jets and giant jets?

- Do runaway electron processes play any role in TLE mechanisms?
- What meteorological conditions favor blue jets, giant jets and upward lightning? Do these phenomena pose any threat to stratospheric aerospace vehicles?
- Do cosmic rays, gravity waves, meteors and meteoric dust influence TLE imitation and dynamics (Zabotin and Wright, 2001; Suszcynsky et al., 1999; Wescott, 2001)?
- Do TLEs occur in the atmosphere of other planets?

Acknowledgments

This work was primarily supported by the National Science Foundation under grant ATM-0221215. We wish to thank our many colleagues who have contributed to this research program over the past decade, including Russ Armstrong, Gar Bering, Bill Boeck, Dennis Boccippio, Steven Cummer, Kenneth Cummins (Vaisala, Inc.), Hiroshi Fukunishi, Martin Füllekrug, Matt Heavner, Gary Huffines, Umran Inan, Stephen Mende, Liv Nordem Lyons, Thomas Nelson, Victor Pasko, Colin Price, Steve Reising, Mitsu Sato, Dave Sentman, Mark Stanley, David Suszcynsky, Yukihiro Takahashi, Mike Taylor, O. H. Vaughan, Earle Williams, Gene Wescott and, especially, the late Prof. John R. Winckler, along with the numerous research faculty and students from around the world who participated in the many sprite Campaigns at the Yucca Ridge Field Station.

Bibliography

American Meteorological Society (2000). *The Glossary of Meteorology*. American Meteorological Society, Boston, second edition. 855 pp.

Armstrong, R. A. and Lyons, W. A. (2000). Satellite and ground-based data exploitation for NUDET discrimination, characterizing atmospheric electrodynamic emissions from lightning, sprites, jets and elves. Final Report, DOE Contract #DE-AC04-98AL79469, 213 pp.

Armstrong, R. A., Shorter, J. A., Taylor, M. J., Suszcynsky, D. M., Lyons, W. A., and Jeong, L. S. (1998). Photometric measurements in the SPRITES'95 & '96 campaigns of nitrogen second positive (399.8 nm) and first negative (427.8 nm) emissions. *J. Atmos. Sol.-Terr. Phys*, 60:787–800.

Barrington-Leigh, C. P., Inan, U. S., and Stanley, M. (2001). Identification of sprites and elves with intensified video and broadband array photometry. *J. Geophys. Res.*, 101:1741–1750.

Barrington-Leigh, C. P., Inan, U. S., Stanley, M., and Cummer, S. A. (1999). Sprites directly triggered by negative lightning discharges. *Geophys. Res. Lett.*, 26:3605–3608.

Bedard Jr., A. J., Lyons, W. A., Armstrong, R. A., Nelson, T. E., Hill, B., and Gallagher, S. (1999). A search for low-frequency atmospheric acoustic waves associated with sprites, blue jets, elves and storm electrical activity. *EOS Trans. AGU, Fall Meet. Suppl.*, 80(46). Abstract.

Berger, K., Anderson, R. B., and Kroninger, H. (1975). Parameters of lightning flashes. *Electra*, 80:223–2237.

Bering, E. A. (1997). The global circuit, global thermometer, weather-by-product, or climate modulator. *Rev. Geophys. Res. Suppl.*, pages 845–862.

Bering, E. A. III, Benbrook, J. R., Bhusal, L., Garrett, J. A., Paredes, A. M., Wescott, E. M., Moudry, D. R., Sentman, D. D., Stenbaek-Nielsen, H. C., and Lyons, W. A. (2004). Observations of transient luminous events (TLEs) associated with negative cloud-to-ground (-CG) lightning strikes. *Geophys. Res. Lett.*, 31:doi:10.1029/2003GL018659.

Bering, E. A., III, Benbrook, J. R., Garrett, J. A., Paredes, A. M., Wescott, E. M., Moudry, D. R., Sentman, D. D., and Stenbaek-Nielsen, H. C. (2002). The electrodynamics of sprites. *Geophys. Res. Lett.*, 29:doi:10.1029/2001GL013267.

Blanc, E., Farges, T., Roche, R., Brebion, D., Hua, T., Labarthe, A., and Melinkov, V. (2004). Nadir observations of sprites from the International Space Station. *J. Geophys, Res.*, 109:doi:10.1029/2003JA009972.

Boccippio, D. J., Williams, E. R., Lyons, W. A., Baker, I., and Boldi, R. (1995). Sprites, ELF transients and positive ground strokes. *Science*, 269:1088–1091.

Boeck, W. L., Jr., O. H. Vaughan, Blakeslee, R., Vonnegut, B., and Brook, M. (1992). Lightning induced brightening in the airglow layer. *Geophys. Res. Lett.*, 19:99–102.

Boeck, W. L., Jr., O. H. Vaughan, Blakeslee, R. J., Vonnegut, B., Brook, M., and McKune, J. (1995). Observations of lightning in the stratosphere. *J. Geophys. Res.*, 100:1465–1475.

Brooks, C. E. P. (1925). The distribution of thunderstorms over the globe. *Geophys. Mem., Air Ministry, Meteorology Office, London*, 24:147–164.

Byers, H. R. and Braham, R. R. (1949). The thunderstorm. *U.S. Weather Bureau, Washington, D.C.*, page 287.

Cho, M. and Rycroft, M. J. (2001). Non-uniform ionization of the upper atmosphere due to the electromagnetic pulse from a horizontal lightning discharge. *J. Atmos. Sol.-Terr. Phys*, 63:559–580.

Christian, H. J., Blakeslee, R. J., Boccippio, D. J., Boeck, W. L., Buechler, D. E., Driscoll, K. T., Goodman, S. J., Hall, J. M., Koshak, W. J., Mach, D. M., and Stewart, M. F. (2003). Global frequency and distribution of lightning as observed from space by the Optical Transient Detector. *J. Geophys. Res.*, 108(D1):doi:10.1029/2002LD002347.

Corliss, W. R. (1977). *Handbook of Unusual Natural Phenomena*. Glen Arm, MD.

Cotton, W. R. and Anthes, R. A. (1989). *Storm and Cloud Dynamics*. Academic Press, New York. 883 pp.

Cummer, S. A. and Inan, U. S. (2000). Modeling ELF radio atmospheric propagation and extracting lightning currents from ELF observations. *Radio Science*, 35:385–394.

Cummer, S. A. and Lyons, W. A. (2004). Lightning charge moment changes in U.S. high plains thunderstorms. *Geophys. Res. Lett.*, 31:doi:10.1029/2003GL019043.

Cummer, S. A. and Lyons, W. A. (2005). Implications of impulse charge moment changes in sprite-producing and non-sprite-producing lightning. *J. Geophys. Res.*, 110(A40304):doi:10.1029/2004JA010812.

Cummer, S. A. and Stanley, M. (1999). Submillisecond resolution lightning currents and sprite development, observations and implications. *Geophys. Res. Lett.*, 26:3205–3208.

Cummins, K. L., Murphy, M. J., Bardo, E. A., Hiscox, W. L., Pyle, R. B., and Pifer, A. E. (1998). A combined TOA/MDF technology upgrade of the U.S. National Lightning Detection Network. *J. Geophys. Res.*, 103:9035–9044.

Davidson, J. E. (1893). Thunderstorms and the auroral phenomena. *Nature*, 47:582.

Davis, C., Atkins, N., Bartels, D., Bosart, L., Coniglio, M., Bryan, G., Cotton, W., Dowell, D., Jewett, B., Johns, R., Jorgensen, D., Knievel, J., Knupp, K., Lee, W. C., McFarquhar, G., Moore, J., Przybylinski, R., Rauber, R., Smull, B., Trapp, R., Trier, S., Wakimoto, R., Weisman, M., and Ziegler, C. (2004). The bow echo and MCV experiment (BAMS), observations and opportunities. *Bull. Am. Met. Soc.*, 85:1075–1093.

Demetriades, N., Murphy, M. M., and Holle, R. L. (2003). The importance of total lightning in the future of weather nowcasting. In *Proceedings, Intl. Lightning Detection Conference*, Helsinki.

Dowden, R. L., Brundell, J. B., and Rodger, C. J. (2002). VLF lightning location by time of group arrival (TOGA) at multiple sites. *J. Atmos. Sol.-Terr. Phys*, 64:817–830.

Fishman, G. J., Bhat, P. N., Mallozzi, R., Horak, J. M., Koshut, T., Kouveliotou, C., Pendleton, G. N., Meegan, C. A., Wilson, R. B., Paciesas, W. S., Goodman, S. J., and Christian, H. J. (1994). Discovery of intense gamma-ray flashes of atmospheric origin. *Science*, 264:1313.

Franz, R. C., Nemzek, R. J., and Winckler, J. R. (1990). Television image of a large upward electrical discharge above a thunderstorm system. *Science*, 249:48–51.

Fujita, T. T. (1992). The mystery of severe storms. wind research laboratory. Technical report, University of Chicago.

Fukunishi, H., Takahashi, Y., Kubota, M., Sakanoi, K., Inan, U. S., and Lyons, W. A. (1996). Elves, lightning-induced transient luminous events in the lower ionosphere. *Geophys. Res. Lett.*, 23:2157–2160.

Fukunishi, H., Takahashi, Y., Uchide, A., Sera, M., and Miyasato, K. R. (1999). Occurrences of sprites and elves above the sea of Japan near Hokuriko in winter. *EOS Supplement*, 80(46):F217. Abstract.

Füllekrug, M. and Constable, S. (2000). Global triangulation of intense lightning discharges. *Geophys. Res. Lett.*, 27:333–336.

Füllekrug, M., Price, C., Yair, Y., and Williams, E. R. (2002). Intense oceanic lightning. *Ann. Geophys.*, 20:133–137.

Gerken, E. A. and Inan, U. S. (2003). Observations of decameter-scale morphologies in sprites. *J. Atmos. Sol.-Terr. Phys.*, 65:567–572.

Gerken, E. A. and Inan, U. S. (2004). Comparison of photometric measurements and charge moment changes in two sprite-producing storms. *Geophys. Res. Lett.*, 31:doi:10.1029/2003GL0118751.

Gerken, E. A., Inan, U. S., and Barrington-Leigh, C. P. (2000). Telescopic imaging of sprites. *Geophys. Res. Lett.*, 27:2637–2640.

Golde, R. H. (1977). *Lightning*, volume 1, Physics of lightning. Academic Press, London. 496 pp.

Hampton, D. L., Heavner, M. J., Wescott, E. M., and Sentman, D. D. (1996). Optical spectra characteristics of sprites. *Geophys. Res. Lett.*, 23:89–92.

Hardman, S. F., Dowden, R. L., Brundell, J. B., Bahr, J. L., Kawasaki, Z., and Rodger, C. J. (2000). Sprite observations in the Northern Territory of Australia. *J. Geophys. Res.*, 105:4689–4697.

Heavner, M. J. (2000). *Optical spectroscopic observations of sprites, blue jets, and elve: Inferred microphysical processes and their macrophysics implications*. Ph.D. dissertation, University of Alaska, Fairbanks.

Holzworth, R. H., McCarthy, M. P., Thomas, J. N., Chin, J., Chinowsky, T. M., Taylor, M. J., and Jr., O. Pinto (2005). Strong electric fields from positive lightning strokes in the stratosphere. *Geophys. Res. Lett.*, 32(doi:10.1029/2004GL021554).

Hsu, R. R., Su, H. T., Chen, A. B., Lee, L. C., Asfur, M., Price, C., and Yair, Y. (2003). Transient luminous events in the vicinity of Taiwan. *J. Atmos. Sol.-Terr. Phys.*, 65:561–566.

Hu, W., Cummer, S., Lyons, W. A., and Nelson, T. E. (2002). Lightning charge moment changes for the initiation of sprites. *Geophys. Res. Lett.*, 29:doi:10.1029/2001GL014593.

Huang, E., Williams, E., Boldi, R., Heckman, S., Lyons, W., Taylor, M., Nelson, T., and Wong, C. (1999). Criteria for sprites and elves based on Schumann resonance observations. *J. Geophys. Res.*, 104:16943–16964.

Inan, U. S., Barrington-Leigh, C., Hansen, S., Glukhov, V. S., Bell, T., and Rairden, R. (1997). Rapid lateral expansion of optical luminosity in lightning-induced ionospheric flashes referred to as 'elves'. *Geophys. Res. Lett.*, 24(5):583–586.

Inan, U. S., Reising, S. C., Fishman, G. J., and Horack, J. M. (1996). On the association of terrestrial gamma-ray bursts with lightning and implications for sprites. *Geophys. Res. Lett.*, 23:1017–1020.

Isrealevich, P. L., Yair, Y., Devir, A., Joseph, J., Levin, Z., Mayo, I., Moalem, M., Price, C., Ziv, B., and Sternlieb, A. (2004). Transient airglow enhancements observed from the space shuttle Columbia during the MEIDEX sprite campaign. *Geophys. Res. Lett.*, 31:doi:10.1029/2003GL019110,2004.

Krehbiel, P. R., Thomas, R. J., Rison, W., Hamlin, T., Harlin, J., and Davis, M. (2000). GPS-based mapping system reveals lightning inside storms. *EOS, Trans. Amer. Geophys. Union*, 81:21–25.

Krider, E. P., Noggle, R. C., Pifer, A. E., and Vance, D. L. (1980). Lightning direction-finding systems for forest fire detection. *Bull. Am. Met. Soc.*, 61:980–986.

Laing, A. G. and Fritsch, J. M. (1997). The global population of mesoscale convective complexes. *Quart. J. Roy. Met. Soc.*, 123:389–405.

Lang, T, Miller, L. J., Weisman, M., Rutledge, S. A., III, L. J. Barker, Bringi, V. N., Chandrasekar, V., Detwiler, A., Doesken, N., Helsdon, J., Knight, C., Krehbiel, P., Lyons, W. A., MacGorman, D., Rasmussen, E., Rison, W., Rust, W. D., and Thomas, R. J. (2004). The Severe Thunderstorm Electrification and Precipitation Study (STEPS). *Bull. Am. Met. Soc.*, 85:1107–1125.

Lehtinen, N. G., Inan, U. S., and Bell, T. F. (2001). Effects of thunderstorm-driven runaway electrons in the conjugate hemisphere, purple sprites, ionization enhancements and gamma rays. *J. Geophys. Res.*, 106:28841–28856.

Lyons, W. A. (1994a). Characteristics of luminous structures in the stratosphere above thunderstorms as imaged by low-light video. *Geophys. Res. Lett.*, 21:875–878.

Lyons, W. A. (1994b). Low-light video observations of frequent luminous structures in the stratosphere above thunderstorms. *Mon. Wea. Rev.*, 122:1940–1946.

Lyons, W. A. (1996a). Sensor system to monitor cloud-to-stratosphere electrical discharges. Final report, NASA contract nas10-12113, Kennedy Space Center.

Lyons, W. A. (1996b). Sprite observations above the U.S. High Plains in relation to their parent thunderstorm systems. *J. Geophys. Res.*, 101:29641–29652.

Lyons, W. A. and Armstrong, R. A. (1997). NO_x production within and above thunderstorms: The contribution of lightning and sprites. In *3rd Conf. on Atmospheric Chemistry*, Boston. American Meteorological Society. Preprint.

Lyons, W. A. and Armstrong, R. A. (2004). A review of electrical and turbulence effects of convective storms on the overlying stratosphere and mesosphere. In *AMS Symposium on Space Weather - Annual Meeting*, Boston. American Meteorological Society.

Lyons, W. A., Armstrong, R. A., III, E. A. Bering, and Williams, E. R. (2000). The hundred year hunt for the sprite. *EOS*, 81:373–377.

Lyons, W. A., Bauer, K. G., Eustis, A. C., Moon, D. A., Petit, N. J., and Schuh, J. A. (1989). R·SCAN's National Lightning Detection Network. In *Fifth Intl. Conf. on Interactive Information and Processing Systems for Meteor., Ocean. and Hydrology*, Boston. American Meteorological Society. The first year progress report - Preprint.

Lyons, W. A. and Cummer, S. A. (2004). Lightning, sprites and supercells. In *Proc. 22nd Conf. Severe Local Storms*, Boston. American Meteorological Society.

Lyons, W. A. and Cummer, S. A. (2005). Lightning characteristics of the aurora, NE record hail stone producing supercell of 22-23 June 2003 BAMEX. In *Conf. on the Applications of Lightning Data (AMS)*. Preprint.

Lyons, W. A. and Keen, C. S. (1994). Observations of lightning in convective supercells within tropical storms and hurricanes. *Mon. Wea. Rev.*, 122:1897–1916.

Lyons, W. A., Nelson, T., Williams, E. R., Cummer, S. A., and Stanley, M. A. (2003a). Characteristics of sprite-producing positive cloud-to-ground lightning during the 19 July STEPS mesoscale convective systems. *Mon. Wea. Rev.*, 131:2417–2427.

Lyons, W. A. and Nelson, T. E. (1998). Electrical activity on a late season high plains tornadic squall line associated with sprites and elves. In *19th Conf. on Severe Local Storms (AMS)*, Boston. American Meteorological Society. Preprint.

Lyons, W. A., Nelson, T. E., Armstrong, R. A., Pasko, V. P., and Stanley, M. (2003b). Upward electrical discharges from the tops of thunderstorms. *Bull. Am. Met. Soc.*, 84:445–454.

Lyons, W. A., Uliasz, M., and Nelson, T. E. (1998). Climatology of large peak current cloud-to-ground lightning flashes in the contiguous United States. *Mon. Wea. Rev.*, 126:2217–2233.

Lyons, W. A. and Williams, E. R. (1993). Preliminary investigations of the phenomenology of cloud-to-stratosphere lightning discharges. In *Conference on Atmospheric Electricity*, pages 725–732, Boston. American Meteorological Society. Preprint.

MacGorman, D. R. and Rust, W. D. (1998). *The Electrical Nature of Storms*. Oxford University Press.

MacKenzie, T. and Toynbee, H. (1886). Meteorological phenomena. *Nature*, 33:26.

Maddox, R. A. (1980). Mesoscale convective complexes. *Bull. Am. Met. Soc.*, 61:1374–1387.

Malan, D. (1937). Sur les decharges orageuses dans la haut atmosphere. In *Academie des Sciences*. Third session.

Marshall, R. A. and Inan, U. S. (2005). High-speed telescopic imaging of sprites. *Geophys. Res. Lett.*, (in press).

Mazur, V., Shao, X.-M., and Krehbiel, P. R. (1998). "Spider" lightning in intracloud and positive cloud-to-ground flashes. *J. Geophys. Res.*, 103:19811–19822.

Mende, S. N., Rairden, R. L., Swenson, G. R., and Swenson, W. A. (1995). Sprite spectra, N_2 first positive band identification. *Geophys. Res. Lett.*, 22:2633–2636.

Miyasato, R., Fukunishi, H., Fukunishi, Y., and Taylor, M. J. (2003). Energy estimation of electrons producing sprite halos using array photometer data. *J. Atmos. Sol.-Terr. Phys.*, 65:573–581.

Moudry, D., Stenbaek-Nielsen, H., Sentman, D., and Wescott, E. (2003). Imaging of elves, halos and sprite initiation at 1 ms time resolution. *J. Atmos. Sol.-Terr. Phys*, 65:509–518.

NASA (2003). Potential for space/atmospheric environmental effects in the Columbia Shuttle Orbiter Disaster. Report of the Space/Atmospheric Environment Scientist Panel to Space Shuttle Vehicle Engineering Office, NASA/JSC.

National Research Council (1986). *The Earth's Electrical Environment.* Studies in Geophysics. National Academy Press, Washington, DC. 263 pp.

Neubert, T., Allin, T. H., Stenbaek-Nielsen, H., and Blanc, E. (2001). Sprites over Europe. *Geophys. Res. Lett.*, 28:3585–3588.

Orville, R. A. and Henderson, R. W. (1986). Global distribution of midnight lightning, Sept. 1977 to August 1978. *Mon. Wea. Rev.*, 114:2640–2653.

Orville, R. E., Huffines, G. R., Burrows, W. R., Holle, R. L., and Cummins, K. L. (2002). The North American Lightning Detection Network (NALDN) – first results, 1998-2000. *Mon. Wea. Rev.*, 130:2098–2109.

Pasko, V. P., Inan, U. S., and Bell, T. F. (1996). Sprites as luminous columns of ionization produced by quasi-electrostatic thunderstorm fields. *Geophys. Res. Lett.*, 23:649–652.

Pasko, V. P., Inan, U. S., and Bell, T. F. (2000). Fractal structure of sprites. *Geophys. Res. Lett.*, 27:497–500.

Pasko, V. P., Inan, U. S., and Bell, T. F. (2001). Mesosphere-troposphere coupling due to sprites. *Geophys. Res. Lett.*, 28:3821–3824.

Pasko, V. P., Inan, U. S., and Taranenko, Y. N. (1997). Sprites produced by quasi-electrostatic heating and ionization in the lower ionosphere. *Journal of Geophysical Research*, 102(3):p. 4529.

Pasko, V. P., Inan, U. S., Taranenko, Y. N., and Bell, T. F. (1995). Heating, ionization and upward discharges in the mesosphere due to intense quasi-static thundercloud fields. *Geophys. Res. Lett.*, 22:365–368.

Pasko, V. P., Stanley, M. A., Mathews, J. D., Inan, U. S., and Woods, T. G. (2002). Electrical discharge from a thunderstorm top to the lower ionosphere. *Nature*, 416:152–154.

Pasko, V.P., Inan, U.S., Bell, T.F., and Reising, S.C. (1998). Mechanism of ELF radiation from sprites. *Journal of Geophysical Research*, 25(18):3493.

Pielke Sr., R. A. (2002). *Mesoscale Meteorological Modeling*. Academic Press, Orlando.

Price, C., Burrows, W., and King, P. (2002a). The likelihood of winter sprites over the Gulf Stream. *Geophys. Res. Lett.*, 29:doi:10.1029/2002GL015571.

Price, C., Greenberg, E., Yair, Y., Satori, G., Bor, J., Fukunishi, H., Sato, M., Israelevich, P., Moalem, M., Devir, A., Levin, Z., Joseph, J. H., Mayo, I., Ziv, B., and Sternlieb, A. (2004). Ground-based detection of TLE-producing intense lightning during the MEIDEX mission on board the space shuttle Columbia. *Geophys. Res. Lett.*, 31(doi:10.1029/2004GL020711).

Price, C. P., Asfur, M., Lyons, W., and Nelson, T. (2002b). An improved ELF/VLF method for globally geolocating sprite-producing lightning. *Geophys. Res. Lett.*, 29:doi:10.1029/2001GL013519.

Rakov, V. A. and Uman, M. A. (2003). *Lightning, Physics and Effects*. Cambridge University Press. 687 pp.

Ray, P. J., editor (1986). *Mesoscale Meteorology and Forecasting*. American Meteorological Society. 793 pp.

Rodger, C. J. (1999). Red sprites, upward lightning, and VLF perturbations. *Rev. of Geophys.*, 37:317–336.

Roussel-Dupré, R. and Gurevich, A. V. (1996). On runaway breakdown and upward propagating discharges. *J. Geophys. Res.*, 101:2297–2310.

Rowland, H. L. (1998). Theories and simulations of elves, sprites and blue jets. *J. Atmos. Sol.-Terr. Phys.*, 60:831–844.

Rycroft, M. J., Israelsson, S., and Price, C. (2000). The global electrical circuit, solar activity and climate change. *J. Atmos. Sol.-Terr. Phys.*, 62:1563–1576.

Salanave, L. E. (1980). *Lightning and its spectrum*. University of Arizona Press, Tucson. 136 pp.

Sato, M., Fukunishi, H., Kikuchi, M., Yamagishi, H., and Lyons, W. A. (2003). Validation of sprite location based on ELF observations at Syowa station in Antarctica. *J. Atmos. Sol.-Terr. Phys.*, 65:609–616.

Sentman, D. D. and Wescott, E. M. (1993). Observations of upper atmospheric optical flashes recorded from an aircraft. *Geophys. Res. Lett.*, 20(24):2857–2860.

Sentman, D. D., Wescott, E. M., Osborne, D. L., Hampton, D. L., and Heavner, M. J. (1995). Preliminary results from the sprites 94 aircraft campaign 1. Red sprites. *Geophys. Res. Lett.*, 22:1205–1208.

Smith, D. A., Eack, K. B., Harlin, J., Heavner, M., Jacobson, A., Massey, R., Shao, X. M., and Wiens, K. C. (2002). The Los Alamos Sferic Array: A research tool for lightning investigations. *J. Geophys. Res.*, 107:doi:10/1029/2001JD000502.

Stanley, M., Brook, M., and Krehbiel, P. (2000). Detection of daytime sprites via a unique sprite VLF signature. *Geophys. Res. Lett.*, 27:871–874.

Stanley, M., Krehbiel, P., Brook, M., Moore, C., and Rison, W. (1999). High speed video of initial sprite development. *Geophys. Res. Lett.*, 26:3201–3204.

Stanley, M. A. (2000). *Sprites and their parent discharges*. Ph.d. dissertation, New Mexico Institute of Mining and Technology. 164 pp.

Stenbaek-Nielsen, H. C., Moudry, D. R., Wescott, E. M., Sentman, D. D., and Sabbas, F. T. Sao (2000). Sprites and possible mesospheric effects. *Geophys. Res. Lett.*, 27:3829–3832.

Su, H. T., Hsu, R. R., Chen, A. B., Wang, Y. C., Hsiao, W. S., Lai, W. C., Lee, L. C., Sato, M., and Fukunishi, H. (2003). Gigantic jets between a thundercloud and the ionosphere. *Nature*, 423:974–976.

Su, H.-T., Hsu, R.-R., Chen, A. B.-C., Lee, Y.-J., and Lee, L.-C. (2002). Observation of sprites over the Asian continent and over oceans around Taiwan. *Geophys. Res. Lett.*, 29(7):doi:10.1029/2001GL013737.

Suszcynsky, D. M., Strabley, R., Roussel-Dupré, R., Symbalisty, E. M., Armstrong, R. A., Lyons, W. A., and Taylor, M. (1999). Video and photometric observations of a sprite in coincidence with a meteor-triggered jet event. *J. Geophys. Res.*, 104:31361–31367.

Taraneko, Y. K., Inan, U. S., and Bell, T. F. (1993). The interaction with the lower ionosphere of electromagnetic pulses from lightning, heating, attachment and ionization. *Geophys. Res. Lett.*, 20:1439–1542.

Turman, B. B (1977). Detection of lightning superbolts. *J. Geophys. Res.*, 82:2566.

Uman, M. A (1987). *The lightning discharge*. Number 39 in International Geophysics Series. Academic Press, Orlando.

Uman, M. A. and Krider, E. P. (1989). Natural and artificially initiated lightning. *Science*, 246:457–464.

Uman, M. A. and Rakov, V. A. (2003). The interaction of lightning with airborne vehicles. *Progress in Aerospace Sciences*, 39:61–81.

Vaughan Jr., O. H., Blakeslee, R., Boeck, W. L., Vonnegut, B., Brook, M., and Jr., J. McKune (1992). A cloud-to-space lightning as recorded by the Space Shuttle payload bay TV cameras. *Mon. Wea. Rev.*, 120:1459–1461.

Vaughan Jr., O. H. and Vonnegut, B. (1989). Recent observations of lightning discharges from the top of a thundercloud into the air above. *J. Geophys. Res.*, 94:13179–13182.

Volland, H. (1982). *Handbook of Atmospherics*, volume 1. CRC Press, Boca Raton, FL. 377 pp.

Vonnegut, B. (1980). Cloud-to-stratosphere lightning. *Weather*, 35:59–60.

Wescott, E. M. (2001). Triangulation of sprites, associated halos and their possible relation to causative lightning and micrometeors. *J. Geophys. Res.*, 106:10467–10477.

Wescott, E. M., Sentman, D. D., Heavner, M. J., Hampton, D. L., Lyons, W. A., and Nelson, T. (1998). Observations of 'columniform' sprites. *J. Atmos. Sol.-Terr. Phys.*, 60:733–740.

Wescott, E. M., Sentman, D. D., Heavner, M. J., Hampton, D. L., Osborne, D. L., and Jr., O. H. Vaughan (1996). Blue starters, brief upward discharges from an intense Arkansas thunderstorm. *Geophys. Res. Lett.*, 23:2153–2156.

Wescott, E. M., Sentman, D. D., Osborne, D., Hampton, D., and Heavner, M. (1995). Preliminary results from the Sprites 94 aircraft campaign: Blue jets. *Geophys. Res. Lett.*, 22:1209–1212.

Wescott, E. M., Sentman, D. D., Stenbaek-Nielsen, H. C., Huet, P., Heavner, M. J., and Moudry, D. R. (2001). New evidence for the brightness and ionization of blue starters and blue jets. *J. Geophys. Res.*, 106:21549–21554.

Williams, E. R. (1988). The electrification of thunderstorms. *Sci. Am.*, 259:88–99.

Williams, E. R. (1992). The Schumann resonance: A global tropical thermometer. *Science*, 256:1184–1186.

Williams, E. R. (1998). The positive charge reservoir for sprite-producing lightning. *J. Atmos. Sol.-Terr. Phys.*, 60:689–692.

Williams, E. R. (2001). Sprites, elves, and glow discharge tubes. *Phys. Today*, 54:41–47.

Wilson, C. T. R. (1925). The electric field of a thunderstorm and some of its effects. *Proc. Phys. Soc. London*, 37:32D–37D.

Wilson, C. T. R. (1956). A theory of thundercloud electricity. *Proc. Roy. Soc. Lond., Series A*, 236:297–317.

Yair, Y., Isrealevich, P., Devir, A. D., Moalem, M., Price, C., Joseph, J. H., Levin, Z., Ziv, B., Sternlieb, A., and Teller, A. (2004). New observations of sprites from the space shuttle. *J. Geophys. Res.*, 109:doi:10.1029/2003JD004497.

Zabotin, N. A. and Wright, J. W. (2001). Role of meteoric dust in sprite formation. *Geophys. Res. Lett.*, 28:2593–2596.

THE MICROPHYSICAL AND ELECTRICAL PROPERTIES OF SPRITE-PRODUCING THUNDERSTORMS

Earle Williams [1] and Y. Yair [2]

[1] *Parsons Laboratory, Massachusetts Institute of Technology, Cambridge, MA 02139, USA.*

[2] *Department of Life and Natural Sciences, The Open University of Israel, Ra'anana, Israel.*

Abstract Sprites and elves are caused by exceptional lightning flashes in storms distinguishable in both size and structure from ordinary thunderstorms. An ice-based mechanism for charge separation known as the non-inductive process is capable of explaining macroscopic features of both ordinary and sprite-producing thunderstorms. Accordingly, the microphysical basis for this mechanism is reviewed. A key distinction for the sprite-producing storm in summer is the development of a large and laterally-extended reservoir of positive charge in the lower portion of the cloud, favorable to the occurrence of positive ground flashes with exceptional charge moments (>500 C·km). The electrical structure of sprite-producing storms in summertime is presently better resolved than wintertime scenarios.

3.1   Introduction

This chapter is concerned with the microphysical and electrical properties of sprite-producing thunderstorms. The vast observational record relates the appearance of sprites to an exceptional magnitude of vertical charge moment in the causal lightning flash, and ultimately to the magnitude of the charge transferred to the ground. It has already been shown that the overwhelming majority of these ground flashes are of a positive polarity. Thus, we shall focus the discussion on the electrical processes operating within clouds, leading to charge separation and the creation of charge centers that will eventually enable the formation of this type of flash. The electrical configuration of sprite-producing clouds departs from the usual positive-dipole structure of ordinary thunderclouds and this aspect is given considerable emphasis.

3.2 The Non-Inductive Charging Process in Thunderclouds

Thunderstorm electrification is an intensely-studied field with thousands of published works encompassing observational, laboratory, theoretical and modeling studies. It is beyond the scope of this chapter to cover the detailed aspects of the synergistic processes occurring within thunderstorms, and we shall therefore give a succinct overview of only the mechanism that is thought to be the dominant one for charge separation. A number of comprehensive reviews document the various mechanisms that were invoked to explain the electrical structure observed in thunderstorms, and we refer the interested reader to the extensive, updated and detailed summaries published as chapters in the books by MacGorman and Rust (1998), Rakov and Uman (2003) and Jayarante (2003).

While there are many electrical processes and particle interactions that take place within the dynamic and turbulent environment of a thundercloud, it is widely recognized that the most potent one (vis-a-vis charge separation) is the non-inductive interaction between graupel particles and ice crystals. In its most basic description, this process calls for collisions between particles and the selective transfer of a distinct polarity to the larger particle. In ordinary thunderclouds, the smaller ice crystals are charged positively and then carried aloft, while the larger graupel particles charge negatively and descend relative to the smaller particles. Measurements of the charge transfer during collisions of vapor-grown ice crystals and a riming (graupel) target have found that the sign of the charge transferred depends on the ambient temperature and the local liquid water content (LWC). It is convenient to represent this dependence in the LWC-T phase space (Figure 1). Clearly, the sign on the rimer target (representing the graupel particle) is positive for high temperatures for both low and high cloud water contents, while negative charging occurs at lower temperature and intermediate water contents. The results of the University of Manchester (UMIST) group (Jayarante et al., 1983; Saunders et al., 1991; Brooks and Saunders, 1994), who also ran laboratory simulation experiments, introduced a factor termed the effective liquid water content (EW). The laboratory set-up of these researchers was different from Takahashi (1978), and consequently important differences are evident in the sign of the charge on the graupel particle, most notably a negative charging regime at high temperatures and low effective water contents (denoted **a** in Figure 2) and positive at high EW. An attempt to account for differences between Takahashi (1978) and the UMIST results was performed recently by the University of Cordoba in Argentina. Pereyra et al. (2000) studied the charge transfer between ice crystals rebounding from a riming target, where the ice crystals were mixed into a stream of supercooled droplets (as in Takahashi, 1978), and not grown

MICROPHYSICS OF SPRITE STORMS 59

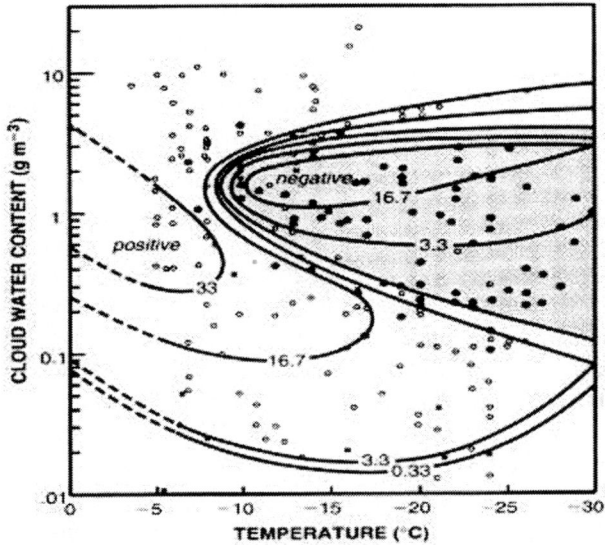

Figure 1. The polarity and magnitude of charges acquired by a riming graupel target colliding with ice particles, presented in the cloud Liquid Water Content (LWC) -Temperature phase space. Based on the experiments of Takahashi (1978), used with permission.

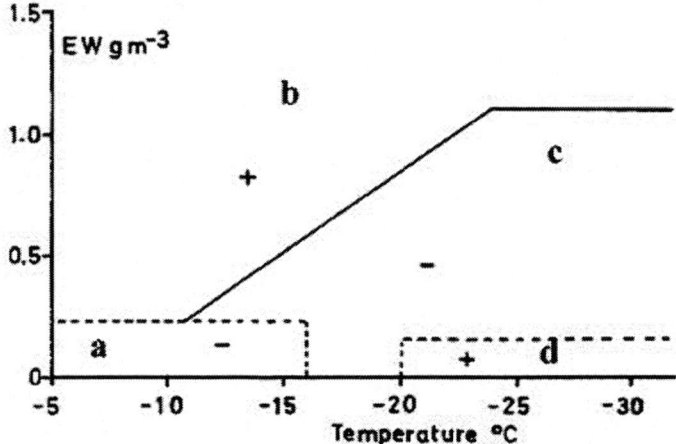

Figure 2. Charging regimes in the Effective Water Content (EW)-Temperature phase space, based on the results of laboratory experiments conducted by the UMIST group (Saunders et al., 1991, with permission).

in the same cloud as the droplets that were used to rime the graupel target (as in the UMIST experiments). The Argentina results showed better agreement with Takahashi (1978) than with the earlier UMIST results (Saunders et al., 1991). The rimer was found to charge negatively at higher values of EW. The mixing effect also leads to the positive charging of graupel at low EW and high temperatures, in agreement with Takahashi (1978). This effect probably arises from enhanced vapor deposition to the riming graupel surface, a process which favors positive charging.

More recent laboratory work by Berdeklis and List (2001) made use of a sophisticated wind tunnel with accurate control on liquid water content and with ice crystals grown in a separate chamber and subsequently introduced into the wind tunnel flow. In the water-saturated conditions that all laboratory workers seek to achieve as representative of the thunderstorm updraft, the charging results in this study agree more closely with those of Takahashi (1978) than with Saunders et al. (1991). In particular, for an effective liquid water content of 0.5 g/m^3, the simulated graupel particle acquired negative charge for temperatures lower than -10 C, but positive charge for higher temperatures.

The detailed physical basis for the non-inductive collisional process remains poorly understood. Nevertheless, a phenomenological hypothesis involving the growth conditions of the interacting ice particles in mixed phase conditions was introduced more than 60 years ago (Findeisen, 1940; Findeisen and Findeisen, 1943), and has been profitably reactivated more recently (Jayarante et al., 1983; Baker et al., 1987; Williams et al., 1991). The basic hypothesis, as currently formulated, states that the more rapidly growing of the two colliding particles charges positively. In this context, sublimation (induced by warming in the riming process) is a state of negative growth.

A more microscopic description of the collisional encounter between graupel and ice was suggested by Baker and Dash (1994), who stated that ice in equilibrium with vapor has a disordered quasi-liquid layer at the ice-vapor interface. The thickness of the layer increases with temperature and growth rate by deposition. The exact nature of the growth rate of the impacting particles at the point of contact determines the direction of mass transfer between their quasi-liquid layers. This exchanged mass carries with it charge from the upper part of the electrical double layer and thus determines the polarity of the charge transfer between the particles. The particle with the higher temperature (and thicker quasi-liquid layer) would lose mass and negative charge to the colder (and less thick quasi-liquid layer) particle, thereby leaving the warmer particle positively charged. Thus, the faster growing particles (by vapor transfer) acquire positive charge (Baker et al., 1987). Similar conclusions were reported earlier by Williams et al. (1991).

The role of surface defects and growth regimes was further discussed by Dash et al. (2001). The charging process has three stages: growth before collision, impact and separation. A full description of this process requires the accurate characterization of particle surfaces, collisions and ambient conditions where ice and graupel interact. The qualitative aspects of mass and charge transfer are such that the faster a particle grows, the larger its surface density of negative ions. Collisions create melting at the surfaces of both colliding particles, releasing negative ions into the liquid. Upon separation, each particle takes approximately equal amounts of mass of the liquid. The faster growing particle, having lost more negative ions, becomes positively charged. Liquid-like mass exchange accompanies charging, with the colder particle gaining mass. This theory however is relevant to growth from the vapor of both particles, but does not extend to rimed surfaces that are common in thunderclouds. Also, it is not entirely clear how this description fits with the changes in saturation (with respect to water and ice) encountered in different regions and in different phases of the thundercloud.

If we accept the validity of the laboratory results presented by Takahashi (1978) and Pereyra et al. (2000) to adequately represent the conditions within thunderclouds, we can postulate that charge transfers during encounters of ice crystals and graupel pellets in the prevalent conditions of temperature, liquid water content and mixing in thunderstorms will lead to the normal polarity (positive-over- negative charge) usually found in observations. A variant of these conditions will shift the charging action to other regions in the LWC-T phase space and can account for a reversed polarity, such as is found in the stratiform region of MCS (Williams et al., 1994, discussed in the next section).

3.3 Cloud Scale Charge Structure Possible with the Non-Inductive Mechanism

The non-inductive ice-ice charging mechanism discussed above, and characterized by the laboratory stimulations discussed earlier, has been shown to be capable of creating a variety of cloud-scale charge structures (Williams et al., 1991, 1994; Williams, 1995, 2001a; Mansell et al., 2005). Four distinct prototypical storm configurations are illustrated in Figure 3. The dominant laboratory parameter (and by inference, the dominant cloud parameter) influencing the mixed phase microphysics and the large-scale electrical structure is the supercooled water content. This critical parameter is modulated in turn by meteorological conditions at larger scales – the lifting condition of moist air and the depletion of cloud water by the Bergeron process, by precipitation formation, and by mixing of drier environmental air.

Following Williams (1995), negative-charging of graupel particles and normal positive dipole structures (Figure 3b) are favored by intermediate values of liquid water content (0.5-2.0 g/m^3), whereas positive-charging of graupel particles and inverted dipole structures are favored by both low (<0.5 g/m^3) (Figure 3a) and high (>2 g/m^3) (Figure 3d) values. It is useful to consider the relative prevalence of these three distinct situations. By storm count on a global basis, the ordinary thunderstorms with intermediate liquid water content (Figure 3b) are strongly dominant. Lifting generally originates several kilometers below the 0°C isotherm, an essential condition for adiabatic cloud water contents of several g/m^3 (Ludlam, 1980; Johnson, 1993), and thereby the 'intermediate' condition of ultimate liquid water content. The vertical air speeds are often of the order of the precipitation particle fallspeeds and so depletion of cloud water is a dominant process. The lower negative charge on larger, faster-falling (graupel) particles appear to be favorable for the dominance of ground flashes with negative polarity in this condition. The existence of vertical wind shear is capable of 'tilting' the normally upright positive dipole (Figure 3c), a condition believed to be favorable for the occurrence of ground discharges from the upper positive pole (Brook et al., 1982; Rust et al., 1981; Curran and Rust, 1992; Brannick and Doswell III, 1992; Levin et al., 1996). The presently available evidence however supports the predominance of negative polarity lightning with a dominant positive dipole (Figure 3b). Given that sprites are almost exclusively caused by positive ground flashes (Boccippio et al., 1995; Barrington-Leigh et al., 1999; Williams, 2001b), the positive dipole (Figure 3b) would appear to be an unfavorable configuration for sprites.

At the high end of the scale of cloud water content, essentially all laboratory results (Takahashi, 1978; Saunders et al., 1991; Pereyra et al., 2000) show that the larger ice particles acquire positive charge in collisions with smaller ice particles and crystals, thereby setting up conditions for a large-scale dipole with inverted electrical polarity (Figure 3d). Meteorological conditions favorable to large cloud water contents are also a lifting process from relatively low levels, the existence of broad, undiluted updrafts, and sufficiently unstable conditions that the updraft speeds are so large that cloud water depletion by both the riming process and warm rain coalescence is suppressed. Such conditions were identified in Colorado during STEPS (Rust and MacGorman, 2002; Lang et al., 2004). Based on the association between an inverted polarity structure in the mixed phase region (Figure 3d) and the predominance of clustered ground flashes with positive polarity, it now appears that storms studied a decade ago by Knapp (1994), Stolzenburg (1994), Brannick and Doswell III (1992), Seimon (1993) and MacGorman and Burgess (1994), also possessed an inverted polarity of the main cloud dipole. The ultra-large hail produced by storms in this cloud water regime (Polsten, 1996) is solid evidence for extraordinary updraft speeds (Williams, 2001a). Given the need for positive ground flashes for sprites, this regime would also appear to be a good candidate as a sprite producer, but the presently available observations (Lang et al., 2004; Cummer and Lyons, 2005) do not show sprites over such supercells. The explanation probably lies in the small vertical charge moments associated with these positive ground flashes. Like the main negative charge in ordinary thunderclouds, the positive charge reservoirs are expected to be relatively confined laterally in supercells (Figure 3d).

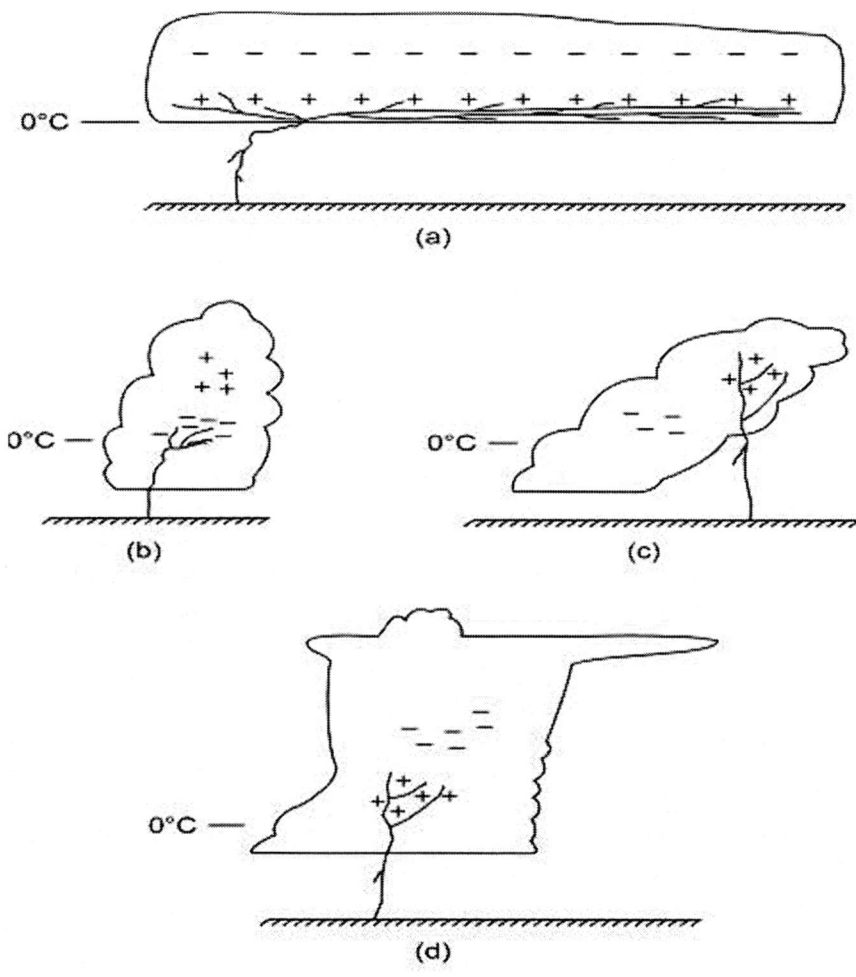

Figure 3. Prototypical charge structures and cloud-to-ground lightning, explicable by the non-inductive charging process: (a) stratiform region of a mesoscale convective system (b) ordinary thundercloud (c) tilted thundercloud (by vertical wind shear) (d) supercell thunderstorm with inverted polarity.

At the other extreme of the range of liquid water content, the lifting conditions are quite different – low speed (~1 m/s), laterally extensive, and originating within 1 km of the 0°C isotherm. The latter condition guarantees maximum adiabatic water contents which are less than 0.5 g/m^3. These conditions are prevalent during the dissipation and the End-of-Storm-Oscillation (EOSO) (Moore and Vonnegut, 1977; Livingston and Krider, 1978) stage of ordinary thunderstorms, and in the stratiform precipitation region of mesoscale convective systems (Houze, 1989; Rutledge and Petersen, 1994). The dominant electrostatic structure is often a simple inverted dipole within the mixed phase region (Figure 3a) (Shepherd et al., 1996), but can also be a more complicated, multi-layered charge structure. Even in these conditions, the positive charge near 0°C is frequently a dominant feature (Williams, 1998). The prevalent lightning type during this stage are laterally extensive 'spiders' (Mazur et al., 1995; Boccippio et al., 1995; Boccippio, 1996; Williams, 2001b; Lyons et al., 2003), often connecting to Earth in an energetic positive ground flash. These flashes appear to be the dominant 'parents' for sprites in summer storms.

3.4 The Electrical Structure Inside Sprite-Producing Storms in Summertime

Having identified a summertime meteorological regime with low water content and gentle ascent that appears favorable for sprite lightning, we can now explore this general regime in greater detail. The 'general regime' is comprised of two categories – the End-of-Storm-Oscillation (EOSO) and the stratiform region of the mesoscale convective system (MCS). The summertime MCS receives much attention in the literature because of the extensive work on sprites in the Great Plains of the United States (Sentman et al., 1995; Lyons, 1996; Huang et al., 1999; Hu et al., 2002) and this circumstance has left an impression that an MCS is a meteorological requirement for sprites. Such is not the case, as the MCS is presently defined in the literature, and the evidence for this is abundant (Williams, 1998; Hayakawa et al., 2004; Pinto Jr. et al., 2004; Yair et al., 2004; Adachi et al., 2004). One long-standing obstacle to understanding here is the arbitrary threshold assigned to an MCS – 20,000 km^2 (Zipser, 1982; Mohr and Zipser, 1996a,b). This threshold is a rather arbitrary one and has no physical basis as a break in the physics with changing scale. Nor is any physical basis attached to the 100,000 km^2 threshold for the largest MCS called Mesoscale Convective Complex (MCC) (Laing and Fritsch, 1997), the most sprite-productive storm on the planet (Lyons et al., 2003). Williams and Boccippio (1993) and Boccippio (1996) studied nighttime thunderstorms in central Florida that had all the characteristics of the stratiform precipitation region of MCSs – radar bright band, rear inflow jet, extended lifetime following afternoon deep convection, positive ground flashes with laterally extensive spider lightning, but which often did not attain the formal MCS area threshold (Zipser, 1982; Mohr and Zipser, 1996a,b).

The EOSO is the electrical manifestation of a smaller-scale phenomenon that is essentially the dissipation phase of convective scale thunderstorms (Moore and Vonnegut, 1977; Livingston and Krider, 1978; Williams and Boccippio, 1993). The horizontal scale of the EOSO is typically larger than the convective scale ($\sim$10 km) because multiple afternoon thunderstorms are involved in the dissipation stage. Because mesoscale features are collections of convective scale elements, the transition from the convective scale to the mesoscale is almost invariably a continuum, and no physically-based threshold scale can be identified.

Mesoscale phenomena are often migratory however, and convective scale phenomena (e.g., air mass thunderstorms) are often stationary. In this context, the trailing stratiform region of an MCS can be viewed as the EOSO of migratory storms in dissipation. The trailing stratiform region is the dissipative residue of the leading deep convection.

This general picture is supported by the common physical characteristics in the EOSO and the stratiform precipitation region of the MCS: an occurrence late in the diurnal cycle of moist convection, weak vertical air motion, laterally extensive radar bright band, elevated cloud base height, a dominant reservoir of positive charge, and a predominance of ground flashes with positive polarity with horizontally extensive 'spider' lightning propagation visible near cloud base height (Figure 3a). The common occurrence of the bright band is evidence for snowfall with aggregation of dendritic ice crystals just above the melting level. This precipitation type and the associated microphysics are quite distinct from the graupel and hail present in regimes of larger liquid water content.

It is important to note that if the number of thunderstorms contributing by dissipation to the EOSO is sufficiently small, then no lightning may be present during the period of inverted electric field polarity at the ground. Isolated thunderstorms can decay with no indication of a radar bright band. Storms can also decay and produce an EOSO, but not produce any positive ground flashes during the period of inverted electric field. Such is typical of the small mountain thunderstorms at Langmuir Laboratory in New Mexico (Moore and Vonnegut, 1977). The EOSO beneath the large island thunderstorms ('Hector') near Darwin, Australia is more likely to produce positive ground flashes (L. Carey, personal communication, 2004) and in Florida, positive ground flashes with spider lightning are quite prevalent during this phase (Williams and Boccippio, 1993; Boccippio et al., 1995; Mazur et al., 1995; Boccippio, 1996, pp. 234).

Detailed soundings of the electric field have disclosed the electrical structure of stratiform precipitation accompanying the EOSO and the MCS, though the larger-scale MCS has been explored more thoroughly in this regard (Marshall and Rust, 1993; Stolzenburg, 1994; Shepherd et al., 1996). The lateral uniformity in these systems often justifies a simple one-dimensional interpretation of the vertical profiles of electric field. A vertically confined layer of positive electric charge near the 0°C isotherm is a dominant feature, with a compensating negative layer above as shown by observations in Figure 4 (Shepherd et al., 1996). Hence the gross vertical charge structure in the mixed phase region is opposite to that in ordinary thunderclouds. The dominant lower layer of positive charge just above the radar bright band is consistent with the pronounced downward-directed electric field at the Earth's surface during the EOSO (Moore and Vonnegut, 1977; Livingston and Krider, 1978). The occurrence of a vertical dipole charge structure entirely within the mixed phase region is evidence that ice particle collisions by the non-inductive mechanism are causing the dipole by differential particle motions under gravity. Melting cannot be a contribution to this charge separation process because the positive charge layer is frequently observed above the radar bright band where the in situ temperature is <0°C. This point is well illustrated by Figure 3 showing the location of the positive charge layer and the radar bright band. The observations in Figure 4 can be compared with the idealized prototype in Figure 3a.

Charge structures more complicated than the simple two-layer inverted dipole are frequently observed in stratiform precipitation (Marshall and Rust, 1993; Stolzenburg, 1994; Shepherd et al., 1996; Marshall et al., 1996; Schurr and Rutledge, 2000a,b). Yet the lower layer appears to be the dominant reservoir for the positive ground flashes causal to sprites, displayed in Figure 3a (Williams, 1998; Lyons et al., 2003). The key evidence for this claim comes from studies during the STEPS (Severe Thunderstorm Electrification and Precipitation Study) experiment in 2000 in which simultaneous electrostatic, electromagnetic and optical measurements were undertaken of sprite-producing storms (Lyons et al., 2003). Detailed mapping of the 'spider' lightning structures of sprite-producing ground flashes with VLF methods showed preferential propagation just above the 0°C isotherm. The vertical charge moments of these energetic flashes were determined with ELF methods (Huang et al., 1999; Hu et al., 2002; Lyons et al., 2003). The large charge moments were also confirmed by electrostatic measurements. These large charge moments were attributed to the laterally extensive charge reservoirs present in this regime of stratiform precipitation (Lyons et al., 2003; Nelson, 1997, pp. 110). The total charge transfers in sprite-producing lightning are found to be of the order of 100 C or more, and an order of magnitude larger than in typical cloud-to-ground flashes in ordinary, more compact, thunderstorms.

Though the best electrical documentation of sprite-producing storms is found at mid-latitude sites, it is important to point out that tropical EOSO and MCS stratiform behavior is similar to mid- latitude. L. Carey (personal communication, 2004) has explored the dissipating behavior of island thunderstorms in

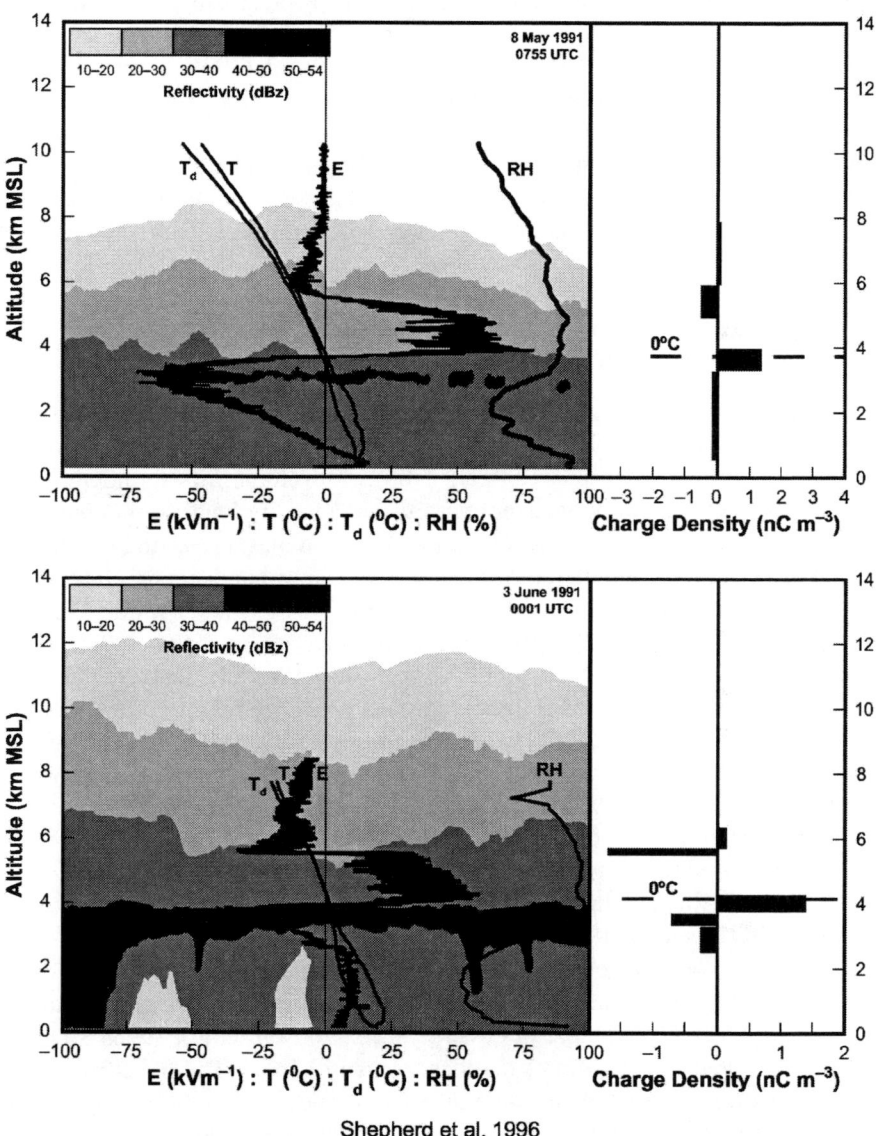

Figure 4. The location of radar bright band (left hand panels) and the lower positive charge layer (right hand panels) in two mesoscale convective systems probed with radar and electric field soundings. In both cases, the dominant positive layer is (partially) compensated by a negative charge layer at higher altitude, and within the 'mixed phase region'.

tropical Australia. Willams and Eckland (1992) and Rutledge et al. (1993) have investigated the trailing stratiform regions of a tropical squall line. Chauzy et al. (1985) made soundings of the electric field in tropical African squall lines that show features similar to the ones in Shepherd et al. (1996) at mid-latitude, within the regions of stratiform precipitation. One of the authors has explored convection within the Amazon basin of Brazil (Williams, 2002). Here it was found that stratiform precipitation with radar bright band and spider lightning was a relatively scarce phenomenon (one clear case in perhaps 40 days of local observations), but when it did occur, the general features of electrical structure were also similar to those at mid-latitude.

3.5 The Electrical Structure inside Sprite-Producing Storms in Wintertime

Winter thunderstorms are different from their summer-time counterparts described above mostly in their dynamical evolution (Kitagawa and Michimoto, 1994) and vertical extent (Krehhbiel, 1986). Developing in a colder atmosphere, they tend to exhibit weaker updrafts and a lower cloud top, nonetheless maintaining the existence of the mixed-phase region that is essential for charge separation. Globally, thunderstorms occur in northern-hemisphere winter only in very specific regions: the coastal areas of the Sea of Japan, over the Gulf of Mexico and along the Gulf Stream in the Atlantic, and over the coastal areas of the Mediterranean Sea in southern Europe and the Middle East (we refer the interested readers to the global lightning maps produced by the NASA/GHCC team at Marshall Space Flight Center, at: http://thunder.msfc.nasa.gov/data/OTDsummaries/). So far, sprites above winter thunderstorms have been observed almost exclusively in Japan (Fukunishi et al., 1999; Hayakawa et al., 2004; Takahashi et al., 2003; Hobara et al., 2001; Adachi et al., 2004), with the single exception of the 1999 observation of sprites and elves above a winter storm over Albania, during the Leonid-meteor shower airborne observation campaign in November 1998 (Yano et al., 2001).

Hence we choose to focus our discussion only on the conditions prevailing in Japan winter thunderstorms. Clearly similar features prevail in other locations as well (see for example in Levin et al., 1996; Yair et al., 1998; Altaratz et al., 2003, who studied winter thunderstorms in Israel). The meteorological conditions that are conducive to the development of winter thunderstorms in Japan, and the temporal evolution of these storms were studied by Kitagawa and Michimoto (1994) and are reviewed in Chapter 8 in Rakov and Uman (2003). Kobayashi et al. (1996) suggested a classification to three types based on mesoscale features: frontal, winter-monsoon and cold air-mass thunderclouds. Frontal thunderclouds are formed with the passage of a synoptic cold front, and are typified by a relatively high flash rate (> 1.5 per minute). In the winter monsoon regime, the thunderclouds usually develop within cold air-masses that move in an unstable atmosphere above the warm Tsushima current west of Japan, and are carried inland by the prevailing north-westerly winds from the Asian continent (Siberian air-mass). Bands of cumulus clouds, sometimes as shallow as 5 km, approach the coastline in both longitudinal and transversal snow bands (depending on the wind-shear strength) and produce intensive snow storms. Active cumulonimbus clouds are embedded within these bands. These thunderclouds exhibit an intermediate flash rate (<1 per minute). Cold air-mass thunderclouds are least active electrically with only a few flashes per hour (sometime named "one flash" thunderstorms (Michimoto, 1993)). Surface temperatures in these conditions are $\sim+5°C$ during early and late winter and $\sim0°C$ in midwinter, ensuring the presence of the mixed-phase region within the cloud and the operation of the non-inductive charge separation mechanism. Lightning occurs mostly over the sea and near the coast, and no more than 20-30 km inland (Hojo et al., 1989). The location of the -10°C isotherm above ground level was found to be a decisive factor for lightning activity (Michimoto, 1993), and in situations when it is at or below 1.4 km MSL, no lightning is observed. This finding is probably due to the weak updrafts that prohibit effective charge separation by the non-inductive mechanism. The monthly average height of the -10°C isotherm in Japan changes from 1.9 km MSL in January to 2.8 km MSL in March. Cloud tops in these situations are found to be lower than 8 km, and seldom below the -30°C isotherm and thus the total vertical extent of these clouds is in the range of 5-7 km.

Japan winter thunderstorms exhibit a relatively high percentage of positive ground flashes compared to sprite producing summer thunderstorms (see previous paragraph). However, not all these positive ground flashes have a large enough charge moment to generate sprites. For example, Hayakawa et al. (2004) report that on the night they imaged sprites, there were 477 cloud-to-ground discharges, out of which $\sim48\%$ were positive, a value that exceeds the Japan winter average of 33% (Miyake et al., 1992). This feature of winter storms was attributed to the existence of a strong wind-shear, which may create a "tilted dipole" (Figure 3c) and thus offset the upper positive charge near the anvil from the lower negative center (Brook et al., 1982). However, other works have not found any correlation between the percentage of positive ground flashes and the strength of the shear (Kitagawa, 1992), and an alternative view was suggested in which a positive monopole structure exists throughout the cloud, providing a charge source for almost exclusively positive ground flashes (Kitagawa and Michimoto, 1994). In-situ balloon measurements conducted by Takahashi et al. (1999) support a tripole electrical structure in winter thunderstorms, where positively charged ice crystals reside near the -20°C isotherm, above a negatively charged region of falling graupel around the -10°C isotherm. At lower levels, the weakly negatively charged graupel becomes positively charged due to the higher temperature, creating small regions of positive charge.

Radar studies (Fukao et al., 1991) and in-situ microphysical measurements (Takahashi et al., 1999) indicate that lightning in winter is associated with the contact and rebound of graupel and ice crystals or snowflakes. Radar work by Maekawa et al. (1993) showed that the threshold for lightning (though not generally the lightning responsible for sprites) was in reflectivity cores with values >40 dBZ, in a storm structure with graupel particles in the middle section and ice crystals near cloud top. The graupel particles were descending, and according to Kitagawa (1992) removed negative charge, leaving the cloud a positive monopole. This view is partially supported by recent in-situ measurements (T. Takahashi, personal communication, 2004) which suggest that the depletion of negative charge by sedimentation leads to the creation of very large horizontal positive charge reservoirs that may eventually generate the large charge moment needed for sprites. If this is the case, a regime with enhanced cloud water content (Takahashi, 1978) should predominate at the higher levels within these clouds. Additional in situ measurements are needed to establish this regime.

Ishii et al. (2003) studied the VHF sources of 68 ground flashes in winter thunderstorms occurring over the Sea of Japan. They differentiate between three types of flashes based on the vertical location of the VHF emissions. In Type A, the positive charge layers are located around the -30°C isotherm between 4-6 km MSL, while in Type B (further divided to B1 and B2) the charges reside at higher temperatures $\sim$-10°C and lower altitudes around 1-2 km MSL. A detailed analysis of 9 +CGs showed that the majority of the studied flashes originated within a low positive charge layer, typical of type B. It should be pointed out that these 9 flashes were not studied in conjunction with the appearance of sprites, and exhibited medium peak currents (<88 kA) and a low total charge (<70C). Thus, they probably indicate the existence of a reservoir of positive charge at lower levels, but not of the magnitude believed to be essential for sprite production. Ishii et al. (2003) conclude that this charge structure differs from the tilted dipole (Figure 3c) because their cells were inland and east of the coast, while those of Brook et al. (1982) were all over the sea surface.

Takahashi et al. (2003), who studied sprites produced by thunderstorms in the Hokoriku region, reported a cell size $\sim$30 km, based on IR satellite images, and estimated their height to be $\sim$7 km. Hayakawa et al. (2004) integrated radar data with the Japanese Lightning Detection Network (JLDN) and with measurements in the ELF band of sprite-producing winter lightning in the same region. Based on the total area of radar reflectivity of the storm they conclude that an MCS is not a necessary condition for the production of sprites in Japan winter thunderstorms (at least compared to the Mohr and Zipser (1996a) and Mohr and Zipser (1996b) definition of an MCS), as the lightning has a sufficiently large charge moment even in considerably smaller-scale storms. However, their size is still considered to be larger than ordinary summer thunderstorms in that region and Figure 5 (Hayakawa et al., 2004) clearly shows areas in excess of 20×20 km with high reflectivity where the sprite-producing lightning occurred. The vertical extent of these clouds, based on radar cross sections (RHI – Radar Height Indicator), show high reflectivity (>30 dBZ) extending up to 6 km and the clouds are described as "flat-topped stratiform, extending extensively horizontally, but extending to ~ 6 km altitude". The radar data reported in this study do not indicate the existence of a melting-layer associated bright band, a characteristic that would have indicated a similarity to the charge structure assumed to be the positive charge reservoir in continental MCSs (Williams, 1998). Other Japanese researchers also report only weak and temporary echoes of bright-band in the RHI cross section of winter thunderstorms (K. Michimoto, personal communication) probably because cloud base is very often colder than $0°C$. Hayakawa et al. (2004) also show that winter sprites appear solely in conjunction with strong positive ground flashes, which had charge moments (Qds) above a threshold level of 200-300 C·km (with an accuracy of 10%), a value that is comparable to those determined by Huang et al. (1999), Hu et al. (2002) and Cummer (2003) for summer thunderstorms. This finding is further supported by the prevalence of long duration, very large lightning currents (ms scale), sometimes 2-3 orders of magnitude larger than regular currents, in many winter storms. For example, Miyake et al. (1992) reported charge transfers larger than 10^3 C. Such large discharges are reminiscent of the giant spider lightning which are often found to precede sprites that drain the vast shallow regions of positive charge in summer storms (Williams, 1998; Lyons et al., 2003). The surprisingly large charge accumulations documented in Japan winter storms may also find explanation in the fact that charge regions in winter are located in the lower, denser regions of the atmosphere where the dielectric strength of air is larger, thereby enabling larger charge accumulations. The density scale height (~ 7 km) in the atmosphere is of the same order as the difference in altitude between major charge reservoirs in summer and in winter.

The emerging picture of sprite-producing thunderstorms in Japan is that of vertically limited clouds which develop within a larger, stratiform-like region, not unlike the large continental summer MCS. Two possible generating mechanisms for +CG lightning seem to be active in Japanese winter storms: (a) the formation of large areas of positive charge near the $0°C$ isotherm, and (b) the tilting of the upper, positively charged cloud top (anvil) by the strong wind shear. Which of the two mechanisms is responsible for the generation of the strong positive ground flashes, with a sufficiently large charge moment to spawn sprites, requires additional study. It is clear that a combination of radar, in-situ and remote measurements of cloud structure should be performed in tandem with sprite observations, to clarify the causal relations between the cloud life-cycle, charge structure and the sprite-producing cloud-to-ground lightning flashes. In the opinion of the authors of this chapter, the understanding of cloud electrification in general will be much improved when the behavior of Japan winter storms is clarified.

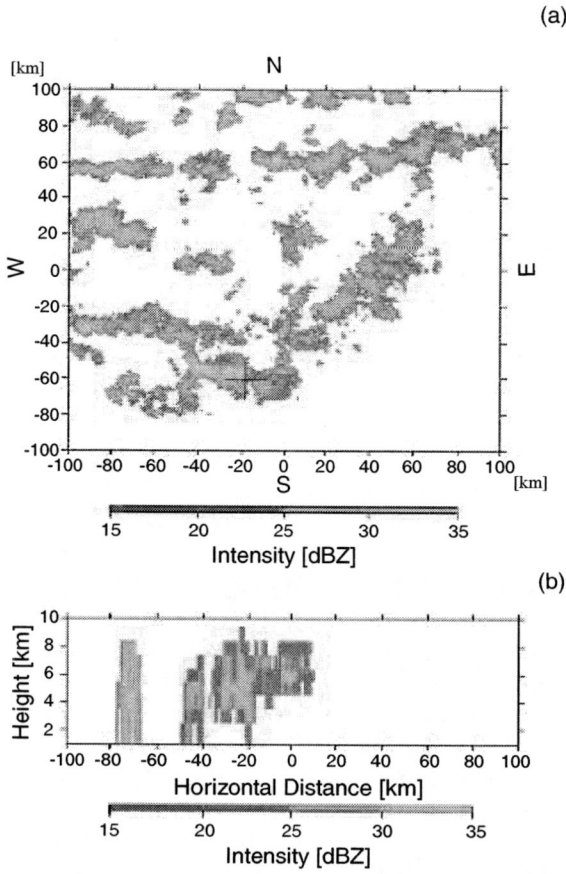

Figure 5. Radar observations for a sprite producing winter thunderstorm, taken from Hayakawa et al. (2004) (a) Radar CAPPI at a height of 5 km at 1640–1782 UT on 14 December. A '+' symbol indicates the position of the sprite. (b) Vertical cross section of radar image along the EW plane crossing the position ('+' symbol in Figure 4a) of the lightning discharge leading to that sprite. (American Geophysical Union, used with permission).

3.6 Gaps in Knowledge and Future Needs

While the microphysical, electrical and meteorological setting for sprites is mapped in sufficient detail for continental summer storms in the U.S., there seems to be a lack of a combined, integrated picture of these conditions in sprite-producing storms elsewhere on Earth. There is now ample evidence for the occurrence of sprites on a global scale (Sentman et al., 1995; Lyons, 1996; Hardman et al., 2000; Su et al., 2002; Neubert et al., 2001; Füllekrug and Price, 2002; Hayakawa et al., 2004; Pinto Jr. et al., 2004), both above the continents and the oceans, based on optical observations from the ISS (Blanc et al., 2004), the Space Shuttle (Yair et al., 2004) and the FORMOSAT satellite (Mende et al., 2004). The parent lightning for these optically observed sprites appears in storm systems which differ in size and dynamics from continental summer MCSs, and requires further study. While some data exist on the charge structure and lightning generation in tropical and winter storms, there seems to be lacking a comprehensive view of these processes in relation to sprite formation. Clearly, local lightning networks have a limited coverage, and thus we need to study the properties of lightning based on remote ELF and VLF methods and rely on global networks that involve stations in several, remotely located, countries. On top of the electromagnetic data, we need radar and satellite data to help define the meteorological conditions prevailing within those storm systems. This will enable us to typify the charge structure which eventually gives rise to the exceptional ground flashes that affect the mesosphere.

The almost-exclusive association of sprites with ground flashes of positive polarity remains an outstanding scientific problem (Williams et al., 2004). The threshold condition for dielectric breakdown in the upper atmosphere by lightning in the troposphere, put forward by C. T. R. Wilson nearly 80 years ago (Wilson, 1925), provides a viable condition for sprite initiation (Pasko et al., 1995; Huang et al., 1999; Hu et al., 2002). The critical lightning source property is the vertical charge moment – the product of total charge transfer and the height above ground from which the charge is removed. The problem is that this threshold charge moment is theoretically independent of polarity. Electromagnetic observations in the ELF region (Williams et al., 2004) show that roughly 10% of supercritical charge moments on a global basis are from negative ground flashes, yet less than 0.1% of optically-observed sprites have been attributed to negative ground flashes (Barrington-Leigh et al., 1999).

One possible explanation for this paradox is the known physical difference between positive and negative ground flashes. Large positive flashes tend to have longer durations of charge transfer than negative ones (Williams et al., 2004). This difference may be related to the sizes and shapes of the respective charge reservoirs. Laterally extended positive charge reservoirs in proximity to the ground may be more common than negative reservoirs because of the water-content-related ice microphysics discussed earlier in Section 3.2. A long duration of charge transfer may maintain an electrical stress on the mesosphere that will sustain the growth and luminosity of a sprite, thereby guaranteeing its detectability in video camera observations. A shorter period of stress by a negative flash may produce a halo (Bering et al., 2004) which is more elusive to detection by ground-based observations

Acknowledgments

We wish to deeply thank our colleagues Tsutumo Takahashi, Masaru Ishii, Zen Kawasaki, Kohichiro Michimoto, Yasuhide Hobara, Larry Carey, Colin Price and Martin Füllekrug for their helpful comments and suggestions in preparing this manuscript.

Bibliography

Adachi, T., Fukunishi, H., Takahashi, Y., and Sato, M. (2004). Roles of the EMP and QE field in the generation of columniform sprites. *Geophys. Res. Lett.*, 31(4):doi:10.1029/2003GL019081.

Altaratz, O., Levin, Z., Yair, Y., and Ziv, B. (2003). Lightning activity over land and sea on the eastern coast of the Mediterranean. *Mon. Wea. Rev.*, 131:2060–2070.

Baker, B., Baker, M. B., Jayaratne, E. R., Latham, J., and Saunders, C. P. R. (1987). The influence of diffusional growth rates on the charge transfer accompanying rebounding collisions between ice crystals and soft hailstones. *Quart. J. Roy. Met. Soc.*, 113:1193–1215.

Baker, M. B. and Dash, J. G. (1994). Mechanisms of charge-transfer between colliding ice particles in thunderstorms. *Journal of Geophysical Research-Atmospheres*, 99(D5):10621–10626.

Barrington-Leigh, C. P., Inan, U. S., Stanley, M., and Cummer, S. A. (1999). Sprites directly triggered by negative lightning discharges. *Geophys. Res. Lett.*, 26:3605–3608.

Berdeklis, T. and List, R. (2001). The ice crystal-graupel collision charging mechanism of thunderstorm electrification. *Jour. Atmos. Sci*, 58:2751–2770.

Bering, E. A., Wescott, E., Bhusal, L., Benbrook, J. R., Jackson, A., Moudry, D., Sentman, D. D., Nielsen-Stenbaek, H., Garrett, J., and Lyons, W. A. (2004). Observations of transient luminous events (TLEs) associated with negative cloud to ground (-CG) lightning strokes. *Geophys. Res. Lett.*, 31:L05104.

Blanc, E., Farges, T., Roche, R., Brebion, D., Hua, T., Labarthe, A., and Melnikov, V. (2004). Nadir observations of sprites from the International Space Station. *J. Geophys. Res.*, 109(A2):doi:10.1029/2003JA009972.

Boccippio, D. (1996). *The electrification of stratiform anvils*. PhD thesis, Massachusetts Institute of Technology.

Boccippio, D. J., Williams, E., Heckman, S. J., Lyons, W. A., Baker, I., and Boldi, R. (1995). Sprites, ELF transients and positive ground strokes. *Science*, 269:1088–1091.

Brannick, M. L. and Doswell III, C. A. (1992). An observation of the relationship between supercell structure and lightning ground strike polarity. *Wea. Forec.*, 7:143–149.

Brook, M., Nakano, M., Krehbiel, P., and Takeuti, T. (1982). The electrical structure of the Hokuriku winter thunderstorms. *J. Geophys. Res.*, 87(NC2):1207–1215.

Brooks, I. M. and Saunders, C. P. R. (1994). An experimental investigation of the inductive mechanisms of thunderstorm electrification. *JGR*, 99:10627–10632.

Chauzy, S., Chonbg, M., Dellanoy, A., and Despiau, S. (1985). The June 22 tropical squall line observed during the COPT '81 experiment: Electrical signature associated with dynamical structure and precipitation. *J. Geophys. Res.*, 90:6091–6098.

Cummer, S. A. (2003). Current moment in sprite-producing lightning. *J. Atmos. Sol.-Terr. Phys.*, 65:499–508.

Cummer, S.A. and Lyons, W.A. (2005). Implications of lightning charge moment changes for sprite initiation. *J. Geophys. Res.*, 110(A4):Art. no. 04304.

Curran, E. B. and Rust, W. D. (1992). Positive ground flashes produced by low-precipitation thunderstorms in Oklahoma on 26 April 1984. *Mon. Wea. Rev.*, 120:544–553.

Dash, J. B., Mason, B. L., and Wettlhufer, J. S. (2001). Theory of collisional charging of ice: Microphysics of thunderstorm electrification. *J. Geophys. Res.*, 106:20395–20402.

Findeisen, W. (1940). On the origin of storm electricity. *Meteor. Z.*, 57:6. (in German).

Findeisen, W. and Findeisen, E. (1943). Investigation on the ice splinter formation on rime layers, (A contribution to the origin of storm electricity and the microstructure of cumulonimbi). *Meteor. Z.*, 60(5). (in German).

Fukao, S., Meakwa, Y., Sonoi, Y., and Yoshino, F. (1991). Dual polarization radar observation of thunderclouds on the coast of the Sea of Japan in the winter season. *Geophys. Res. Lett.*, 18:179–182.

Fukunishi, H., Takahashi, Y., Uchida, A., Sera, M., Adachi, K., and Miyasato, R. (1999). Occurrences of sprites and elves above the sea of Japan near Hokuriku in winter. *EOS, AGU Fall Meeting*, 80(46):F217.

Füllekrug, M. and Price, C. (2002). Estimation of sprite occurrences in Central Africa. *Meteor. Z.*, 11:99.

Hardman, S. F., Dowden, R. L., Brundell, J. B., Bahr, J. L., Kawasaki, Z.-I., and Rodger, C. J. (2000). Sprite observations in the Northern Territory of Australia. *J. Geophys. Res.*, 105:4689–4697.

Hayakawa, M., Nakamura, T., Hobara, Y., and Williams, E. (2004). Observation of sprites over the Sea of Japan and conditions for lightning-induced sprites in winter. *J. Geophys. Res.*, 109(A01312):doi:10.1029/-2003JA0099905.

Hobara, Y., Iwasaki, N., Hayashida, T., Hayakawa, M., Ohta, K., and Fukunishi, H. (2001). Interrelation between ELF transients and ionospheric disturbances in association with sprites and elves. *Geophys. Res. Lett.*, 28:935–938.

Hojo, J., Ishii, M., Kawamura, T., Suzuki, F., Komuro, H., and Shiogama, M. (1989). Seasonal variation of cloud-to-ground lightning flash characteristics in the coastal area of the Sea of Japan. *J. Geophys. Res.*, 94(13):207–212.

Houze, R. A. (1989). Observed structure of mesoscale convective systems and implications for large-scale heating. *Quart. J. Roy. Met. Soc.*, 115(487):425–461.

Hu, W., Cummer, S., Lyons, W. A., and Nelson, T. E. (2002). Lightning charge moment changes for the initiation of sprites. *Geophys. Res. Lett.*, 29:doi:10.1029/2001GL014593.

Huang, E., Williams, E., Boldi, R., Heckman, S., Lyons, W., Taylor, M., Nelson, T., and Wong, C. (1999). Criteria for sprites and elves based on Schumann resonance observations. *J. Geophys. Res.*, 104:16943–16964.

Ishii, M., Saito, M., Hojo, J., and Kami, K. (2003). Location of charges associated with positive CG flashes in winter. In *Proceedings of the XII ICAE meeting*, pages 151–154, Versailles, France.

Jayarante, E. R., Saunders, C. P. R., and Hallet, J. (1983). Laboratory studies of the charging of soft-hail during ice crystal interactions. *Q. J. R. Meteorol. Soc.*, 109:609–630.

Jayarante, R. (2003). Thunderstorm electrification mechanisms. In Cooray, V., editor, *The Lightning Flash*, pages 17–44. The Institution of Electrical Engineers.

Johnson, D. (1993). The onset of effective coalescence growth in convective clouds. *Quart. J. Roy. Met. Soc.*, 119:925–933.

Kitagawa, N. (1992). Charge distribution of winter thunderclouds. *Res. Lett. Atmos. Electr.*, 12:143–153.

Kitagawa, N. and Michimoto, K. (1994). Meteorological and electrical aspects of winter thunderclouds. *J. Geophys. Res.*, 99(D5):10713–10721.

Knapp, D. I. (1994). Using cloud-to-ground lightning data to identify tornadic thunderstorm signatures and nowcast severe weather. *Natl. Wea. Dig.*, 19:35–42.

Kobayashi, F., Shimura, T., Wada, A., and Sakai, T. (1996). Lightning activities of winter thundercloud systems around the Hokuriku coast of Japan. *Proc. 10th ICAE*, pages 560–563.

Krehhbiel, P. R. (1986). *The electrical structure of thunderstorms*, pages 90–113. National Academy Press.

Laing, A. G. and Fritsch, J. M. (1997). The global population of Mesoscale Convective Complexes. *Quart. J. Roy. Met. Soc.*, 123:389–405.

Lang, T. J., Miller, L. J., Weisman, M., Rutledge, S. A., Barker III, L. J., Bringi, V. N., Chandrasekar, V., Detwiler, A., Doeskin, N., Helsdon, J., Knight, C., Krehbiel, P., W. A. Lyons, CCM, MacGorman, D., Rasmussen, E., Rison, W., Rust, W. D., and Thomas, R. J. (2004). The Severe Thunderstorm Electrification and Precipitation Study (STEPS). *Bull. Am. Met. Soc.*, 85(8):1107.

Levin, Z., Yair, Y., and Ziv, B. (1996). Positive cloud-to-ground flashes and wind shear in Tel-Aviv thunderstorms. *Geophys. Res. Lett.*, 23(17):2231–2234.

Livingston, J. M. and Krider, E. P. (1978). Electric fields produced by Florida thunderstorms. *J. Geophys. Res.*, 83:385–401.

Ludlam, F. H. (1980). *Clouds and storms*, chapter The behavior and effects of water in the atmosphere. The Pennsylvania State University Press.

Lyons, W. A. (1996). Sprite observations above the U.S. High Plains in relation to their parent thunderstorm systems. *J. Geophys. Res.*, 101:29641–29652.

Lyons, W. A., Nelson, T. E., Williams, E. R., Cummer, S. A., and Stanley, M. A. (2003). Characteristics of sprite-producing positive cloud-to-ground lightning during the 19 July 2000 STEPS mesoscale convective systems. *Mon. Wea. Rev.*, 131:2417–2427.

MacGorman, D. R. and Burgess, D. W. (1994). Positive cloud-to-ground lightning in tornadic storms and hailstorms. *Mon. Wea. Rev.*, 122:1671–1697.

MacGorman, D. R. and Rust, W. D. (1998). *The electrical nature of storms*. Oxford University Press. 421 pp.

Maekawa, Y., Fukao, S., Sonoi, Y., and Yoshimo, F. (1993). Distribution of ice particles in wintertime thunderclouds detected by a C band dual polarization radar: a case study. *J. Geophys. Res.*, 98:16613–16622.

Mansell, E. R., MacGorman, D. R., Ziegler, C. L., and Straka, J. M. (2005). Charge structure in a simulated multicell thunderstorm. *J. Geophys. Res.*, 100(D12101):doi:10.1029/2004JD005287.

Marshall, T. C. and Rust, W. D. (1993). Two types of vertical electric structures in stratiform precipitation regions of mesoscale convective systems. *Bull. Am. Met. Soc.*, 74:2159–2170.

Marshall, T. C., Stolzenberg, M., and Rust, W. D. (1996). Electric field measurements above mesoscale convective systems. *J. Geophys. Res.*, 101:6979–6996.

Mazur, V., Krehbiel, P. R., and Shao, X.-M. (1995). Correlated high speed video and interferometric observations of a cloud-to-ground lightning flash. *J. Geophys. Res.*, 100:25731–25753.

Mende, S. B., Frey, H., Hsu, R., Su, H., Chen, A., Lee, L., Fukunishi, H., and Takahashi, Y. (2004). Sprite imaging results from the ROCSAT2 ISUAL instrument. *EOS, AGU Fall Meeting*, 85(47):AE51A–02.

Michimoto, K. (1993). A study of radar echoes and their relation to lightning discharges of thunderclouds in the Hokuriku District, II: Observations and and analysis of "single flash" thunderclouds in midwinter. *J. Meteor. Soc. Japan*, 71:195–204.

Miyake, K., Suzuki, T., and Shinjou, K. (1992). Characteristics of winter lightning current in Japan Sea coast. *IEEE Trans. Pow. Del.*, 7:1450–1456.

Mohr, K. I. and Zipser, E. J. (1996a). Defining mesoscale convective systems by their 85-GHz ice-scattering signatures. *Bull. Am. Met. Soc.*, 77(6):1179–1189.

Mohr, K. I. and Zipser, E. J. (1996b). Mesoscale convective systems defined by their 85-GHz ice scattering signature: Size and intensity comparison over tropical oceans and continents. *Mon. Wea. Rev.*, 124(11):2417–2437.

Moore, C. B. and Vonnegut, B. (1977). Lightning. In Golde, R. H., editor, *The Thundercloud*, volume 1, pages 51–98. Academic Press.

Nelson, T. A. (1997). Spatial relationships between radar reflectivity and sprite- and elve-producing cloud-to-ground lightning strokes. Master's thesis, Mankato State University, Minnesota.

Neubert, T., Allin, T., Stenbaek-Nielsen, H., and Blanc, E. (2001). Sprites over Europe. *Geophys. Res. Lett.*, 28:3585–3588.

Pasko, V. P., Inan, U. S., Taranenko, Y. N., and Bell, T. F. (1995). Heating, ionization and upward discharges in the mesosphere due to intense quasi-electrostatic thundercloud fields. *Geophys. Res. Lett.*, 22:365–368.

Pereyra, R. G., Avila, E. E., Castellano, N. E., and Saunders, C. P. R. (2000). A laboratory study of graupel charging. *J. Geophys. Res.*, 105:20803–20813.

Pinto Jr., O., Saba, M. M. F., Pinto, I. R. C. A., Tavares, F. S. S., Solorzano, N. N., Naccarato, K. P., Taylor, M., Pautet, P. D., and Holzworth, R. H. (2004). Thunderstorm and lightning characteristics associated with sprites in Brazil. *Geophys. Res. Lett.*, 31(13):13103–13106.

Polsten, K. L. (1996). Synoptic patterns and environmental conditions associated with very large (4" and greater) hail events. In *18th Conf. on Severe Local Storms*, pages 349–356, Indianapolis, IN. American Meteorological Society.

Rakov, V. A. and Uman, M. A. (2003). *Lightning, Physics and Effects*. Cambridge University Press.

Rust, W. D. and MacGorman, D. R. (2002). Possibly inverted-polarity electrical structures in thunderstorms during STEPS. *Geophys. Res. Lett.*, 29(12):doi:10.1029/2001GL014303.

Rust, W. D., MacGorman, D. R., and Arnold, R. T. (1981). Positive cloud-to-ground lightning flashes in severe storms. *Geophys. Res. Lett.*, 8:791–794.

Rutledge, S. A. and Petersen, W. A. (1994). Vertical radar reflectivity structure and cloud-to-ground lightning in the stratiform region of MCSs: Further evidence for in situ charging in the stratiform region. *Mon. Wea. Rev.*, 122:1760–1776.

Rutledge, S. A., Williams, E. R., and Petersen, W. A. (1993). Lightning and electrical structure of mesoscale convective systems. *Atmos. Res.*, 29:27–53.

Saunders, C. P.R., Keith, W. D., and Mitzeva, R. P. (1991). The effect of liquid water content on thunderstorm charging. *J. Geophys. Res.*, 96:11007–11017.

Schurr, T. and Rutledge, S. (2000a). Electrification of stratiform regions in mesoscale convective systems, Part I, an observational comparison of symmetric and asymmetric MCSs. *J. Atmos. Sci.*, 57:1961–1982.

Schurr, T. and Rutledge, S. (2000b). Electrification of stratiform regions in mesoscale convective systems, Part II, two dimentional numerical simulations of a symmetric MCS. *J. Atmos. Sci.*, 57:1961–1982.

Seimon, A. (1993). Anomalous cloud-to-ground lightning in an F5 tornado-producing supercell thunderstorm on 28 August 1990. *Bull. Am. Met. Soc.*, 74:189–203.

Sentman, D. D., Wescott, E. M., Osborne, D. L., Hampton, D. L., and Heavner, M. J. (1995). Preliminary results from the Sprites 94 aircraft campaign: 1. Red sprites. *Geophys. Res. Lett.*, 22:1205–1208.

Shepherd, T. R., Rust, W. D., and Marshall, T. C. (1996). Electric fields and charges near 0°C in stratiform clouds. *Mon. Wea. Rev.*, 124(5):919–938.

Stolzenburg, M. (1994). Observations of high ground flash densities of positive lightning in summertime thunderstorms. *Mon. Wea. Rev.*, 122:1740–1750.

Su, H. T., Hsu, R. R., Chen, A. B. C., Lee, Y. J., and Lee, L. C. (2002). Observation of sprites over the Asian continent and oceans around Taiwan. *Geophys. Res. Lett.*, 29(4):doi:10.1029/2001GL013737.

Takahashi, T. (1978). Riming electrification as a charge generation mechanism in thunderstorms. *J. Atmos. Sci.*, 35:1536–1548.

Takahashi, T., Tajiri, T., and Sonoi, Y. (1999). Charges on graupel and snow crystals and the electrical structure of winter thunderstorms. *J. Atmos. Sci.*, 56:1561–1578.

Takahashi, Y., Miyasato, R., Adachi, T., Adachi, K., Sera, M., Uchida, A., and Fukunishi, H. (2003). Activities of sprites and elves in the winter season, Japan. *J. Atmos. Sol.-Terr. Phys.*, 65:551–560.

Willams, E. R. and Eckland, W. (1992). 50 MHz profiler observations of trailing stratiform precipitation: Constraints on cloud microphysics and charge separation. In *Int'l Conf. on Cloud Physics and Precipitation*, Montreal, PQ, Canada. Preprint.

Williams, E. R. (1995). Comment on "Thunderstorm electrification laboratory experiments and charging mechanisms" by C. P. R. Saunders. *J. Geophys. Res.*, 100:1503–1505.

Williams, E. R. (1998). The positive charge reservoir for sprite-producing lightning. *J. Atmos. Sol.-Terr. Phys.*, 60:689–692.

Williams, E. R. (2001a). The electrification of severe storms. In Doswell, C. A., editor, *Severe Convective Storms*, volume III, pages 527–561. American Meteorological Society.

Williams, E. R. (2001b). Sprites, elves and glow discharge tubes. *Physics Today*, pages 41–47.

Williams, E. R. and Boccippio, D. J. (1993). Dependence of cloud microphysics and electrification on mesoscale vertical air motions in stratiform precipitation. In *Conference on Atmospheric Electricity*, pages 825–831, St. Louis, MO. AMS. Preprints.

Williams, E. R., Downes, E., Boldi, R., Lyons, W., and Heckman, S. (2004). The polarity asymmetry of sprite-producing lightning: A paradox, Preprint volume. In *Conference on Atmospheric Electricity*, pages 9–11, Belo Horizonte, Brazil. Brazilian Society for Electrical Protection.

Williams, E. R., Zhang, R., and Boccippio, D. (1994). Microphysical growth state of ice particles and large-scale electrical structure of clouds. *J. Geophys. Res.*, 99:10787–10792.

Williams, E. R., Zhang, R., and Rydock, J. (1991). Mixed-phase microphysics and cloud electrification. *J. Atmos. Sci.*, 48:2195–2203.

Williams, E. R., et. al. (2002). Contrasting convective regimes over the Amazon: Implications for cloud electrification. *J. Geophys. Res. - LBA Special Issue*, 107(D20):doi:10.1029/2001JD000380.

Wilson, C. T. R (1925). The electric field of a thundercloud and some of its effects. *Proc. Roy. Soc. London*, 37(32D).

Yair, Y., Israelevich, P., Devir, A. D., Moalem, M., Price, C., Joseph, J. H., Levin, Z., Ziv, B., Sternlieb, A., and Teller, A. (2004). New observations of sprites from the Space Shuttle. *J. Geophys. Res.*, 109(D15201):doi:-10.1029/2003JD004497.

Yair, Y., Levin, Z., and Altaratz, O. (1998). Lightning phenomenology in the Tel-Aviv area from 1989 to 1996. *J. Geophys. Res.*, 103(D8):9015–9025.

Yano, H., Abe, S., and Takahashi, Y. (2001). High-definition TV imagery of elves and sprites over the Mediterranean Sea during the 1999 Leonid meteor shower peak. In *Proceeding of the 1st AP-RASC Meeting*, Tokyo, Japan.

Zipser, E. J. (1982). Use of a conceptual model of the life-cycle of mesoscale convective systems to improve very-short-range forecasts. In Browning, K., editor, *Nowcasting*, pages 191–204. Academic Press, London and New York.

GLOBAL THUNDERSTORM ACTIVITY

Colin Price
Department of Geophysics and Planetary Science, Tel Aviv University, Ramat Aviv, Israel.

Abstract   Sprites, elves and other transient luminous events (TLEs) are known to exist only above thunderstorms. It is therefore important to know where these thunderstorms occur around the globe, and how their distribution varies temporally and spatially. The majority of thunderstorms on Earth occur within the tropical regions between ±30° latitude of the equator (∼50% of the surface area of the globe). This is due to the maximum solar heating in the tropics, and the atmospheric general circulation patterns between the tropics and the subtropics (Hadley Circulation). Along the thermal equator, which migrates with the seasons, air masses from the northern and southern hemispheres converge along the intertropical convergence zone (ITCZ). This is the latitudinal position of the majority of the globe's rainfall and thunderstorm activity. However, in the tropics these thunderstorms are concentrated mainly over the continental regions (Americas, Africa and southeast Asia) with little thunderstorm activity observed over the oceans. The reason for the preference of thunderstorms to continental regions is likely related to the larger daily surface heating over land as compared with the oceans. In the extra-tropical regions thunderstorms form along the polar front, the boundary between warm moist air from the tropics, and cool dry air from polar regions. Recent satellite measurements of lightning indicate a mean global rate of ∼45 flashes/second. In fair weather regions the integrated effect of global thunderstorms and other electrified clouds can be observed via the atmospheric global electric circuit. The global thunderstorms charge the Earth's surface negatively with a mean charge of 500,000 Coulombs, and a mean potential between the ionosphere (∼80 km) and the Earth's surface of 250 kV. The diurnal variation of the atmospheric electric circuit (and global thunderstorms) has a maximum around 18 UT and a minimum around 03 UT known as the Carnegie Curve.

4.1   The Earth's Energy Balance

The Earth is located at a mean distance of 150 million km from our Sun, and at this distance we receive on average 1367 W/m² of energy, known as the solar constant (S_o). However, since we live on a sphere, the solar radiation does not intersect with the surface at the same angle at all latitudes. When the solar rays are perpendicular (90°) to the equator (at noon), the rays are tangential (0° or just above the horizon) at the north and south poles (Figure 1a). Hence as we move to higher latitudes, the same energy from the sun is spread over larger areas, resulting in less energy absorbed per unit area. This is the prime reason why the tropics are warmer than the higher latitudes (Figure 1b). When we compare the incoming absorption of energy from the sun (short wave radiation) relative to the outgoing terrestrial radiation (long wave radiation) we notice a latitudinal imbalance (Figure 1c). The latitudinal gradient of incoming energy (black curve) is much larger than the latitudinal gradient in the outgoing energy (grey curve). The climate system tries to adjust for this imbalance by transporting excess heat from the tropics to high latitudes. This transport is done via the atmosphere and the oceans. Ocean currents transport warm waters away from the equator, while bringing cooler waters equatorward, helping to

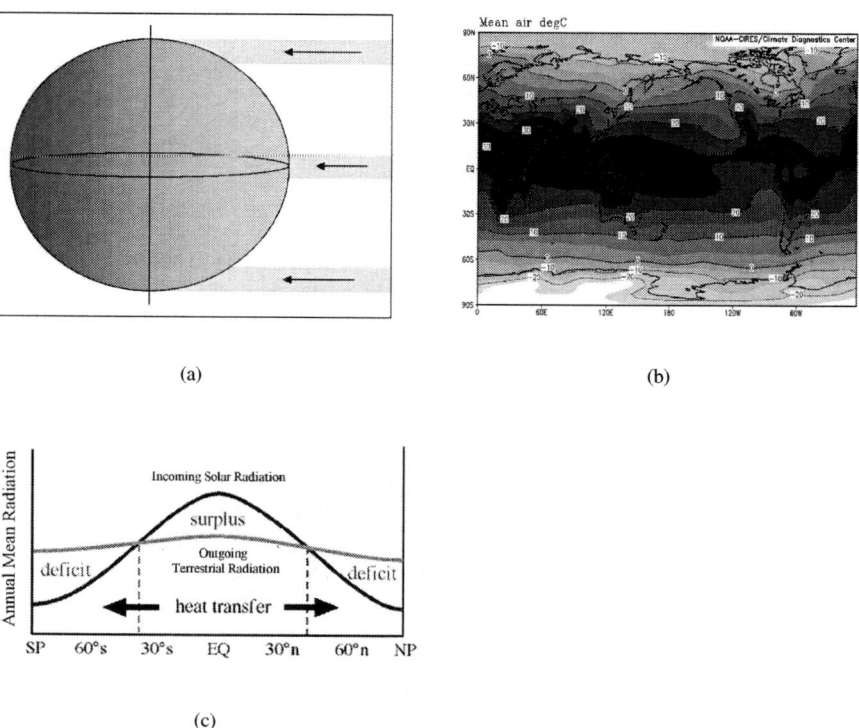

Figure 1. (a) The dependence of absorbed solar radiation on latitude; (b) The annual mean surface temperatures showing the warmest areas in the tropics (http://www.cdc.noaa.gov/cdc/reanalysis/reanalysis.shtml); c) Latitudinal energy balance of the Earth showing the incoming energy from the sun (black curve) relative to the outgoing radiation from the Earth (grey curve). The surplus of energy in the tropics results in large poleward transports of heat via the atmosphere and oceans.

reduce this imbalance. In the atmosphere, the transfer of heat poleward occurs primarily through the meridional general circulation of the atmosphere, where thunderstorms have a major role in the transport of this energy both away form the warm surface, and then away from the warm tropics.

4.2   The General Circulation of the Atmosphere

The maximum heating at the surface in the tropics results in rising thermals and convection in the atmosphere. The resulting region of rising air along the thermal equator is know as the intertropical convergence zone (ITCZ) due to the convergence of air masses from the northern and southern hemispheres along this boundary. The thermal equator migrates north and south of the geographic equator according to the seasons, where the thermal equator is furthest north in June-August during the northern hemisphere summer, and furthest south in December-February during the southern hemisphere summer.

Although the region of maximum heating by the sun covers a broad region of thousands of kilometers, the ITCZ is actually very narrow and compact ($\sim$100 km over the oceans). This narrow region of convergence is due to the extra driving force of the convection along the ITCZ resulting from the release of latent heat as water vapor is condensed to form cloud droplets as the air rises and cools in the atmosphere. The release of latent heat adds additional heat to the cloud parcel and enhances the buoyancy in the deep convective clouds, increasing the vertical updrafts.

In general, a significant difference in updraft intensity is observed between oceanic and continental convection in the tropics (Lemone and Zipser, 1980; Jorgenson and Lemone, 1989). The reason for this difference is still a topic of research, but updraft velocities in oceanic thunderstorms may reach 10 m/s, while over continental regions the updrafts may reach 50 m/s (Price and Rind, 1992; Williams and Stanfill, 2002; Williams et al., 2004). Since updraft intensity plays a major role in thunderstorm electrification and lightning frequencies (Baker et al., 1995, 1999), this dramatic difference in thunderstorm dynamics results in the lightning activity over the oceans being an order of magnitude less than over the continents. In fact the boundary between land and ocean is very clearly seen in satellite images of tropical lightning activity, for example in Figure 2 (Christian et al., 2003). However, not all tropical continental thunderstorms are intense lightning generators. The tropical monsoon periods are characterized by the seasonal onshore flow of moist oceanic air, resulting in heavy rainfall in continental thunderstorms, however with low lightning rates (Petersen et al., 2002; Williams et al., 2002). This occurs in the Indian Monsoon, the African Monsoon, the Brazilian Monsoon and the Australian Monsoon. Intense lightning activity prefers a somewhat dry environment, which may explain the difference between African and South American lightning ac-

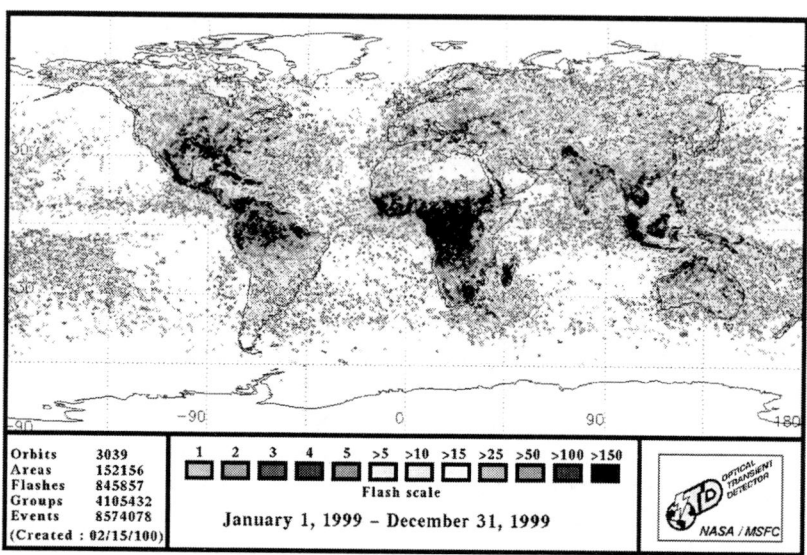

Figure 2. Global lightning activity for 1999 from the satellite OTD sensor showing lightning activity over the tropical land areas (http://thunder.msfc.nasa.gov/otd).

tivity (Williams and Satori, 2004). A field campaign to observe sprites during the monsoon season is not recommended.

The air that rises within the tropical convective storms eventually reaches the tropopause and the stable stratosphere between 15-20 km altitude, and then is forced to flow north and south away from the equator. Large amounts of water vapor are deposited in the upper troposphere resulting in the moistening of the upper tropospheric environment (Price, 2000). This poleward moving air is influenced by the Coriolis force, resulting in a deviation of the winds towards the east in both hemispheres. This air continues to radiate heat to space and hence cools and sinks as it is transported northeast/southeast away from the tropics (Figure 3). The air continues to subside up until 30° north and south, heating adiabatically as it sinks. These regions of subsidence in the sub-tropics define the regions of our planet where deserts are found. Both the subsidence (resulting in high pressure at the surface) and the heating of the air as it sinks (resulting in low relative humidity) result in minimal precipitation in these regions. This meridional circulation pattern, with rising motion along the ITCZ and sinking motion in the subtopics, is know as the Hadley circulation (Figure 4), and is extremely important in the redistribution of heat, moisture and momentum on the planet. Near the surface, the sub-tropical regions and the ITCZ are connected via the north-easterly (northern hemisphere) and south-easterly (southern hemisphere) trade winds that blow between ±30°

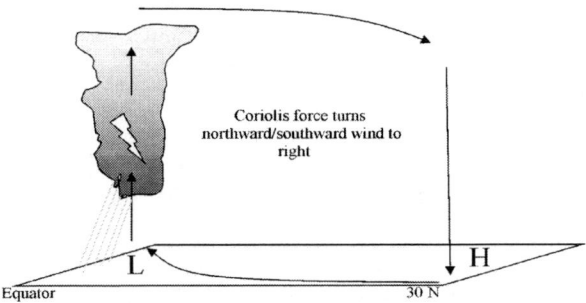

Figure 3. Schematic description of the Hadley circulation between the continental tropics and subtropics in the northern hemisphere.

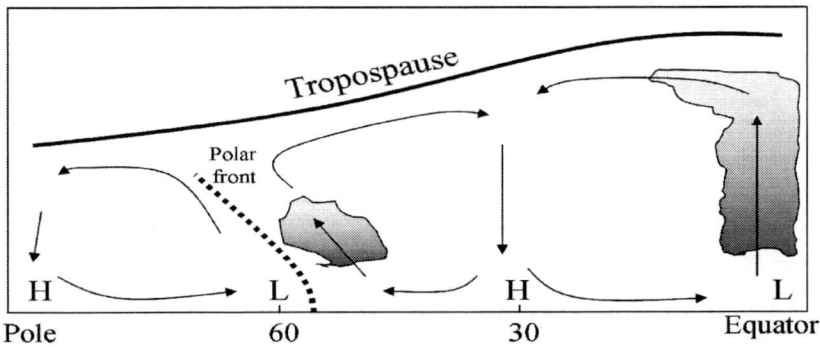

Figure 4. Schematic representation of the general circulation of the atmosphere showing 3 cells in each hemisphere, with 2 regions of convective activity in each hemisphere along the ITCZ and the polar front.

latitude. In addition to the return flow towards the equator at the surface in the Hadley circulation, part of the tropical air reaching the surface around 30° moves poleward, where it meets cold dry air arriving from the Arctic (northern hemisphere) and Antarctic (southern hemisphere) (Figure 4). This additional region of convergence, known as the polar front, is another region of forced uplift and hence the development of thunderstorms, lightning, and transient luminous events (TLEs). The polar front also migrates with the seasons from around 50-60° latitude in the summer, to 30-40° in the winter. These are the latitude bands of the midlatitude synoptic storm systems, associated with cold and warm fronts. These regions are associated primarily with summer thunderstorm activity over the mid-latitude continents, although oceanic winter thunderstorm activity and sprite formation also occurs in these regions. Sprites have been observed above winter thunderstorms in Japan and the former Yugoslavia (Takahashi et al., 2003; Jenniskens et al., 2000). As we move further poleward we encounter another region of subsidence in the polar regions, where again few thunderstorms are observed. Regions of surface convergence are normally associated with low surface pressure (L) while surface subsidence results in high atmospheric pressure at the surface (H).

4.3   Frontal Thunderstorms in Mid-Latitude Regions

Frontal thunderstorms occur at the boundary (front) between different air masses, normally cold-dry polar air meeting warm-moist tropical air. The greater the differences between the air masses (temperature and humidity) the greater the atmospheric instabilities that develop, and the greater the intensity of these storms. The intensity appears visibly as frequent lightning discharges, and sometimes sprite activity. Mid-latitude storms generally rotate around a region of low pressure (anti-clockwise in the northern hemisphere (cyclonic rotation) and clockwise (anti-cyclonic) in the southern hemisphere), while simultaneously propagating eastward around the globe with the general westerly flow in midlatitudes (30-60° latitude). Each storm has two fronts that separate the cold polar air from the warm tropical air (Figure 5a). The warm front (represented by red semi-circles) is defined by the forward motion of warm air over colder air, while the cold front (represented by blue triangles) is defined by the forward motion of cold air into regions of warm air (Figure 5a). The symbols point in the direction of frontal motion. Since cold-dry air is denser than warm-moist air, the cold air behind the cold front digs under the warm moist air, forcing it upward over a short distance (Figure 5b). The rapid uplift of air results in strong convective storms often associated with intense lightning activity. The line of thunderstorms that often form along the cold front is called a squall line (Figure 5c). On the other hand, along the warm front the warm air slowly rises over the colder denser air, resulting in a broad region of weak rainfall and showers and weak electrical activity. Therefore, for sprite observations in regions of mid-latitude storms, the cold front regions would provide the best conditions for intense lightning activity, and perhaps TLE observations. Large instabilities in the atmosphere can also occur along frontal zones that divide warm-moist air from cold-dry air around high pressure cells (e.g. The Bermuda High). The flow around a High Pressure Cell is clockwise in the northern hemisphere, opposite to the direction of flow around a low pressure system (Figure 5a). The air is stable at the center of the high pressure resulting in clear, hot and humid weather in the summer months. At the edges of the high pressure region, the mixing of cool dry air and warm moist air along the polar front allows afternoon convection to be initiated. This afternoon convection often produces a ring of thunderstorms around the high pressure center which will also rotate in a clockwise manner around the high pressure cell. This often produces the common "ring of fire" thunderstorm pattern which often develop into mesoscale convective systems (MCSs) which can produce complexes of thunderstorms called mesoscale convective complexes (MCCs). These huge thunderstorm complexes are prolific lightning producers, with many of the U.S. sprite observations occurring above these storms (Lyons et al., 2003).

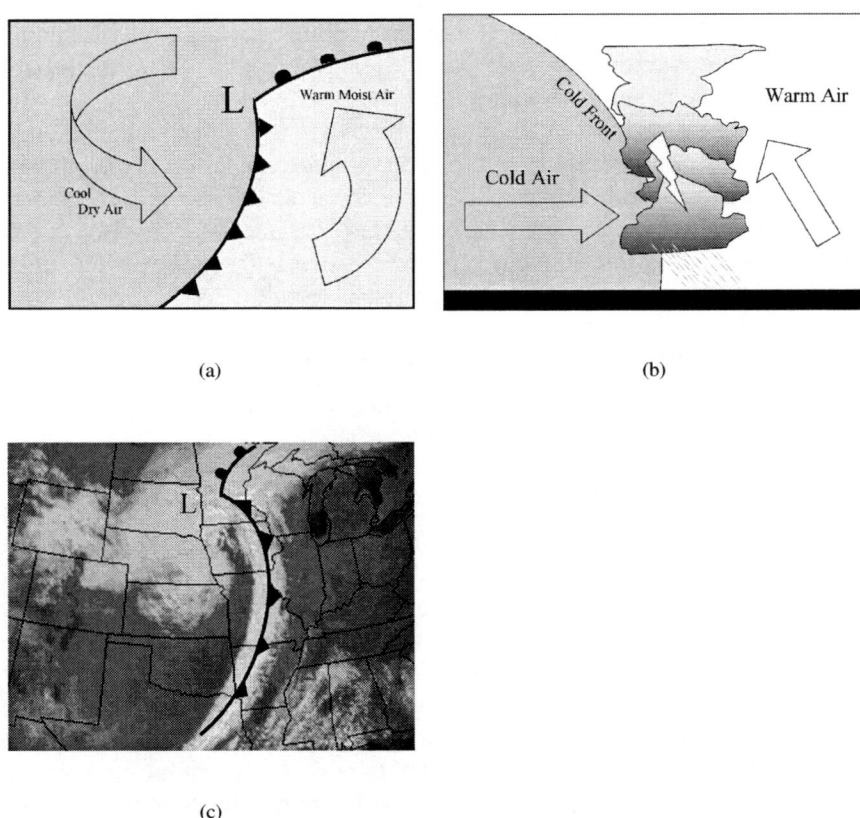

Figure 5. (a) Schematic representation of a midlatitude low pressure system with the cold and warm fronts shown with triangles and semi-circles respectively. In the northern hemisphere the rotation is anticlockwise around the low pressure center; (b) a cross section through a cold front showing the thunderstorm development along the cold front; and (c) a satellite image showing the cloud formation along the cold front.

Global thunderstorms can therefore occur in two very different environments. First, the tropical airmass type thunderstorms resulting from the diurnal heating of the surface of the Earth. These late afternoon thunderstorms occur mainly in the tropics but can also occur during summer months in mid-latitudes where instabilities (static instabilities) can develop in the afternoons on hot summer days. Second, the frontal thunderstorms occur primarily in mid- to high latitudes where different types of air masses interact and result in instabilities (baroclinic instabilities) along cold, warm and stationary fronts. Frontal thunderstorms can occur at any hour of the day, over continent and ocean, and during summer and winter. What is needed for these thunderstorms to develop is a strong density gradient between adjacent air masses. These density gradients can be caused either by temperature differences, humidity differences, or a combination of both. In general these instabilities are largest in the summer months in mid latitudes, however over the relatively warm oceans the instabilities can also be large in the winter months.

It should be noted that in addition to the above ways of producing thunderstorms we also observe thunderstorms due to orographic forcing (uplift over topography). Mountain ranges and islands force air to flow upwards and can initiate instabilities that trigger the formation of thunderstorms. Hence locations to the south of the Himalayas have very intense lightning activity due to the forced uplift of moist air penetrating inland from the Indian Ocean.

4.4 Global Observations of Lightning

In the last decade we have acquired a great deal of knowledge about global lightning and thunderstorms from satellite observations. The two primary sensors used were the Optical Transient Detector (OTD) and the Lightning Imaging Sensor (LIS) (Christian et al., 2003; Williams, 2005). The OTD sensor obtained data over a 5-year period from 1995-2000, while the LIS sensor was launched in December 1997 aboard the Tropical Rain Measuring Mission (TRMM) satellite, and is still working to this day (http://thunder.msfc.nasa.gov).

The satellite data show the annual migration of global lightning into the northern hemisphere during the northern hemisphere summer, and then southward into the southern hemisphere during their summer, in agreement with the seasonal migration of the ITCZ and the atmospheric circulation patterns. During spring and fall the distribution of lightning is fairly symmetric about the equator. Approximately 90% of the global lightning observed from space is over the continental regions, while located in the summer hemisphere. The oceanic lightning that does exists is primarily over the relatively warm oceans in the winter hemispheres along the polar front (southeast of South America, Africa and Australia during JJA and north Atlantic, Mediterranean Sea and Japan Sea during DJF). Since lightning is mainly a continental phenomenon, which occurs more often in the summer hemisphere, the northern hemisphere has more lightning activity than the southern hemisphere summer. Hence on a global basis, the Earth's lightning activity peaks in July-August, with a minimum activity in January- February (Figure 6). This asymmetry in mid-latitude land area between the hemispheres, which affects the global lightning, is not seen in the tropical lightning activity. In fact, in the tropics there is slightly more lightning in January than in June (Figure 6). This could be due to the eccentricity of the Earth's orbit around the Sun, resulting in the maximum/minimum solar radiation in the tropics in January/July. Although the satellite detectors sample only a fraction of the global lightning, being in a polar orbit, it is estimated that the global flash rate is approximately 45 flashes/second (Christian et al., 2003). This is less than 50% of the long standing estimate of 100 flashes/second (Brooks, 1925).

On a diurnal basis, tropical thunderstorms are generally active in the late afternoons, and into the evening hours as a result of the solar forcing as the sun heats up the surface during the day (Figure 7). The satellite data show that the continental thunderstorms peak between 1600-1700 local time (dominated by the tropical thunderstorms), with a minimum activity in the early morning hours (06:00-10:00 local time). Over the oceans the thunderstorms are equally distributed during the day, since the ocean temperatures are fairly constant throughout the day. While solar radiation is absorbed only within a few millimeters of the land surface during the day, the same radiation over the oceans is absorbed within 10-100 meters of the ocean surface. Hence the diurnal temperature range over the continents is much greater than over the oceans, with direct impacts on the instabilities that develop on a daily basis, influencing the diurnal thunderstorm activity. Due to the large diurnal variability, the global mean of 45 flashes/second can vary from less than 10 flashes/second, to more than 80 flashes/second. It should be noted that only part of the globe is found at local afternoon at any one time.

4.5 The Global Atmospheric Electric Circuit

If we consider the universal time variations of thunderstorm activity, we find three maxima during each 24 hour period, displayed in Figure 8a (Whipple, 1929). These three maxima correspond to the three tropical thunderstorm regions (Figure 2) that each peak in the late afternoon hours (Figure 7) as a result of solar heating of the surface during the day. The first peak at 09 UT is due to thunderstorm activity in southeast Asia, the second peak at 14 UT is due to lightning activity in Africa, while the third peak at 20 UT is a result of thunderstorm activity in South America. Plotting the global thunderstorm activity as a function of universal time (Figure 8a "world") shows a minimum in thunderstorm activity near 03 UT, when the sun is over the Pacific Ocean,

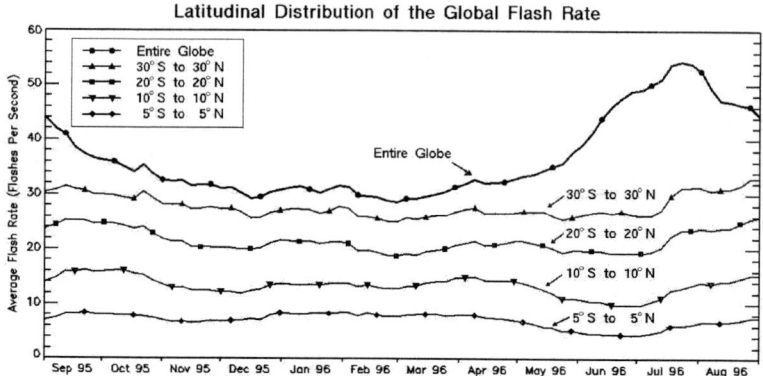

Figure 6. The monthly variations of lightning flash rate for different latitude bands (http://thunder.msfc.nasa.gov)

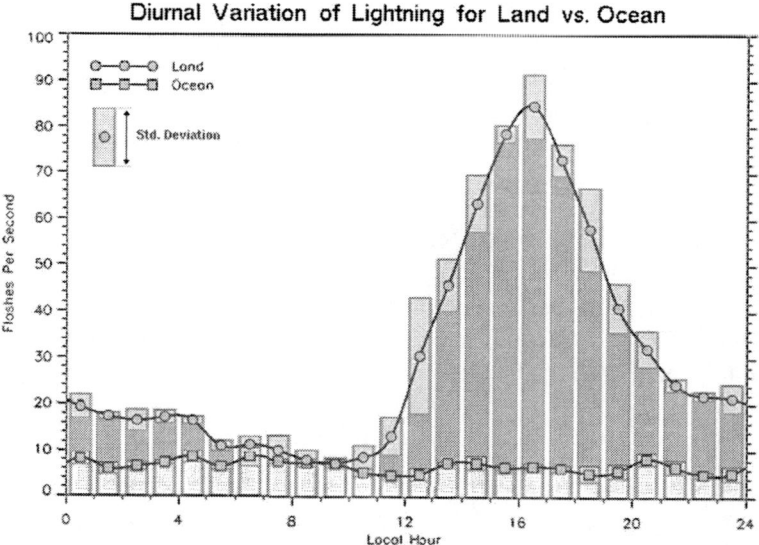

Figure 7. Local hour variations of global lightning activity over land and ocean (http://thunder.msfc.nasa.gov)

and a maximum in thunderstorm activity between 14 UT and 19 UT. Over many years researchers have measured the electrical potential gradient close to the Earth's surface in fair weather regions (no clouds, rain or precipitation nearby, and low pollution levels) and have noticed that often the electric field (mean value of 100 V/m) shows a similar diurnal variation as described above (Mauchly, 1923; Hoffman, 1923). Measurements aboard the Carnegie research vessel in the 1920s showed that the mean electric potential at the surface in clear-sky conditions varies in a way very similar to global thunderstorm activity (Figure 8b). Furthermore, integrating the vertical electric potential with height (using balloons or aircraft) gives the potential difference between the ionosphere and the Earth, know as the ionospheric potential, that has a mean value of 250 kV (Markson, 1985). Since the electric field drops rapidly with altitude, a good estimate of the ionospheric potential can be obtained by integrating the field within the troposphere (up to altitudes of 20 km). The ionospheric potential has also been shown to exhibit a diurnal cycle similar to the Carnegie Curve with a maximum near 18 UT and a minimum around 03 UT displayed in Figure 8b (Markson, 1986). It is therefore believed that the Earth-atmosphere system acts like a global electric circuit (Bering et al., 1998; Rycroft et al., 2000). The thunderstorms and regions of electrified storms act as the generators (batteries) of the circuit. Conduction currents flow upward above the storms to the ionosphere. The ionosphere is close to an equipotential surface and therefore the currents flow horizontally around the globe, and return to the Earth's surface in the fair weather regions. These currents can be measured at the surface and are of the magnitude of 2 pA/m^2 (1 pA = 10^{-12} Amperes). The Earth is also highly conductive and hence the currents flow back within the Earth to the regions of thunderstorm activity to close the electric circuit. The currents observed in the atmosphere are due to the finite conductivity of

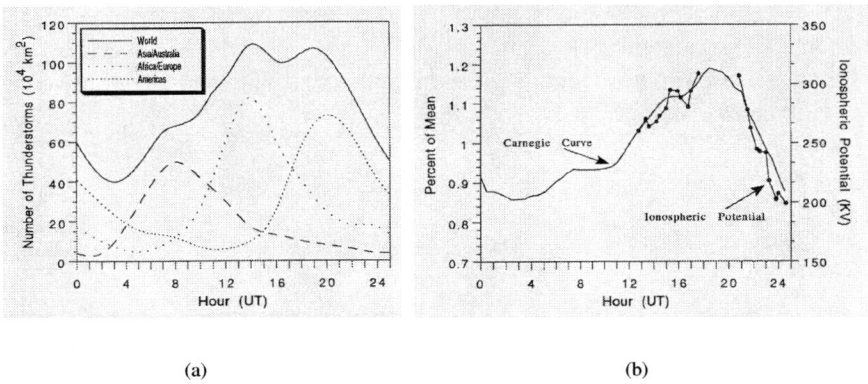

Figure 8. (a) The diurnal variations of thunderstorm area in the three tropical continental regions (adapted from Whipple, 1929), and (b) The diurnal variations of the fair weather potential gradient PG about the mean value of 1, together with measurements of the ionospheric potential (Markson, 1986).

the atmospheric dielectric. It is easy to show that if all thunderstorm activity were to cease, the global electric circuit currents, fields, and charges would decay and disappear within 10 minutes (Bering et al., 1998). Hence the constant charge on the Earth (500,000 Coulombs), the fair weather fields and currents are maintained by the never ending thunderstorm activity around the globe.

4.6 Future Directions

As has been described above, there are different types of thunderstorms around the globe, however, the global studies of sprites are limited to a small number of locations. Only now are we starting to use space platforms to investigate global sprite distributions (Yair et al., 2004; Chern et al., 2003). One recommendation would be to expand field experiments to additional regions of the globe (Lyons, Chapter 2). Most research has actually focused in mid-latitude thunderstorms, while the large majority of thunderstorms occur in the tropics. How does tropical sprite activity differ from mid-latitude sprite activity above thunderstorms (Mende, Chapter 6)? How do the parent lightning flashes that trigger sprites differ (if at all) between regions. Winter thunderstorms are generally much smaller physically than summer thunderstorms, yet still produce sprites. What is special about the winter storms that produce sprites? Most sprites are produced by positive polarity lightning. Is this true in all sprite-producing thunderstorms? Are sprites over the oceans also produced by positive lightning? Studies show many large (charge moment) negative discharges over the oceans (Füllekrug et al., 2002; Greenberg and Price, 2004), although few sprite observations have been targeted above oceanic thunderstorms. Do the physical characteristics of sprites differ according to the type of thunderstorm, region, season and lightning type? And finally, how important are sprites and their associated lightning to the global atmospheric electric circuit ? Are sprite-producing storms the main contributors to the global atmospheric electric circuit ?

Bibliography

Baker, M. B., Blyth, A. M., Christian, H. J., Latham, J., Miller, K. L., and Gadian, A. M. (1999). Relationship between lightning activity and various thundercloud parameters: Satellite and modeling studies. *Atmos. Res.*, 51:221–236.

Baker, M. B., Christian, H. J., and Latham, J. (1995). A computational study of the relationships linking lightning frequency and other thundercloud parameters. *Quart. J. Roy. Met. Soc*, 121:1525–1548.

Bering, E. A. III, Few, A. A., and Benbrook, J. R. (1998). The global electric circuit. *Phys. Today*, 51(10):24–30.

Brooks, C. E. P. (1925). The distribution of thunderstorms over the globe. *Geophys. Mem.*, 3(24):147–164.

Chern, J. L., Hsu, R. R., Su, H. T., Lee, L. C., Mende, S. B., Fukunishi, H., and Takahashi, Y. (2003). Global survey of upper atmospheric Transient Luminous Events on the ROCSAT-2 satellite. *J. Atmos. Sol.-Terr. Phys.*, 65(5):647–659.

Christian, H. J., Blakeslee, R. J., Boccippio, D. J., Boeck, W. L., et al. (2003). Global frequency and distribution of lightning as observed from space by the Optical Transient Detector. *J. Geophys. Res.*, 108(4005):doi:10.1029/2002JD002347.

Füllekrug, M., Price, C., Yair, Y., and Williams, E. R. (2002). Intense oceanic lightning. *Ann. Geophys.*, 20:133.

Greenberg, E. and Price, C. (2004). A global lightning location algorithm based on the electromagnetic signature in the Schumann resonance band. *J. Geophys. Res.*, 109(D21111):doi:10.1029/2004JD004845.

Hoffman, K. (1923). Bericht über die in Ebeltofthafen auf Spitzbergen in den Jahren 1913/4 durchgeführten luftelektrischen Messungen. *Beitr. Phys. Atmos.*, 11(1). Leipzig.

Jenniskens, P., Butow, S., and Fonda, M. (2000). The 1999 Leonid multi-instrument aircraft campaign – an early review. *Earth, Moon and Planets*, 82-83:1–26.

Jorgenson, D. P. and Lemone, M. A. (1989). Vertical velocity in oceanic convection off tropical Australia. *J. Atmos. Sci.*, 51:3183–3193.

Lemone, M. A. and Zipser, E. J. (1980). Cumulonimbus vertical velocity events in GATE. Part I: Diameter, intensity and mass flux. *J. Atmos. Sci.*, 37:2444–2457.

Lyons, W. A., Nelson, T. E., Williams, E. R., Cummer, S. A., and Stanley, M. A. (2003). Characteristics of sprite-producing positive cloud-to-ground lightning during the 19 July 2000 STEPS mesoscale convective system. *Mon. Wea. Rev.*, 131:2417–2427.

Markson, R. (1985). Aircraft measurements of the atmospheric electric global circuit during the period 1971–1984. *J. Geophys. Res.*, 90:5967–5977.

Markson, R. (1986). Tropospheric convection, ionospheric potential and global circuit variations. *Nature*, 320:588–594.

Mauchly, S. J. (1923). Diurnal variations of the potential gradient of atmospheric electricity. *Terr. Magn. Atmos. Electr.*, 28:61–81.

Petersen, W. A., Nesbitt, S. W., Blakeslee, R. J., Cifeli, R., Hein, P., and Rutledge, S. A. (2002). TRMM observations of intraseasonal variability in convective regimes over the Amazon. *J. Clim.*, 15:1278–1294.

Price, C. (2000). Evidence for a link between global lightning activity and upper tropospheric water vapor. *Nature*, 406:290–293.

Price, C. and Rind, D. (1992). A simple lightning parameterization for calculating global lightning distributions. *J. Geophys. Res.*, 97:9919–9933.

Rycroft, M. J., Israelsson, S., and Price, C. (2000). The global atmospheric electric circuit, solar activity and climate change. *J. Atmos. Sol.-Terr. Phys.*, 62:1563–1576.

Takahashi, Y., Miyasato, R., Adachi, T., Adachi, K., Sera, M., Uchida, A., and Fukunishi, H. (2003). Activities of sprites and elves in the winter season, Japan. *J. Atmos. Sol.-Terr. Phys.*, 65:551–560.

Whipple, F. J. W. (1929). On the association of the diurnal variation of the electrical potential gradient in fine weather with the distribution of thunderstorms over the globe. *Quart. J. Roy. Met. Soc.*, 55:1–17.

Williams, E., Chan, T., and Boccippio, D. (2004). Islands as miniature continents: Another look at the land-ocean lightning contrast. *J. Geophys. Res.*, 109(D16206):doi:10.1029/2003JD003833.

Williams, E. and Stanfill, S. (2002). The physical origin of the land-ocean contrast in lightning activity. *Compt. Rend. Phys.*, 3:1277–1292.

Williams, E. R. (2005). Lightning and climate: A review. *Atmos. Res.*, 75:272–287.

Williams, E. R. et al. (2002). Contrasting convective regimes over the Amazon: Implications for cloud electrification. *J. Geophys. Res. - LBA Special Issue*, 107(D20-8082):doi:10.1029/2001JD000380.

Williams, E. W. and Satori, G. (2004). Thermodynamic and hydrological comparison of the two tropical continental chimneys. *J. Atmos. Sol.-Terr. Phys.*, 66:1213–1231.

Yair, Y., Israelevich, P., Dvir, A. D., Moalem, M., Price, C., Joseph, J. H., Levin, Z., Ziv, B., Sternlieb, A., and Teller, A. (2004). New observations of sprites from the Space Shuttle. *J. Geophys. Res.*, 109(D15201):doi:-10.1029/2003JD004497.

IMAGING SYSTEMS IN TLE RESEARCH

Thomas H. Allin [1], T. Neubert [2] and S. Laursen [2]

[1] *Measurement and Instrumentation Systems, Oersted DTU, Technical University of Denmark.*

[2] *Danish Space Research Institute, Copenhagen, Denmark.*

Abstract Almost since first documented in 1989, it has been clear that red sprites, and in general Transient Luminous Events (TLEs) are not temporally resolved at video rates. Still, much can be learned by imagery of sprites at video rates – and, at least, the occurrence of sprites and TLEs is very well documented using the right equipment in the right way.

This chapter provides an introduction to the concepts of low light imagers, and how they can be successfully applied in TLE research. As examples, we describe the 2003 and 2004 Spritewatch systems, which integrate low-light cameras with a digital processing system, and is controllable over the internet.

5.1 Introduction to Low Light Imaging

Low light conditions generally refer to situations in which the human eye, due to darkness, is incapable of providing useful information for orientation. A more standardised definition would be when the ambient light flux is below 1 lux, corresponding roughly to the illumination at the Earth surface from the full moon. With typical optical intensities in the 1-100 mlux range, Transient Luminous Events (TLEs) are low light phenomena.

Imaging in low light conditions is non-trivial for several reasons. Electronic photodetector devices, such as Charge Coupled Device (CCD) sensors, exhibit non-linear behaviour at low light conditions. Simply stated, at low light levels, the noise level on the output of the photodetector circuit becomes comparable to the contribution from the photonic signal. This means that the image quality becomes poor, or even absent.

The reasons for an inadequate photonic signal for high-quality imaging may be many. A common denominator in the description of imaging systems is the *exposure* – the number of photons reaching a detector per unit area per unit time. Hence, at low light levels, the following conditions are all limiting image quality, considered equivalent to the signal-to-noise ratio (SNR) of the electronic sensor:

- Weak photonic signal at sensor surface – source too dim for sensor exposure to ever exceed noise level of sensor, or optical entrance too narrow.
- Too short exposure time – insufficient time for significant buildup of photo-generated charge in sensor pixels.
- Too small pixel size – insufficient area for collecting photons.

From an observer's point of view, there is little to do about the source, unless the experiment takes place in a controlled environment. Most astronomical or geophysical imaging experiments fall outside this category. Hence, to achieve the best possible image of faint sources, it is mandatory to use large aperture optics, and large pixel dimensions. The latter requirement of course limits the achievable spatial resolution.

5.1.1 Optics

The effect of entrance optics on the image intensity is described by the following equation, relating the emitted radiance E from a source to the received irradiance L at the sensor surface, projected by an ideal spherical lens of diameter d and focal length f:

$$L = \frac{\pi}{4} \left(\frac{d}{f}\right)^2 E \cos^4 \alpha \qquad (5.1)$$

– where α is the direction to the object from the sensor surface normal. Radiance is defined as the amount of light energy passing a unit surface per unit time per unit solid angle, measured in W/m²/sr. Hence, equation (5.1) only accounts for the part of the radiance that reaches the lens. Irradiance is the integral over solid angle of the radiance, measured in W/m² and thus the most natural measure of light intensity at the surface of an imaging sensor.

From equation (5.1), it is seen that the optical throughput of a lens scales with the ratio $1/F^2 = (d/f)^2$, where F is the F-number of the lens. Increasing the aperture of a lens while maintaining the focal length results in larger image irradiance. But the image size remains unaltered, as long as the focal length is unaltered. So, if the angular field of view is to remain constant, the sensor size and focal length should be scaled in proportion.

5.1.2 Electronic Imaging Sensors

Most electronic cameras today are based on the Charge Coupled Device (CCD). Other sensor types, such as Active Pixel Sensors, are emerging, and will probably dominate the market for consumer devices before long. The functionality of a CCD is based on change of conduction properties of semiconductor materials when exposed to light in the visible and near-infrared parts

IMAGING SPRITES

of the spectrum. A CCD is typically made of a rectangular silicon substrate, of some 10-100 microns of thickness, the top of which divided spatially into a rectangular mesh of electrodes and channel stops (Figure 1). When set to positive voltages with respect to the substrate bottom, the electrodes and channel stops confine the motion of electrons in the conduction band of the substrate. The rate at which photons arrive determine how many conduction-band electrons are created per unit time in a pixel. As long as an electrode is kept high with respect to the substrate bottom, the pixel is integrating the photo-generated charge. When the integration completes, nearby electrodes exchange electronic charge by switching on and off in a well-defined pattern – this is the technique of *charge coupling*. This is repeated until the electrons reach a horizontally oriented readout register, through which they are transferred until reaching the output gate of the CCD. The output gate is typically a combination of a sense capacitor, inverting the electronic signal, and an amplifier (sometimes referred to as a *charge to voltage converter*). The *sensitivity* of a CCD sensor is sometimes quoted as the increase in output level of this amplifier per photoelectron, in units of μV/electron. The *full well capacity* is the number of photoelectrons that will saturate the output amplifier. Along with the sensitivity, this defines the *dynamic range* of the sensor:

$$\text{Dynamic range} = \frac{\text{Full well capacity}}{\text{sensitivity}} \qquad (5.2)$$

CCD sensors are often categorised according to their physical configuration of imaging and optional charge storage sections. Figure 2 shows the three most common types, and how photocharge is transferred towards the output gate for each. Evidently, the Frame Transfer (FT) CCD has the largest physical foot-

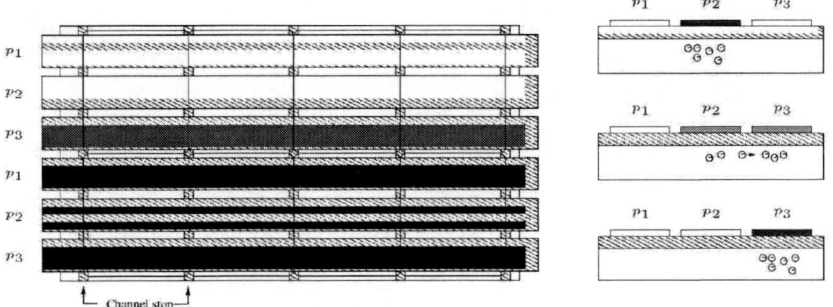

Figure 1. Left: A CCD example layout. Rounded squares mark the subdivision of the CCD into pixels. The clock phases p_1, p_2, p_3 control the vertical movement of electrons in the substrate, while horizontal movement is prohibited by the vertical channel stops. Right: Principle of charge coupling. A shift of charge position is obtained by sequencing the clock phases as shown.

print, while the Full Frame Transfer (FFT) CCD and the Inter-Line Transfer (ILT) CCDs have comparable dimensions. The FFT and the FT CCDs both expose the pixels during readout, and have the largest possible sensitive area (fill factor), while the ILT CCD sacrifices some of its sensitive area to provide a shielded charge storage region as close to the pixels as possible. This means that the ILT CCD has excellent snapshot capabililties, and very short exposure times can be obtained – down to tens of microseconds. The FFT CCD is typically used in high-end, large-format digital Single Lens Reflector (SLR) cameras used by professional photographers, or packed in the focal plane of large astronomical telescopes, while the FT and ILT sensors are more common in video cameras.

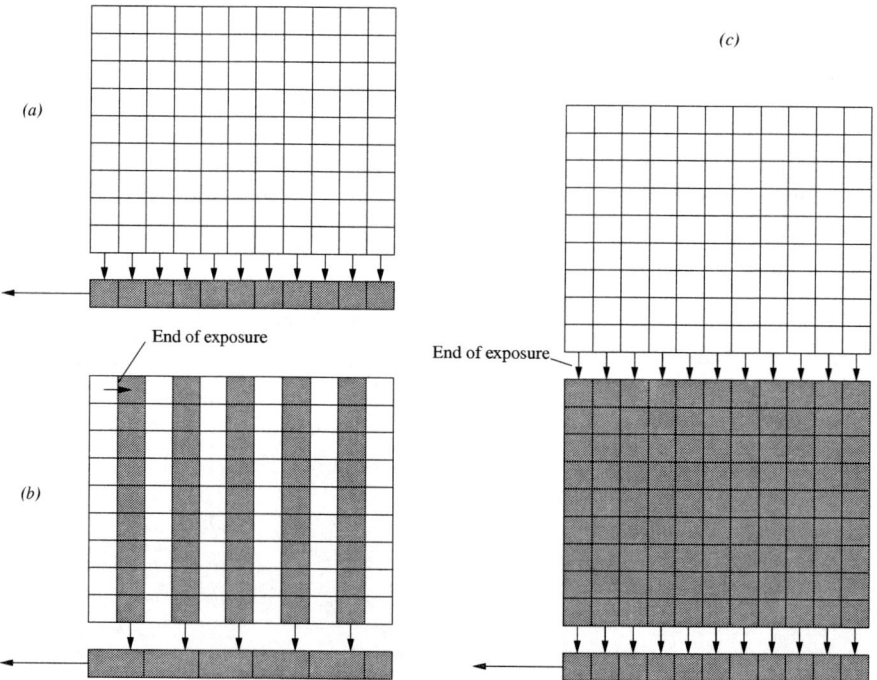

Figure 2. CCD types and readout. (a) Full Frame Transfer (FFT) CCD, (b) Inter-Line Transfer (ILT) CCD, and (c) Frame Transfer (FT) CCD.

5.1.3 Noise and Dynamic range

CCD sensors contain a number of intrinsic noise sources, which add to the photonic signal in a random fashion. These include dark current noise, clock noise, output amplifier noise and reset noise (Holst, 1998). Also the photonic signal fluctuates, dictated by the Poisson distribution: If on average $\overline{m}$ photons are detected in a fixed time interval, then the standard deviation in a count series will be given by $\sigma_m^2 = \overline{m}$. If the photonic fluctuations are independent from circuit noise σ_c, then the following expression provides an estimate of the signal-to-noise ratio at a certain signal level:

$$\text{SNR} = \frac{\eta \overline{m}}{\sqrt{\eta \overline{m} + \sigma_c^2}}, \qquad (5.3)$$

where η is the *quantum efficiency* of the sensor – the probability that a photon will contribute to the electronic signal in a pixel. In terms of electrons, the circuit noise typically contributes $10 - 100$ electrons per pixel rms. Hence, to achieve a signal-to-noise ratio of 100, at a wavelength of 550 nm ($\eta = 0.4$), about 30,000 photons per pixel are needed at the sensor surface. If the effective pixel area is 100 μm^2, and the exposure time 20 ms, this corresponds to a surface irradiance of about 1.5 MR, where 1 R=10^6 photons/cm^2/s.

The dynamic range of an imaging system is the range of input signals over which the system responds linearly. This can be illustrated by the *photon transfer curve*, showing the Root Mean Square (RMS) noise of the output signal versus the average of the output signal over time. Such a curve is shown in Figure 3. The first flat section corresponds to the part where circuit noise dominates. The curve breaks at about 100 electrons, defining the *limiting sensitivity* of the camera. The second part, where the slope of the log-log curve is 1/2, corresponds to the linear range of the sensor – or where the photon noise dominates over circuit noise. The last part of the curve shows sensor saturation, where the output signal no longer increases due to an increase in the photonic signal. For the device in question, the dynamic range covers from about 100 to 100,000 photoelectrons (full well), representable by $\log_2(100,000/100) \simeq 10$ bits.

5.1.4 Intensified versus Non-intensified Imaging

In some cases, it may be advantageous to amplify the photonic signal prior to detection by e.g. a CCD sensor. If the photonic signal is too faint to dominate over the circuit noise in the camera, the signal-to-noise ratio will become poor, perhaps not even exceeding 1. Image intensification is, however, also a noisy and electromechanically bulky affair. Today, mostly micro-channel plate intensifiers (Figure 4) are used, in which photons liberate electrons on a coated input window of the device. A high voltage applied over a capillary structure provides small acceleration tunnels, in which the photoelectrons multiply in a random avalanche fashion, until they reach the output window.

Image intensification introduces two types of noise, in addition to the noise sources intrinsic to the imaging sensor: (1) Gain noise, due to the arbitrary

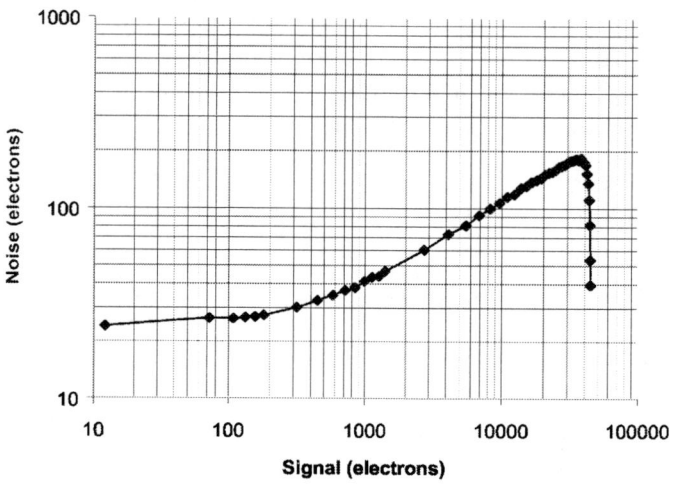

Figure 3. Photon Transfer Curve for Full-Frame Transfer CCD

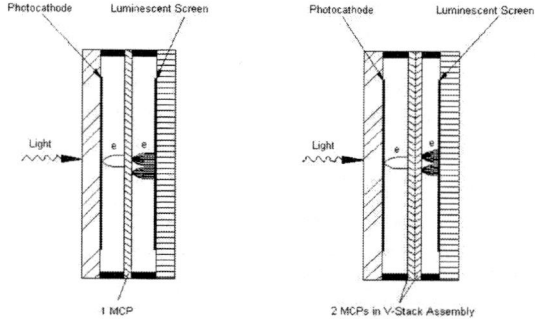

Figure 4. Micro-channel Plate image intensification. Adapted from http://www.proxitronic.de

process of photoliberation of electrons at the input window, the avalanche multiplication process, and finally the conversion of the avalanche back into light at the output window. (2) Thermal electrons and occasionally, ions, leaving the input window unrelated to photon arrival, but still amplified through the high voltage, causes the output window to emit light even in dark conditions.

5.1.5 Summary

Although CCDs are today more sensitive than ever, with bare CCD cameras perfectly able to resolve nearly any kind of TLE (see page 114), there are limiting cases, such as narrow-band spectroscopic imaging, where optical intensification may be necessary. In such applications, the spatial correlation between pixels rather than the individual pixel's signal-to-noise ratio provide the useful information that forms the image. Image intensification will, in the range of applications where imaging with bare CCD cameras is possible, usually not yield better image quality, and is hence only rarely considered worth the extra complexity and cost.

Finally, it is worth mentioning that bare CCD cameras exist, capable of imaging in mlux conditions. Especially, cameras based on the Sony ExView sensor family can be used in night-time conditions, due to their extended sensitivity into the near-infrared. For TLE research, and sprites in particular, this is close to optimal. ExView cameras are available from Watec, JAI, Ikegami and Hitachi, to name a few.

5.2 The Spritewatch Systems

In the summer of 2003, the first Spritewatch system was installed on Pic du Midi de Bigorre, France, 2877 m above sea level. The system was developed in a joint effort between the Danish Space Research Institute, and the Measurement and Instrumentation Systems group at Oersted DTU, Technical University of Denmark. The 2003 Sprite Campaign that included activities conducted by a number of research institutes from Europe, South Africa and the USA, aimed at investigating the effect of transient luminous events on the atmospheric layers in which they occur. Among the new findings during the campaign were the unambiguous detection of infra-sound signatures from red sprites (Farges et al., 2004) and the apparent one-to-one association of red sprites with modifications of the electrical properties of the lower ionosphere (Haldoupis et al., 2004). Also, the campaign team successfully carried out coordinated measurements in geomagnetically conjugate regions, to search for the signature of relativistic electron beams from red sprites in the conjugate hemisphere (Marshall et al., 2004).

5.2.1 The 2003 Spritewatch System

The primary objective of the 2003 Spritewatch system was to provide high-fidelity documentation of the occurrence of sprites and possibly other TLEs, using low-light video imagery. This was to be obtained using a remotely guided camera system, connected to a local processing unit that would do continuous, real-time event detection with no blind periods due to data processing loads. Further, the system should, in real time, report the detected events to a system controller over the internet, and automatically transfer the recorded data to the web server responsible for data distribution to the campaign participants.

A secondary objective was to develop and test a collection of event detection algorithms that would enable the recognition of certain types of TLEs, and an efficient discrimination between this class of events and more common events, such as lightning. The requirements put forth by the science objectives were thus

- Remote guidance of camera unit over the internet.
- Continuous and event-based digital video imagery of highest possible spatial resolution, using two co-aligned cameras.
- Precise (20 ms accuracy) time-tagging of events.
- Real-time data processing and event reporting.
- Continuous system access.

To ensure that these requirements were met, the following issues were addressed during hardware and software selection:

- System ruggedness and stability – should withstand tough weather, including nearby lightning, and recover seamlessly from occasional power- and network outages.
- Real-time availability of uncompressed video data in host memory of local processing unit. Any bottlenecks for the video data would prevent efficient real-time analysis.
- Enough processing capacity in local processing unit for parallel event detection and movie compression.
- Ability to respond to events in rapid succession. Processing, data storage and log reporting should take less time than the field exposure time.
- Prioritising of processes – the system should maintain, as its highest priority, the analysis and local storage of incoming data.

This resulted in the configuration of units as displayed in Figure 5, and described in Table 1.

5.2.2 Hardware in the 2003 system

Rooftop Units The rooftops unit were attached to the metal roof of the supporting building using a mount plate, by means of a star-shaped pattern of space-qualified double-sticking tape. On the mount plate, the guiding unit was fixed – a heavy-duty pan/tilt drive, capable of driving a load of 10 kg to its extreme positions. The Quickset QPT-20 was chosen, due to its proven performance in harsh conditions. Also, the QPT-20 features a microcontroller and is

IMAGING SPRITES 109

operable over a serial RS-232 connection, interfacing easily with control computers. The unit was supplied with 24 V DC from an interface box placed in the building supporting the rooftop mount.

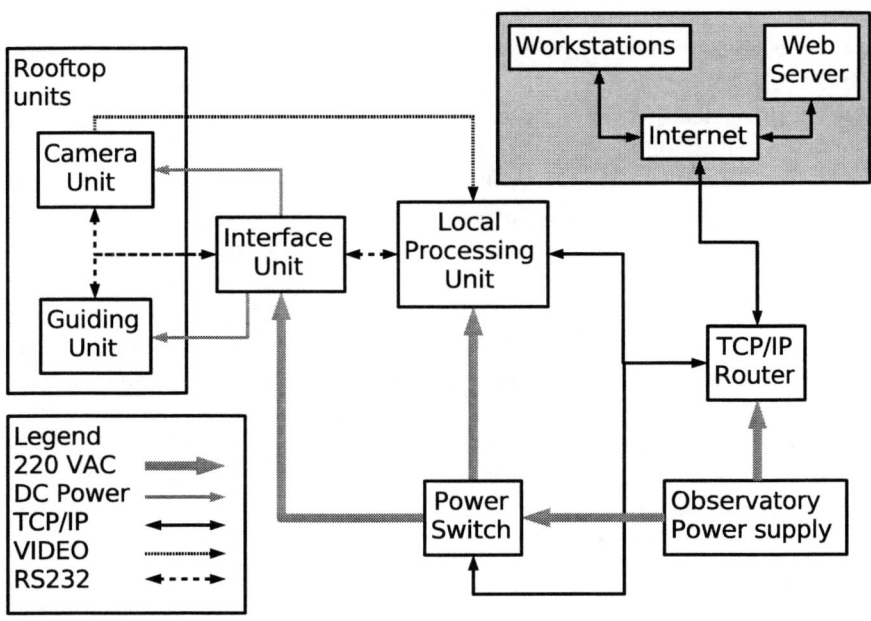

Figure 5. Units in the 2003 Spritewatch system. Units not at Pic du Midi is in the gray box.

Table 1. Table of units in the 2003 Spritewatch system.

Item	Model	Description
Rooftop units		
Guiding unit	Quickset QPT-20	RS-232 I/F
Cameras	JAI CV-S3200	B/W CCD, RS-232 I/F
Indoor units		
Local Processing Unit		PC, P4 Processor
Power Switch	Blackbox	TCP/IP control
Interface unit		Power and RS-232 distribution
Router	D-Link	1/4 up/downlink ports, with NAT & Firewall

A sealed camera housing, containing the two monochrome CCD cameras, was mounted as the load of the pan/tilt unit, as shown in Figure 6. The front entrance to the housing was covered by a high-transmission glass, and heaters inside the housing prevented dew from forming on the inner surfaces. When idle, the camera unit could be parked nose-down, in order to prevent precipitation from reaching the front glass. The cameras chosen were the JAI CV-S3200, and were fitted from the factory with a monochrome type 1/2 Sony ExView HAD CCD sensor. A brief list of properties of these cameras is given in Table 2 – more detailed data sheets can be obtained from http://www.jai.com. The ExView technology allows a relatively higher detection efficiency towards the red and near-infrared parts of the spectrum, which is the primary emission range for the phenomena under investigation. Entrance optics for the cameras were two 16 mm F1.4 monofocal lenses, with a spot size of about 5 μm at 800 nm. The cameras support a variety of integration modes. Long time exposures (up to 128 fields of 20 ms) can be obtained, while a pure 20 ms exposure is readily obtained by turning off the electronic shutter and using the 2:1 interlace readout method. This implies that the horizontal lines in the image are binned in pairs, decimating the vertical resolution by a factor of two. Effectively, the cameras each output a video stream of a spatio-temporal resolution of $752 \times 291 \times 50 = 10,941,600$ pixels per second. The limited sensitivity of the cameras when operated at 20 ms integration is about 1 mlux, established by the number and visual magnitude of stars in the field of view during test observations.

Local Processing Unit The Local Processing Unit (LPU) was a standard personal computer, shown schematically in Figure 7. The purpose of the LPU was to digitise, process and analyse the incoming video streams, and to act as

 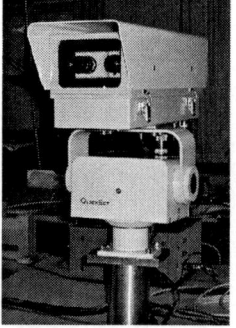

Figure 6. Pic du Midi de Bigorre, 2877 m, 42.9°N,0.01°E, and the 2003 Spritewatch system in the laboratory.

IMAGING SPRITES

Table 2. Properties of the JAI CV-S3200 camera.

Feature	Description
CCD	Sony ICX249AL ExView HAD, type 1/2 (8 mm diagonal)
CCD geometry	752×582 eff. pixels
Pixel size	$8.3 \times 8.6 \mu m^2$
Integration modes	field (20 ms), frame (40 ms), 2^n fields, $n = 1, 2, \ldots, 7$
Video format	CCIR-601, 50 fields per second
Readout method	Progressive, interlace (2:1)
Other	AGC (auto, manual 0/12/24 dB), shutter (off - 1/10000 s)
Sensitivity	1 mlux at 20 ms exposure
Setup	RS-232 I/F
Power	12 V DC / 150 mA (1.8 W)

the link between the system controller and the rooftop units. To realise this, the configuration of devices was chosen as listed in Table 3. The frame grabbers, providing the entry for video data to the LPU, were standard consumer television cards for the PCI bus. The key component on the grabbers is the Conexant BT878A video decoder, which carries out analogue video prefiltering, video digitisation, digital video scaling, digital video format conversion

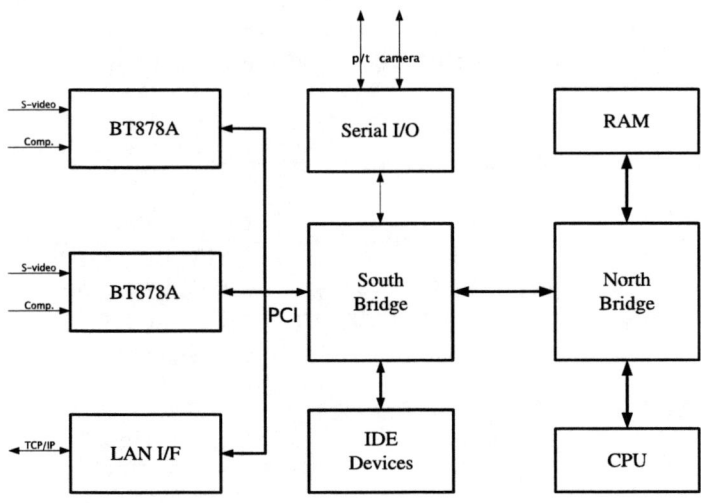

Figure 7. Block diagram of the 2003 Spritewatch local processing unit.

and transfer of digitised video to main memory of the computer in real time. The BT878A supports resolutions up to 768×576×3 bpp[1] (RGB) in the destination format, and has an on-chip FIFO buffer of 76 32-bit words and PCI busmastering capabilities, implying some tolerance towards PCI bus latencies. The BT878A can operate on both, field- and frame basis, depending on the chosen resolution and video format. The BT878A has a built-in line-locking feature, and synchronisation is based on the video source.

Beyond the frame grabbers, most components were standard items. The volatile memory space consisted of 1 GB RAM, and non-volatile memory space was available as an internal IDE hard drive and an external USB hard drive for backup. The main board featured an SIS 645DX chipset, hosting an Intel Pentium IV processorprocessor running at 2.553 GHz. The observatory provided continuous internet connection with a capacity of 2 Mbit/s to be shared between all observatory activities.

5.2.3 Software in the 2003 system

Operating System For the 2003 campaign, the LPU operating system was based on a subset of packages from the Red Hat Linux distribution, version 7.1. Prior to usage in the campaign, all packages were updated to their latest stable versions, or in some cases, recompiled from their sources if no official update was released.

The Linux kernel, containing nearly all hardware-specific libraries in the form of loadable modules, was compiled from its official sources, available from http://www.kernel.org. The most recent stable kernel was chosen, at that time version 2.4.20. In addition, a number of patches were applied to the kernel, in order to make use of the latest driver releases for the frame grabber. Among the patches was the bttv driver, and the latest version of the Video

Table 3. Devices in the 2003 Spritewatch local processing unit.

Device	Model	Description
Processor	Intel P4	2.553 GHz, 533 MHz bus
Mainboard	ASUS P4S533-E	SIS 645DX Chipset
Frame grabbers	Pinnacle PCTV-RAVE	Conexant BT878A video decoder
LAN I/F		On-board
Internal hard drive	Seagate ST3120023A	120 GB
External hard drive	Maxtor	250 GB, USB

[1]Bytes per pixel at 25 Hz = 33,177,600 bytes per second

IMAGING SPRITES   113

4 Linux II video kernel interface layer (the V4L2 API), both available from
http://linux.bytesex.org.

Camera Control Software   No prefabricated control software existed for the
JAI cameras for the Red Hat Linux operating system. Therefore, the utility
cvs3200 was written, with the RS-232 communication protocol kindly made
available by JAI A/S. The program is free software, and it is released under the
GNU General Public License at http://www.allinux.dk.

Guiding Software   No control software existed for using the Quickset QPT-
20 guiding unit with the Linux operating system. Therefore, the utility qt20
was written – a command-line tool, easy to integrate with the system scheduler
in order to generate sequences of target acquisition. The program implements
only a small part of the features and capabilities of the QPT-20 unit – at present,
motion is restricted to 1 dimension (pan or tilt) at a time. The program is
free software, and it is released under the GNU General Public License from
http://www.allinux.dk.

Event Detection Software – the V4L2 API   The Linux video interface layer
and application programming interface, the V4L2 API (Dirk and Schmiek,
2004), is responsible for the control of video acquisition or playback units in a
Linux computer. Basically, when a frame grabber is present in the computer,
and driver modules have successfully been loaded into memory, the frame
grabber will appear as the file /dev/videoN in the file system, where N is
a small integer (at least, no larger than the number of video devices in the system). Any program can now read (capture) and write (play back, if supported)
video data from and to the device, using the primitive read() and write()
system calls, assuming that these calls make sense for the device in use.

Any driver conforming with the V4L2 API must negotiate with the kernel
a memory space allocated when the driver is loaded, to which video data is
transferred. Video data can then be exchanged with applications using this
memory space. The memory management in Linux provides a great advantage
compared to other operating systems. For example, it is possible to lock a large
part (say, 512 of 768 MB) of physical memory (RAM) for this purpose alone,
which is enough for 30-50 seconds of uncompressed video, depending on the
format. Locking the memory implies that it is never swapped to hard disk,
which gives a large performance increase of the event detection software.

The video memory space is divided into buffers, each corresponding to one
video frame. Beyond room for the video data itself, the buffers also hold a
number of useful pieces of information. This includes the buffer size, format,
and timestamp. When the capture device is initialised, the addresses of destination buffers are handled to the frame grabber. From here, the application

controls the buffer usage. The timestamp is inserted by the video driver, immediately after the last DMA transfer of a video image. Thus, the images are timestamped to an accuracy directly comparable to that of the system clock.

For the 2003 Spritewatch system, the initial version of the Spritewatch program had its debut. The program controlled the video data stream from one of the BT878A-based frame grabbers, and maintained an infinite loop of image analysis and event reporting. During 2003 and 2004, the program evolved into a multithreaded program, capable of controlling multiple frame grabbers in parallel – see page 117.

Miscellaneous To provide login access for system controllers, the LPU ran the Secure Shell server, enabling an encrypted communication channel of between control workstations and the LPU. Data were transferred to the campaign server using the File Transfer Protocol (FTP).

To ensure accurate timing, the Network Time Protocol daemon, ntpd, was also installed. Using the travel time of network packets over the internet, along with the reference time from a time server synchronised by an atomic clock, the ntp daemon periodically adjusts the computer clock so that the system time is accurate to within 10 ms from UTC on average. By tuning the ntp daemon, essentially by adjusting the clock more often, and using nearby, high-level time servers, even higher accuracies can be obtained. The 2003 Spritewatch system time was updated every 10 minutes, with nearly linear drifts of about 12 ms in between. Correcting for this drift, the time jitter is close to 1 ms.

5.2.4 Results from 2003

To summarise, the initial results of the effort leading to the 2003 Spritewatch systems were

- Successful demonstration of the applicability of remote agents to advanced imaging problems.

- Verification of robust, parallel real-time processing of multiple, full-resolution video streams on a remote system.

- Successful automatic detection of transient luminous events using the Spritewatch software.

Beyond the technical achievements, the data products from the system were used in multiple research efforts – see e.g. (Farges et al., 2004), (Haldoupis et al., 2004), (Marshall et al., 2004). Figure 8 displays a pair of events, as recorded by the 2003 event detection system. In total 101 sprite events and 2 elves were detected in real time, during the period from July 20 to September 6. More than 150 GB of video files covering the campaign exist, subject to further investigation. The 2003 system is described in more detail in (Allin, 2005).

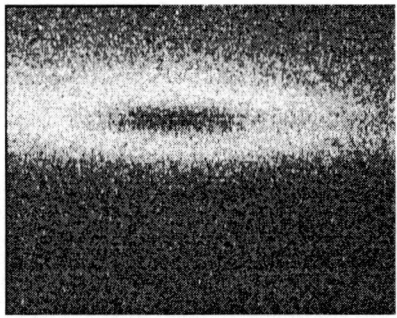 

Figure 8. A couple of events detected during the 2003 campaign – an elve (left) and a sprite (right). The elve image has been contrast enhanced and thus appears to be more grainy than the original.

5.2.5 The 2004 Spritewatch System

In the summer of 2004, an improved version of the Spritewatch system was installed, again at Pic du midi de Bigorre. The 2004 system followed the same design concept as its predecessor, but was improved at a number of points. Firstly, the scope of the system was to provide full-resolution, three-colour video imagery of transient luminous events. Secondly, the system timing was to be based on a 1 pulse per second signal from a GPS receiver, precise to 5 μs from UT, to eliminate the effect of network stability on performance.

5.2.6 Hardware in the 2004 system

As shown in Figure 9, the 2004 Spritewatch system shares quite many features with the 2003 version of the system. Major changes are the GPS receiver, mounted so that it turns with the cameras, but remain vertically oriented. Also, the camera unit now features 4 imagers, of which 3 are equipped with broad-band colour filters, enabling full-resolution colour imagery of optical transients.

The cameras used in the 2004 system were all of the type JAI CV-M50IR, using the same CCD (the Sony ICX249AL) as the CV-S3200 used in the 2003 system. The cameras are simple, monochrome video cameras, controllable by DIP-switches and synchronisation signals. One camera, the survey imager, was equipped with a 16 mm F1.4 monofocal lens, while the filtered imagers were fitted with 50 mm F0.95 monofocal lenses. Using the horizontal and vertical drive output from the survey imager as master synchronisation source, all 4 imagers were synchronised to 20 ms integration using 2:1 interlace scan. The same pan/tilt unit was used, with a slightly improved version of the qt20 control utility. Signal handling was included in the qt20 program in order to support requests from other programs, but this has yet to be fully implemented. The Local Processing Unit (LPU) was enhanced to adopt the four image streams from the camera unit. Four identical image acquisition cards were placed in the PCI slots of the LPU, again based on the BT878A chipset. Otherwise, the LPU configuration was identical to that of the 2003 version. An interface of the GPS receiver with the LPU was built, based on the implementation of a PPS "clock" for the NTP daemon. Rather than using a network server, it is possible to use local hardware with the NTP daemon. Many GPS receivers can be configured to output a TTL signal at the frequency of 1 Hz, locked to the UT second within a few microseconds. Using a level converter, built from the schematic of the 1-PPS Linux kernel interface[2], the TTL signal is converted to RS232 levels, and thus compatible with the serial ports on a PC. Connecting the 1-PPS signal to the DTS pin of the serial port enables the Linux kernel to maintain system time with an accuracy better than 0.1 ms on average. Using the NTP daemon with auxiliary time sources, such as the NMEA data stream from a local GPS, allows any program to use system time and UT interchangeably.

Figure 9. The 2004 Spritewatch system in the DSRI laboratory.

---

[2] Available at http://www.kernel.org/pub/linux/daemons/ntp/pps

5.2.7 The 2004 Spritewatch Software

The Spritewatch also refers to the event detection software of the observation agent described above. The software is free software, released under the GNU General Public License, and is available free of charge from http://www.allinux.dk. From the initial 2003 version, the Spritewatch software has undergone a re-design during 2004. Support for multiple video devices was introduced, and an extensive configuration file now controls the execution of the program.

The program uses *Linux threads* to simulate a parallel execution model, although there is only one CPU in the processing unit. Linux threads allow a Linux program to initiate a number of subprograms-threads – executing in parallel with their parent programs. Threads are preferable to parallel processes (such as created by the `fork()` system call) in a number of ways. Most importantly, they are able to share the memory space assigned to their parent process. This means that it is possible to interchange runtime information between threads by changing values of shared variables, or, as is generally safer on a multitasking system, using signals. Also, the creation of a thread is fast compared to that of a parallel process, since the memory area occupied by the parent process does not need copying.

Structure of the Spritewatch Program The execution of the Spritewatch program is shown schematically in Figure 10. Upon initialisation by the user (or the system scheduler), the program parses its configuration file. In the configuration file, a number of devices (frame grabbers) can be specified. Depending on the number of grabbers declared, a corresponding number of output threads are created. These threads are idle until signalled. When the output threads are running, an analysis thread is created, also idle until signalled. The analysis thread is responsible for event detection, and is typically assigned a dedicated frame grabber (camera). Currently, only one analysis thread is allowed to run at any time. When also the analysis thread is running, the input threads are created. The input threads are each assigned to a specific frame grabber, and are responsible for mapping digitised video data to the host memory. If any of the input devices are marked to be monitored for events, its input thread will further generate a signal to the corresponding analysis thread whenever a new image is ready for processing. The event detection loop continues until the main thread receives a termination signal from the user or system scheduler.

Event Detection. Using image data alone, event detection must be based on changes in the photovisual properties of the scene. As in the 2003 system, image,profiling methods proved to be highly efficient to detect nearly any kind of event. The method is illustrated in Figure 11, which shows two consecutive

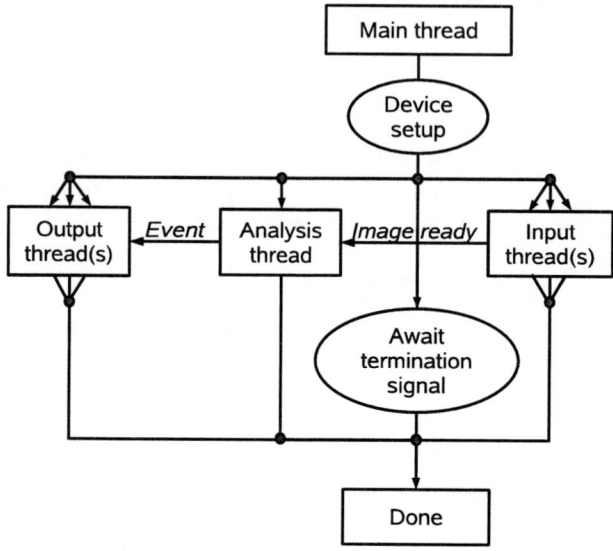

Figure 10. Execution model of the Spritewatch program.

video fields, 20 ms apart. The sprites in the second video field causes the image profiles – the pixel sums along vertical and horizontal directions – to change abruptly, which is easily detected in the algorithms. The need for a quick decision whether an image contains an interesting event or not called for a simple rejection mechanism for non-event images. For the 2003 and 2004 systems, the event size – as based on the profile changes – should exceed a predefined threshold, and the profile change itself should be substantial. Although there is still many non-sprite events getting through the algorithms, it is worth to note that on a typical night of observation, the event detection system will have a sprite detection efficiency higher than 90 %. Compared to a human observer, who may get tired during the night, or need an occasional coffee break, it is safe to say that the Spritewatch provides a large step forward for long-duration TLE campaign instrumentation.

5.3 Conclusions

5.3.1 Summary

A novel concept for transient luminous event observation campaigns has been presented. Using bare CCD cameras and standard computer equipment, it is possible to automate a large part of the observation effort, and create highly advanced and cost-efficient observation systems.

IMAGING SPRITES

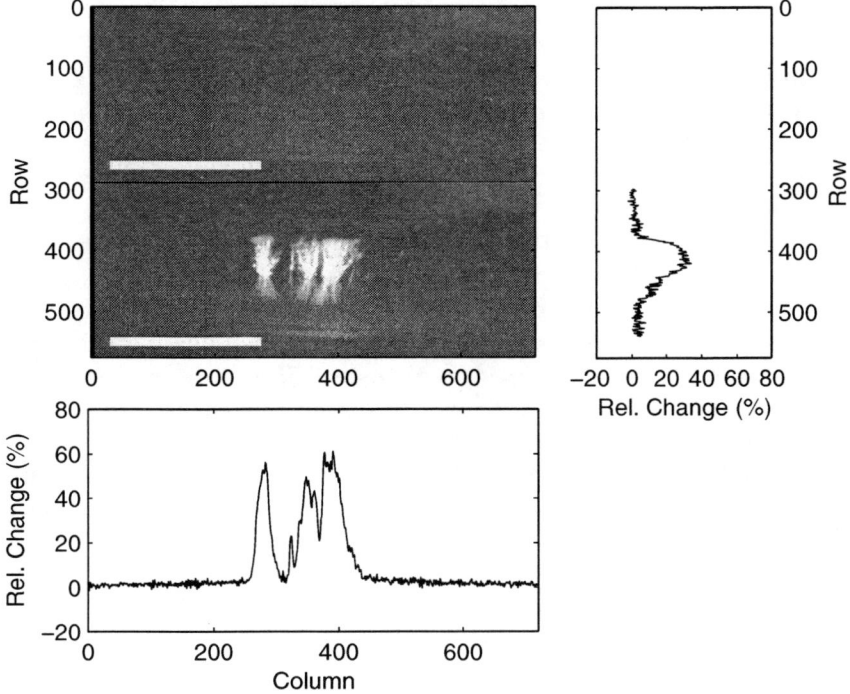

Figure 11. The profiling method of event detection.

The Spritewatch software has developed into a program capable of handling video data from up to four cameras simultaneously, with no data loss. Events are reported with sub-millisecond accuracy of the camera integration window. Future improvements of the Spritewatch software will include a better event classification scheme, and support for digital video interfaces, such as USB or FireWire cameras with higher resolution and/or frame rates.

The choice of cameras for TLE observations should be considered carefully. For sprite observations, the limiting sensitivity should be better than 0.01 lux in order to produce images with good dynamic range, equivalent to contrast. For narrow-band spectral observations, only highly customised bare CCD cameras with large-format sensors and large-aperture optics are applicable. It may be advantageous in such cases to consider using image intensifiers along with standard video cameras.

5.3.2 Future work

The availability of automated observation systems makes large-scale coordinated campaigns much less dependent on manpower availability at possibly remote observation sites. In order to provide plug and play deployment of the Spritewatch systems, the integration of the control computer into the camera head unit using a single-board computer with a digital video interface is currently under investigation.

The concepts and ideas of automated observation systems in TLE research are already applied to space-borne platforms. The ISUAL instrument on board the ROCSAT-II satellite has presented evidence of the truly global occurrence of TLEs, using a combination of photometers and intensified CCD imagers, and the Atmosphere-Space Interaction Monitor (ASIM), to be mounted on the International Space Station in 2008, will carry a miniature multispectral imaging array (MMIA), very similar to the Spritewatch system, along with an X-ray detector. Further development of the Spritewatch agent is on the road map to support such and other missions with important ground-based observations for correlation studies.

Bibliography

Allin, T. H. (2005). *Design Methods for High Speed, High-resolution Digital Multispectral Imagers.* Ph.D. dissertation, Technical University of Denmark

Allin, T. H. *The Spritewatch home page.* http://www.allinux.dk/Spritewatch/

Dirk, B. and M. Schmiek (2004). *The Video For Linux Two API Specification: Draft 0.8.* Available at http://v4l2.bytesex.org.

Farges, T., E. Blanc, A. Le Pichon, T. Neubert, and T. H. Allin (2004). *Identification of infrasound produced by sprites during the Sprite2003 campaign.* Geophysical Research Letters, 32(L01813), doi:10.1029/2004GL021212.

Haldoupis, C., T. Neubert, U. S. Inan, A. Mika, T. H. Allin, and R. A. Marshall (2004). *Subionospheric early VLF signal perturbations observed in one-to-one association with sprites.* Journal of Geophysical Research, 109(A10303), doi:10.1029/2004JA010651.

Holst, G. C. (1998) *CCD Arrays, Cameras, and Displays.* SPIE Optical Engineering Press.

Marshall, R. A., U. S. Inan, T. Neubert, A. Hughes, G. Sátori, J. Bor, A. Collier, and T. H. Allin (2004). *Optical observations geomagnetically conjugate to sprite-producing lightning discharges.* Annales Geophysicae. (in press).

SPACECRAFT BASED STUDIES OF TRANSIENT LUMINOUS EVENTS

Stephen B. Mende [1], Y. S. Chang [2], A. B. Chen [3], H. U. Frey [1], H. Fukunishi [4], S. P. Geller [1], S. Harris [1], H. Heetderks [1], R. R. Hsu [3], L. C. Lee [5], H. T. Su [3] and Y. Takahashi [4]

[1] *Space Science Laboratory, University of California, Berkeley, CA 94720, USA.*

[2] *National Space Program Office, Hsin-Chu, Taiwan.*

[3] *Department of Physics, National Cheng Kung University, Tainan, Taiwan.*

[4] *Department of Geophysics, Tohoku University, Sendai, Japan.*

[5] *National Applied Research Laboratories, Taipei, Taiwan.*

Abstract   The Imager of Sprites and Upper Atmospheric Lightning (ISUAL) is a scientific payload on Taiwan's FORMOSAT-2 (previously known as ROCSAT-2) that provides new observations of transient luminous events (TLEs) from space. The ISUAL project is an international collaboration between the National Cheng Kung University, Taiwan, Tohoku University, Japan and the instrument development team from the University of California, Berkeley. The project was supported by the National Space Program Office in Taiwan. The ISUAL payload includes a visible wavelength intensified CCD imager, a boresighted six wavelength spectrophotometer, and a two channel Array Photometer (AP) with 16 vertically spaced horizontally wide sensitive regions. The imager is equipped with 5 selectable filters on a filter wheel and a 6th open position. The spectrophotometer contains six filter photometer channels, their bandpasses ranging from the far ultraviolet to the near infrared regions. The two channel AP is fitted with broadband blue and red filters. The orbiting platform with this set of instruments will provide the first comprehensive global latitude and longitude survey of TLEs near the midnight local time region. One of the great advantages of spaceborne observations is the lack of the intervening atmosphere between the TLEs and the observer. Ground based observations are often adversely affected by clouds, atmospheric extinction or scattering whereas the space-borne ISUAL instrument measurements provide true emission ratios unobstructed by the variable atmospheric extinction. The channels of the spectrophotometer channels cover the far and mid ultraviolet in addition to channels that respond to various excitation levels of the neutral and ionized nitrogen molecule atmosphere and to emissions from oxygen. The preliminary data shows that the ratio of the emissions is highly variable during the lightning and the associated TLEs. The data is qualitatively consistent with harder characteristic energy electron production in lightning less hard in sprites and even less in elves. The focus of the data analysis will be to solidify these conclusions and to put them on firmer statistical and quantitative basis.

6.1 Introduction

Upward lightning discharges into clear air have been reported by pilots world wide for over a century (e.g., Everett, 1903; Boys, 1926; Vaughan and Vonnegut, 1989). However, scientific investigations of these phenomena did not begin until early 90s. Franz et al. (1990) were the first to record upward electrical discharge events during a storm associated with hurricane Hugo on the night of 22 September 1989 using a low-light-level television camera.

Since then, night-time lightning-induced transient luminous events have been recorded using low-light-level television cameras on the space shuttle (Boeck et al., 1995), on aircrafts (Sentman et al., 1995; Wescott et al., 1995), and on the ground (Rairden and Mende, 1995; Lyons, 1996; Winckler et al., 1996; Stanley et al., 1999; Gerken et al., 2000; Barrington-Leigh et al., 2001; Su et al., 2002). Since then there were observations that indicate other types of TLEs (Lyons et al., 2001; Pasko and George, 2001). Based on ground, aircraft, and space shuttle observations in the past 10 years, TLEs have been classified into several categories including sprites (Sentman et al., 1995), blue jets (Wescott et al., 1995), elves (Fukunishi et al., 1996) and halos (Barrington-Leigh et al., 2001). Systematic observations of TLEs from a free flying spacecraft is expected to produce a much better understanding of TLE processes.

6.2 FORMOSAT-2 Satellite and the ISUAL Instrument

Satellite based studies of upper atmospheric TLE events have several advantages. The most notable one is that global latitude longitude surveys of TLEs can be conducted from satellite orbit. The lack of atmospheric attenuation also provides many advantages such as UV viewing and quantitative interpretation of the measurements regardless of atmospheric conditions or viewing angles. Since TLEs are thunderstorm related phenomena they tend to occur when ground based viewing conditions are relatively unstable. The Imager of Sprites and Upper Atmospheric Lightning (ISUAL) is a scientific instrument package on the Taiwanese FORMOSAT-2 satellite that was launched on May 20, 2004. ISUAL is the first multi-wavelength, quantitative observatory on a free flying satellite primarily dedicated to observing TLEs. The ISUAL payload includes an intensified CCD visible imager with a six position filter wheel, a boresighted six wavelength channel spectrophotometer (SP), and a two channel array photometer (AP) with 16 vertically spaced, parallel, horizontal strip photomultiplier anodes. The imager is equipped with 5 selectable filters on a filter wheel and the 6th filter position is open. The SP six filter channel wavelength band passes range from the far ultraviolet to the near infrared regions (Table 2 in Section 6.2.3). The two AP channels are fitted with broad band filters, one blue and one red. Thus the ISUAL instrument complement provides calibrated emission intensities of sprites and elves, such measurements have been problematic to obtain from the ground.

The high time resolution requirement for TLE studies, produce very high data volumes that is beyond the allocated capacity of most satellite data storage and down link transmission systems. To overcome this difficulty the ISUAL data system operates essentially in a type of burst mode. Data are gathered continuously during the night pass but they are saved only if the on-board hardware control system determines that a significant "trigger" event had occurred. The instrument hardware has several programmable discriminators that produce a trigger when selected SP channel signals exceed a certain threshold. Most triggers are caused by lightning. Without trigger the data is discarded. It is difficult to distinguish the bright lightning from TLEs in real time even with the aid of the SP UV channels that are optimized for TLE detection. In addition, energetic particle events from penetrating space radiation can provide false signals in the phototubes that are indistinguishable from optically induced events. The satellite has a flexible programmable logic system that can be programmed and which will accommodate various "trigger algorithms" and execute them in "real time". The fine tuning of the system is an on going process. At the time of writing of this work we have been able to capture 67 sprites, 580 elves, and 47 halos.

6.2.1 The FORMOSAT-2 Satellite

The ISUAL instrument operates on FORMOSAT-2 (previously known as ROCSAT-2), which is a small-class Taiwanese satellite carrying another instrument, the Remote Sensing Instrument (RSI), that takes high resolution images of the Earth surface under sunlit conditions. The satellite was constructed at the Astrium facility in France and was shipped to Taiwan for system integration and testing, conducted by National Space Program Office (NSPO). It was launched by a Taurus launch vehicle on May 20th, 2004 from the Vandenberg Air Force Base in California.

The primary mission goal of the FORMOSAT-2 program is the RSI observation over the region of Taiwan Island, Taiwan Straits and the remote offshore islands. Therefore frequent revisits of Taiwan and timely availability of the data in Taiwan are a high priority. A sun-synchronous repeating orbit was chosen with overpass at 9:30 and 21:30 local time. The orbit is also a repeating orbit and it traces the same geographic footprint every 24-hour day.

Other relevant information on the FORMOSAT-2 satellite is as follows:

- Weight: approximately 700 kg

- Orbit: polar-orbit and sun-synchronized with altitude of 891 km, repeating orbit i.e. the geographic ground track is constant (centered on Taiwan).
- Orbital plane: 98.99°
- Period: Exactly 14 revolutions per day
- Agility: body rotation ±45° roll, pitch and yaw
- Pointing accuracy <0.7 km
- Pointing knowledge <450 m
- Position knowledge <70 m
- Mission life: 5 years

The ISUAL observing scenario is illustrated in Figure 1a and b. The ISUAL imager is looking essentially horizontally in the direction perpendicular to the orbit plane in the starboard direction as the satellite proceeds northward. Its field of view (FOV) is approximately 20° (1024 pixels) in the horizontal direction (w) and 5 (256 pixels) in the vertical (h). The 4 to 1 aspect ratio of the

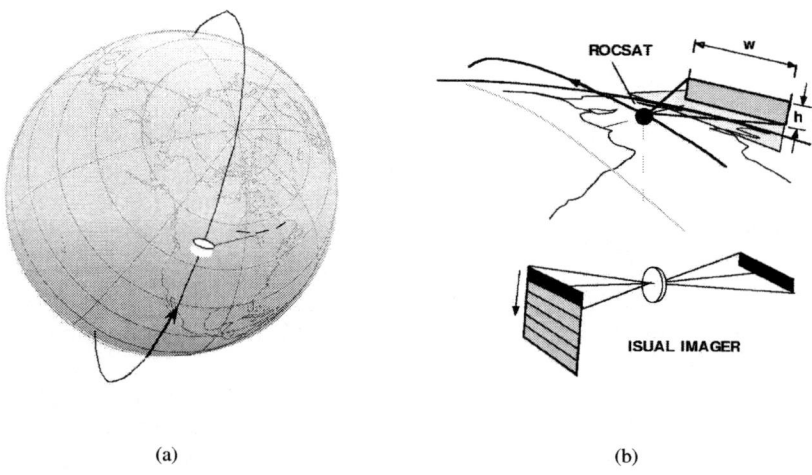

Figure 1. (a) An artist impression of the satellite traveling north while ISUAL is viewing North America. (b) FORMOSAT-2 (shown as ROCSAT) observing scenario showing the satellite orbit track and the approximate view angle. The imager observes a region near the Earth limb perpendicular to the orbit track. The imager takes several horizontally wide images which can be stacked on the CCD.

image allows the storing of 8 images on the 1024×2048 frame transfer CCD. All but one of the image 1024×256 pixel segments are behind an opaque mask. Following each exposure image on the active part of the CCD is transferred down one step as shown by the arrow in the bottom illustration (1b). This technique permits the taking of up to 8 exposures in a quick sequence on the CCD.

The satellite altitude of 891 km defines the straight line viewing distance (range) to the 60 km altitude limb tangent point as 3373 km. The instrument view angle to observe the solid Earth limb is 28.68° below the local horizon at the satellite. The great circle distance along the 60 km altitude layer from the satellite foot point to the tangent point is 3106 km.

The ISUAL instrument block diagram is shown in Figure 2. The instrument consists of 4 packages, the Imager, Spectrophotometer (SP), Array Photometer (AP) and the Associated Electronics Package (AEP). The imager subsystems are the filter wheel, the CCD control electronics, the image intensifier gating and high voltage power supplies (HV). They are all controlled by the digital processing unit (DPU) housed in the AEP. A daylight sensor is included which detects excessive brightness in the field of view of the ISUAL instruments and can shut down the high voltage (HV) supplies by hardware control to protect the instrument from over exposure. The spectrophotometer SP consists of 6 individual photomultiplier units with pre-amplifiers and individually controllable HV supplies each. The Array Photometer is a two-wavelength channel instrument with two 16-channel multi anode photomultipliers and corresponding high voltage supplies. Each sensing instrument has an assigned section of the instrument mass memory and the data is collected at that location. The entire instrument and the data processing sequences are controlled by the DPU. This microprocessor executes the flight software and controls the entire ISUAL instrument operation. In burst mode suitable for TLE observations a lightning flash or a TLE generates photometer signals that are combined by the hardware event detect circuit according to the trigger detection algorithm. This algorithm determines what combination of signals from several photometer channels satisfies the trigger criterion. When the signals satisfy the trigger criteria the hardware processor in the AEP activates the mass memory controller to execute an operating cycle and store the resulting data for downlink transmission. There is a Digital Signal Processor (DSP) attached to the mass memory and it is capable of compressing and repackaging the data for the telemetry format.

6.2.2 The ISUAL Imager

The ISUAL camera is schematically illustrated in Figure 3. There is a light baffle to exclude scattered stray light. Preceding the filters the light goes through a fused silica window, which is used to minimize the radiation exposure of the filters and the following optics. The filters are mounted on a filter wheel driven by a stepper motor under direct control of the DPU.

The imager optics consists of a 62.5 mm focal length F/1.5 lens specially designed for the ISUAL imager by Coastal Optical Incorporated. The lens elements are constructed entirely of space radiation resistant glasses. The image is produced with a vertical dimension 5° by 20° covering a limb region of about 250 km in the vertical and 1000 km in the horizontal direction. The instrument's nominal pointing is 27.5° down from the local horizontal. With this pointing configuration the lower boundary of the field of view intercepts the 60 km altitude region at 2000 km distance from the spacecraft. This view covers a horizontal region of the atmosphere starting at 2000 km from the satellite and extending all the way to the limb at 3106 km perpendicular to the orbit

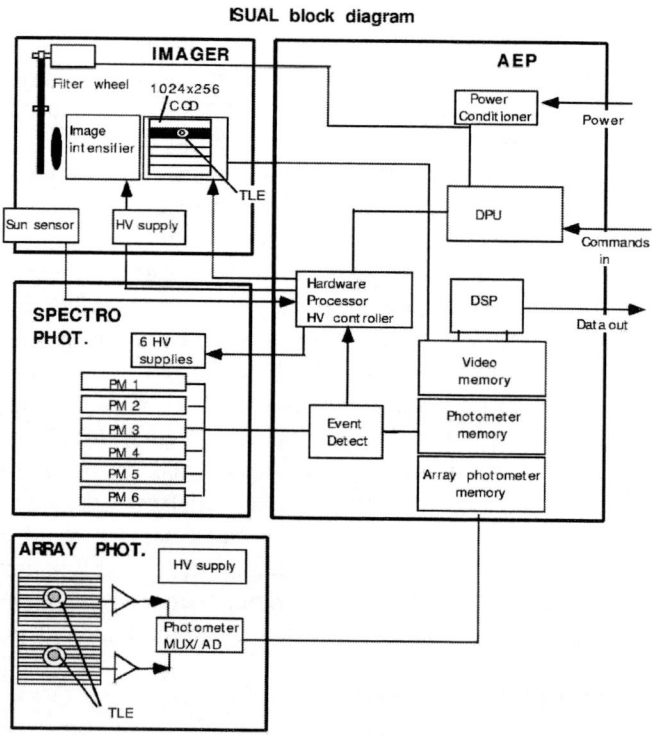

Figure 2. The ISUAL instrument block diagram.

plane. Since the width of the coverage is 1090 km the approximate area of horizontal coverage is $1000 \times (3106\text{-}2000) = 10^6$ km^2.

The detector is an intensified frame transfer CCD of 2048×1024 pixels. The image intensifier (made by Delft Electronic Products) is a proximity focused tube with a red enhanced, so-called S25 photocathode on the input window (Figure 3). The intensifier tube can be operated in a gated configuration by reversing the voltage between the photocathode and the micro-channel plate (MCP). To set the gain of the intensifier, the MCP high voltage is adjusted by the DPU. The electrons from the microchannel plate are accelerated and they impact on the phosphor in the image tube. The image tube phosphor has a separate high voltage supply. A tapered fiber optics (Figure 3) is used to relay the image intensifier output image on to the CCD. The ISUAL imager CCD is used mostly in a 2×2 binned pixel configuration, where each pixel is 12×12 μm and the maximum charge that can be stored on each pixel is 105,000 electrons. One half of the area, (1024×1024 pixels) 512×512 resolution elements, was masked during the manufacturing process and the other half was left open and is light sensitive. In ISUAL there is an additional mask to restrict the light sensitive region to only the top 128×512 resolution element region. The images therefore have 4 to 1 aspect ratio and it is possible to stack eight pictures on the CCD containing 128×512 resolution elements each. The stacking is illustrated at the bottom of Figure 1b. The limb image is focused on the top, unmasked area of the detector. A combination of masks covers the rest of the CCD and this area represents the image storage region for the fast exposure sequences. After an image exposure is completed the entire stack of images is shifted down and the read out electronics picks up the last row of the storage register. The image stacking permits taking a fast sequence of eight exposures containing a sequence of short duration (>1 msec) images. In the standard operating mode, while the system is idle, the detectors take data but only minimal data is recorded. When a flash is detected by the trigger detection hardware driven by pre-selected SP channels, the data capture mode is initiated and a preprogrammed sequence of images is recorded in memory. After each exposure sequence is completed, the CCD storage area is full and no more images are taken until all the images have been completely read out. The duration and repetition rate of the exposures are all programmable. It should be noted that this technique permits the capture of images, which are taken prior to the occurrence of the triggering event. The imager filter chamber can be temperature controlled by regulating heaters located in the filter wheel housing. The filters are of the low temperature coefficient variety with minimal wavelength change due to temperature and the heaters would not be used under normal circumstances. The CCD builds up signal at a slow rate even while it is not exposed to light. This building up is called "dark current". The CCD has a thermoelectric cooler system that can be used to suppress dark current in the long exposure low light level operations that are desirable for airglow and auroral observations. This was also a pre-cautionary design feature that would be important if the spacecraft were to run warmer. The ISUAL instrument can be operated in sprite burst mode to study fast spatial/temporal profiles of TLEs in high time resolution (up to 1000 frame rate i.e. 1 msec exposure time). This was described above and in this mode the photometer trigger algorithm commands the instrument to take up to 8 consecutive exposures and store them on the CCD. After that the instrument data system reads the CCD, which can take up to 100 msec while there is a dead time and the CCD would be unable to take another image even if another trigger were to occur. If continuous trigger readiness and data taking is desirable then the exposure times have to be lengthened to about 20-30 msec. For continuous data taking such as observing aurora and airglow, the ISUAL DPU can be made to issue periodic artificial triggers. There is also a continuous "aurora mode" in which the instrument takes 1 second duration integrations in a continuous sequence.

Examination of the video recordings taken by the space shuttle video cameras (e.g., Vaughan et al., 1992) showed that it could be quite difficult to distinguish TLEs from the accompanying lightning flashes. An intense cloud-to-ground flash usually precedes the sprite event illuminating the thunderstorm cloud above the flash with very high intensity light. A camera viewing from a spacecraft includes both the flash as well as the TLE in the field of view. The

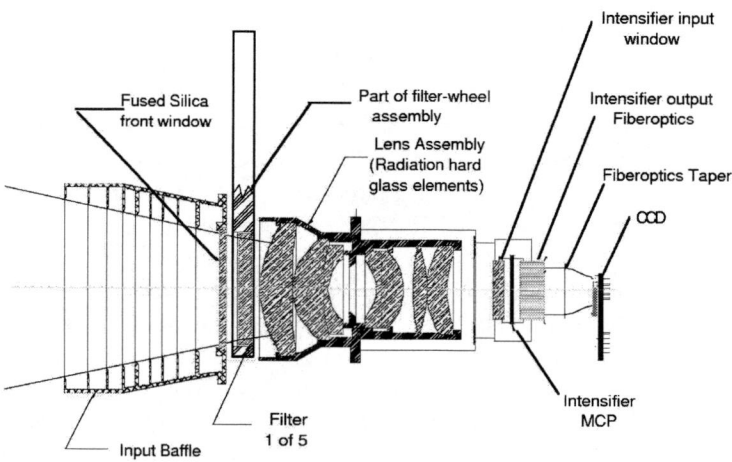

Figure 3. Imager cross sectional view.

intensity of the flash can be so large that a camera that was not designed specifically for the purpose, could be blinded. The ISUAL imager design uses several techniques to minimize the contamination from the parent lightning. ISUAL field of view is directed towards the Earth's limb and flashes near limb have large (altitude) separation between the TLE and the lightning. In some cases the lightning can be completely hidden by the solid Earth's Limb. Appropriate spectral filtering was also used to select specific spectral features (see Table 2 in Section 6.2.3) to maximize the detection of the TLE induced luminosity and reduce the broad continuum signature of lightning. By taking fast exposures it is also possible to maximize the intensity of the short duration TLE compared to the relatively long duration lightning.

A photograph of the ISUAL imager is included as Figure 4. The imager filter properties are shown in Table 1 and Figure 5.

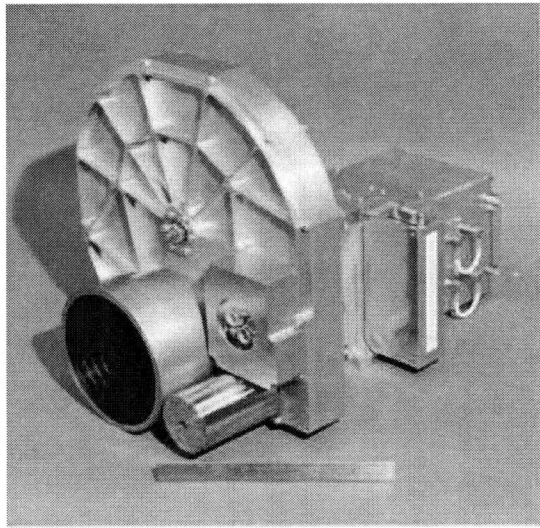

Figure 4. Photograph of the ISUAL imager camera flight unit.

6.2.3 The Spectrophotometer

The ISUAL imager produces an image that is a 2 dimensional luminosity distribution of the TLE in a specified wavelength band with moderate time resolution. For simultaneous measurements of the high time resolution (0.1 ms sampling) temporal profile of the TLE emission in several pass bands, the spectrophotometer (SP) was included into the ISUAL payload. The SP field of view is approximately the same as that of the imager.

Table 1. ISUAL imager filters. Note that the band limits are for paraxial rays only. The filter curves shift towards the blue for any object lying off axis. OI is atomic oxygen.

Imager Filters			
Channel number	Nominal filter	Lower limit 10% (nm)	Upper limit 10% (nm)
1	N_2 1PG	622.8	754
2	760 nm O_2 atmospheric band (airglow, aurora)	757.8	768.8
3	630 nm OI (aurora, airglow)	626.3	636.4
4	557.7 nm OI (aurora, airglow)	554.7	562.8
5	427.8 nm N_2^+ ionized N_2	425.3	431.8
6	Broad band no filter	425	890

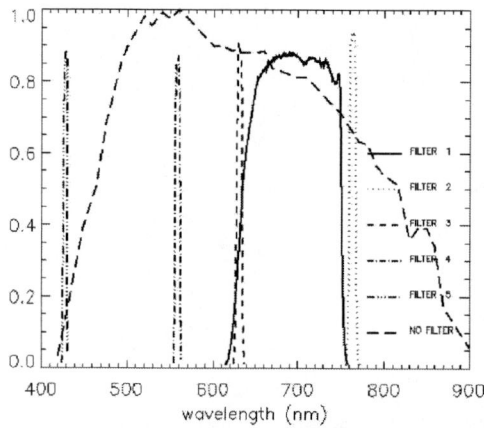

Figure 5. Imager filter profiles for the five filters and the measured responsivity of the unfiltered instrument labeled as filter 6. Its peak is normalized to one.

The spectrophotometer (SP) has six channels that are electrically and mechanically closely similar to each other. A typical channel module is illustrated in Figure 6. The light comes in through a collimator assembly (1) and encounters the filter in the filter assembly (3). The filter is mounted on a special mount that permits the temperature control of the filters by heaters (8). The filter temperature is monitored with a thermostat for telemetry and/or control. The lens, located behind the filter, focuses incoming parallel light on the aperture mask (5). A horizontal slit is cut out of the mask to define the field of view that is similar in size and shape to the imager field of view. The phototubes (6) were rugged photomultipliers procured from EMR in New Jersey. The tube windows and photocathodes were specially selected depending on the wavelength requirement of the channel module. Each visible channel contains a light emitting diode to produce a calibration stimulus for pre-flight and in flight checks on the instrument status. The SP channels operate in an analog mode and the PM tube current is sampled and digitized at a sampling rate of 10 kHz (0.1 ms integration time) in the standard fast mode intended for TLE studies. There is another data rate and the PM tube current is sampled simultaneously at a rate of 1.2 Hz. These data are used to synoptic studies of aurora and airglow. All

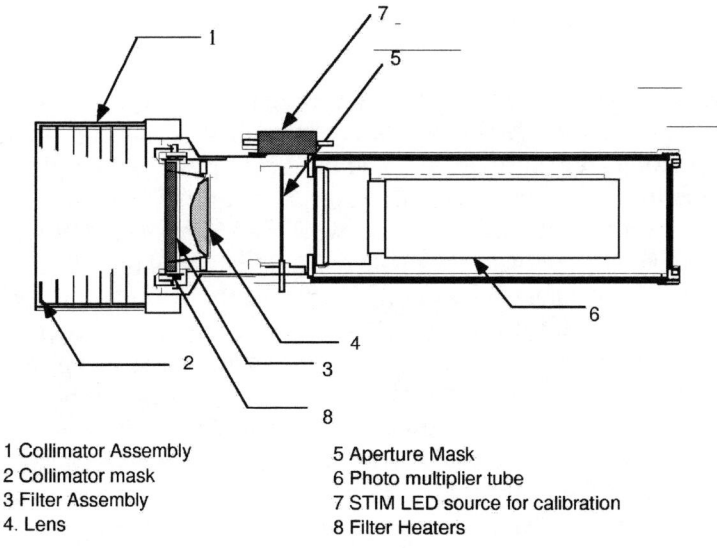

1 Collimator Assembly
2 Collimator mask
3 Filter Assembly
4. Lens
5 Aperture Mask
6 Photo multiplier tube
7 STIM LED source for calibration
8 Filter Heaters

Figure 6. Photometer typical channel cross sectional view.

Figure 7. The ISUAL spectrophotometer.

the photometers are approximately boresighted with the ISUAL imager. The wavelength band selection for these photometers is shown in Table 2 and the band profiles are illustrated in Figure 8. The wavelength selection was guided by the following considerations: It was hoped that the far ultra-violet (FUV) channel 1 of the spectrophotometer would produce a clear UV signature from emissions that originate above the ozone layer and therefore it would not be affected by lightning and that this channel could be used as a TLE trigger for the ISUAL system. Channels 2, 3 and 4 were expected to observe the N_2 2nd positive band at 337 nm, the N_2^+ 1st negative band at 391.4 nm and the N_2 1st positive band to provide information on electron energy distribution of the TLE events. Channel 5 is the permitted transition of O at 777.4 nm, a commonly observed feature in lightning. Channel 6 is a broad band middle UV detector. Channel 4 (mostly N_2 1st positive band) is the most commonly observed spectral feature in TLEs, best understood from numerous ground-based measurements. The baseline design assumed that triggering of the imager would be using channel 4 however other channels could also be programmed to act as triggers. The imager and the photometer were calibrated in the laboratory during pre-flight test. The imager properties, field of view, resolution, absolute sensitivity in all filters, and the variation of sensitivity with MCP high voltage were recorded while the imager was exposed to light sources of known characteristics. The imager absolute sensitivity was obtained with a light source, which was spectrally a continuum and spatially an extended source. By expressing the extended source in equivalent Rayleighs[1] at the peak transmission of the filter an absolute calibration of the imager in digitized CCD signal versus Rayleighs source intensity was obtained. The SP channels were individually calibrated with respect to the size of the field of view, the absolute sensitivity and the variation of sensitivity with high voltage applied to the photomultiplier tube. Since the photometer is sensitive only to photon flux incident on its front lens, the photometer calibration flux is equivalent to the calibration source radiance times the size of the solid angle of the calibration source in units such as Rayleigh times steradians. So in order to find an unknown source intensity in Rayleighs we need to obtain the size of the source i.e. the solid angle that the

Table 2. ISUAL photometer filters. Note that the band limits are for paraxial rays only. The filter curves shift towards the blue for any object lying off axis. OII is singly ionized atomic oxygen.

Spectrophotometer (SP) channels			
Channel number	Nominal filter	Lower limit 10% (nm)	Upper limit 10% (nm)
1	UV Channel	150	280
2	N_2 2PG (0,0) 337	333.5	341.2
3	N_2^+ 391.4 nm (ionized N_2)	387.1	393.6
4	N_2 1PG	608.9	753.4
5	OII 777.4	773.6	784.7
6	Broad band UV	270	410.2

[1] Rayleigh, the unit of intensity of glowing atmospheric gases. It is equivalent to a source producing $10^6/4\pi$ photons per cm^2 per steradian per second at a detector (Chamberlain, 1961, pp. 569). For example a detector will observe a 1 Rayleigh source when viewing end on a column of gas of 1 cm^2 in cross-sectional area, containing 1 million particles where each particle produces 1 photon per second.

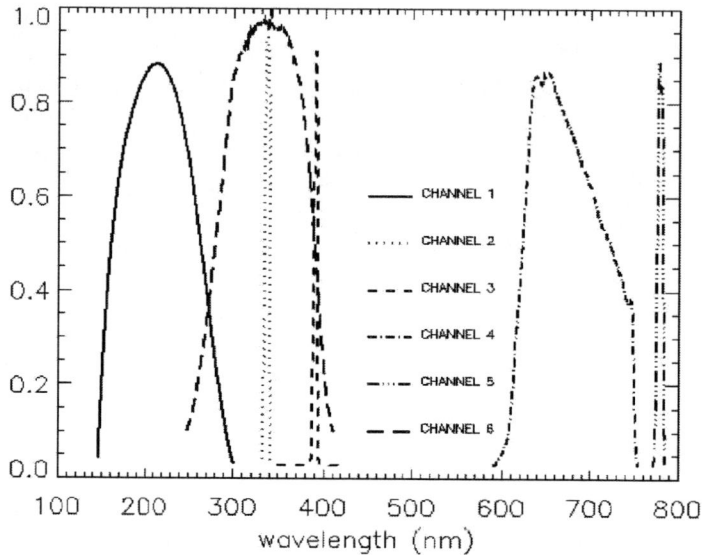

Figure 8. ISUAL photometer channel responsivities in arbitrary units.

unknown source subtended at the photometer. To obtain the unknown source intensity the measured signal has to be divided by the solid angle subtended by the unknown source. The size (solid angle) of the unknown source was derived from the simultaneous imager data.

6.2.4 Data Interpretation of the Spectrophotometer

In order to investigate the energetics of TLE production it would be desirable to relate the measurement to the production rate of a particular band. The SP detects only the fraction of the band emission that is within the photometer filter pass band. We integrated the product of the molecular band components and the wavelength profile of each photometer channel for all major N_2 bands to help us in relating the SP signals to the total molecular band production. In our calculation the channel responsivity maximum for each channel was normalized to unity because the preflight absolute responsivity calibrations (photometer counts per incident photon flux) were obtained at the response peak of each channel. We have adopted Vallance-Jones (1974) intensities for N_2 molecular band components in spite of the fact that the vibrational distribution and rotational temperatures may not be the same as in auroras. The results of these calculations, the fractional contribution of each molecular band of unit intensity for each SP channels is shown in Table 3.

In Table 3, for filter 1 we have included the O_2 absorption as well as the filter response. The absorption was calculated to 80 km altitude and the O_2 cross sections were calculated by a web tool using polynomial coefficients (Minschwaner et al., 1992) including the Herzberg continuum cross section in the range in the wavelength of interest (Yoshino et al., 1988). The O_2 absorption calculation has to be performed on a case-by-case basis because this depends on the geometry of the observation. Filter 1 transmits only in the far and middle UV and it is a very useful channel that can see only those emissions that are produced at altitudes higher than the ozone layer (30 km). The table shows that filter 1 appears to have a strong contribution by components of the Lyman-Birge Hopfield (LBH) band above 190 nm and that there are no other N_2 molecular band contributions. The degree that LBH is emitted depends of course on the lifetime of the $a^1\Pi_g$ state. The $a^1\Pi_g$ state life time was found to be 80 microsec by Shemansky (1969) or >76 but <116 μs by Pilling et al. (1971). Vallance-Jones (1974) states that the quenching coefficient for $N_2(a^1\Pi_g, v' = 0)$ is <3 10^{-10} with a quenching altitude of less than 95 km. However two-photon laser excitation measurements of the quenching rate of the N_2 $a^1\Pi_g$ state for N_2 (Marinelli et al., 1989) shows it to be $2.2\pm0.2\times10^{-11}$cm^3/s. Combining this with the above lifetimes the quenching height should be lowered to 70 km. In summary quenching of LBH is likely to be less important in elves than in sprites or jets.

A good recent reference for other emissions in the FUV (far ultra-violet) and MUV (middle ultra-violet) that are relevant to filters 1 and 2 is the auroral spectra taken by the Midcourse Space Experiment (MSX) satellite (Strickland et al., 2001). For example ISUAL channels 1 and 2 would pass the prominent auroral emission feature seen in the MSX spectra near 214.5 nm (Dick, 1978). The 247 nm OII and 297.2 nm OI emissions have relatively long lifetimes and

Table 3. Table of molecular band contributions into each ISUAL photometer channel.

	N_2 bands 1P	N_2 bands 2P	N_2 LBH > 150 nm	N_2^+ Meinel	N_2^+ 1NG
filter 1	–	–	17%		–
filter 2	–	27.80%	–	–	0.80%
filter 3	–	–	–	–	66%
filter 4	11%	–	–	4.60%	–
filter 5	2.6%	–	–	–	–
filter 6	–	84%	–	–	37%

the atomic O density required for the production of these is minimal at TLE altitudes. Similarly the lifetime of the upper state of the Vegard-Kaplan is also too long to be of significance below 100 km and therefore in TLEs.

The second filter passes the N_2 2P (0,0) band and Table 3 shows that 27.8 % of the emission is within this filter band. Since the filter curves were normalized to unity this number is the same as the (0,0) component value in the band computation of (Vallance-Jones, 1974). The 3rd filter was intended for the measurement of the N_2^+ first negative (0,0) band, which band contains the 66% of all the first negative components. The excitation cross-section for producing the upper state of this by electron impact has been measured by Borst and Zipf (1970). They showed that the ratio of the total ionization cross section to the excitation was nearly constant over the energy range from 30 eV to 10 keV, and had a value of 14.1. This would indicate that where the high energy electrons are dominant i.e. in an aurora the measurement of the 391.4 nm emission of the N_2^+ first negative (0,0) is a direct measurement of the total ion production. However TLEs are likely to produce significant electron fluxes in the less than 30 eV energy range where the ratio between the two cross sections is variable.

Filter 4 is responsive to the N_2 1st positive and N_2^+ Meinel bands with 11 and 4.6 % contribution to each, respectively. These bands were chosen because these bands had been found to be the most intense contributors in sprites and elves in ground based measurements (Mende et al., 1995; Hampton et al., 1996). Channel 5 centered on the 777.4 nm has a strong signal for lightning which is consistent with this feature being used to monitor terrestrial lightning e.g. Christian et al. (2003). It has a minimal response for TLEs in spite of the fact that it is a permitted transition and in this wavelength region the atmosphere is fully transparent. This is because of the lack of atomic oxygen at the altitudes in the mesosphere. On the other hand, lightning processes are sufficiently energetic to dissociate O_2 and excite the atomic O. In the thermosphere, at high altitudes, there is a high density of O and the excitation of this feature is commonly seen in aurora and airglow. Photometer channel 6 covers the broad band in the middle UV with substantial input from the N_2 2nd positive and N_2^+ first negative bands.

6.2.5 The Array Photometers

As we have seen ISUAL has an imager which is capable of producing high spatial resolution images but only with modest time resolution which is insufficient to resolve the temporal variation of the TLE luminosities. There is also the SP, a photometer which has high time resolution but provides no information about the spatial distribution or development of the TLE. A third instrument was incorporated in the ISUAL payload that has the high time resolution characteristic of a photometer but which also provides some spatial resolution in the form of one dimensional vertical imaging. This instrument is called the Array Photometer.

The array photometer (AP) is a two-wavelength channel instrument, each takes data from sixteen separate regions located vertically above each other. This instrument has high temporal and spatial (altitude) resolution for determining the altitude propagation properties of the phenomena.

The ISUAL instrument includes two AP channels. Each channel consists of a bandpass filter, an objective lens and a multi-anode U5900-01-L16 Hamamatsu photomultiplier. One channel has a bandpass filter with a passband between 370-450 nm and the other channel is equipped with a filter of a passband between 530-710 nm. It was highly desirable that the AP have similar wide field of views as the other ISUAL instrument components. The photomultiplier has 16 strip anodes. The total anode area of the photo multiplier tube (PMT) is approximately a square with sides of 16 mm in length. Hence, an objective lens was needed to map the spatially wide image into the square anode area of the PMT. To achieve a wide field of view in the horizontal direction and without degrading the vertical resolution, two cylindrical lenses with different focal lengths were used. The resulting FOV of the array photometer is 22° (horizontal) and 3.6° (vertical). Therefore, the FOV for each anode is 22° by 0.23°, corresponding to ~14 km in height at a range of 3500 km. The two array photometers are bore sighted and the ratio of the measured luminosity can be computed, and the electron energy distribution of the sprites, elves or blue jet-initializing discharge can be estimated (Takahashi et al., 1998; Sera et al., 2001).

To achieve a definitive identification of the different types of sprites, each channel of the AP is sampled at 20 kHz for the first 20 ms after the onset of the causative CG lightning. After that the data is sampled at a lower rate at 2 kHz until ~100 ms after the CG. In the electronic circuitry of the AP, a low pass filter (LPF) with a cut-off frequency of 10 kHz is employed for the faster sampling rate at 20 kHz. While a LPF with a cut-off frequency of 1 kHz is used for the slower sampling rate of 2 kHz. It is known that elves also show an apparent downward propagation due to its geometrical evolution (Inan et al., 1996). This particular property of elves allows us to distinguish elves from the scattered CG flashes, which show no time difference of peak intensities at any elevation (Fukunishi et al., 1999). A sampling rate of 20 kHz with a 10 kHz LPF) is sufficient for measuring the apparent vertical development of elves.

The AP also operates in a burst mode and is triggered by a signal issued by the SP signals and the trigger algorithm. The AP is able to run in pre-triggering or in immediate triggering modes according to the channel of the spectrophotometer used for initializing the triggering signal. Blue jets, aurora and airglow could also be studied by AP at a slow sampling rate of 200 Hz with a 100 Hz low pass filter. This relatively low sampling rate is used for long duration observations where the phenomena do not require the full time resolution of the instrument. One of the unique auroral objectives that could be studies by the AP is the flickering aurora, which is blinking at a rate of up to 100 Hz. In this case the data will be recorded continuously without a triggering signal. The sensitivities of photometers are adjustable by varying the high voltage of the PMTs, so that measurements of luminous emissions can be made with a large intensity range, covering faint airglow to intense filaments of sprite heads.

6.3 Initial Observations with ISUAL

Soon after commissioning the ISUAL instrument it was able to record various types of TLEs. A particularly nice sprite was observed on July 4th 2004 and a colorized version of the sprite is shown in Figure 9a. Underneath, Figure 9b shows the geographic position of the field of view of the imager observing the sprite. Another nice event was observed on the 18th of July 2004. The event is illustrated in Figure 10 where we present the image sequence of all 6 images (29 msec exposures each) taken by the ISUAL imager. The instrument operating mode saved one image prior to the trigger and therefore triggering occurred in the second frame. From the first frame we can see that there was a substantial lightning activity in progress before the trigger. The second frame shows the sprite in full bloom. There is a faint residue in the third frame. In the fourth frame we see another sprite which is located to the right of the one previously observed. The fifth frame is relatively dim but a new sprite is seen on the sixth frame. Whether this is a new sprite or a re-ignition of the one seen in the fourth frame is open to question. The corresponding photometer data is presented in Figure 11. Channel 1 (150-280 nm) is the far ultraviolet channel responding mostly to the LBH emission. Channel 2 (337 nm) is the (0,0) group of the N_2 2nd positive band containing theoretically 27% of the entire band. The 3rd channel is N_2^+ 1st negative showing the presence of ionized N_2. Channel 4 has a filter that passes part of the N_2 1st positive band and some of the N_2^+ Meinel band. It is a broad band filter (see Figure 8) for simplicity it was labeled "N_2 1P". Channel 5 is the 777.4 nm line of atomic O and is a relatively good representation of the total power output of the lightning. Channel 6 (250-390 nm) is a broad band UV filter passing a mixture of N_2 and N_2^+ bands. Note that lightning has a strong continuum emission that would be a contaminant in all TLE channels except in channel 1. The trigger occurred at 21:30:15.316. According to the image (N_2 1P) there was substantial emission in the frame prior to the one containing the trigger. Channel 4, the 1st positive SP filter shows that the baseline was elevated even during the first exposure. Most of the other channels show two distinct peaks between ~308 and 320 msec. The first peak seems to have been caused by the light-

(a) July 4, 2004/21:31:15.451

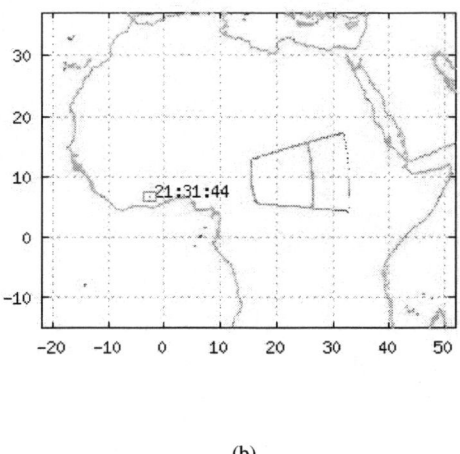

(b)

Figure 9. (a) Sprite event over Africa. (b) The map shows the satellite position in local time and the field of view of ISUAL observing the sprite.

ning. We believe that the second, the larger peak in the first two channels, is caused by the sprite. Channel 1 is the far ultraviolet channel, which is seeing emissions (LBH) that are emitted by the atmosphere above 30 km. There is a very nice strong signal in the second channel which is the N_2 2nd positive. The quenching height of this emission is about 30 km and we would expect minimal contribution from lightning other than the continuum. There is a strong response in channel 4 (at 357 ms) which appears to be a data glitch or penetrating particle event because it is not repeated in any other channel. There are a couple of small flashes at about 372 and 382 ms, the peak that occurs at 372 msec is most likely to be caused by the lightning flash seen in frame 4. The second set of peaks that obey the same pattern is at 433 and 438 ms.

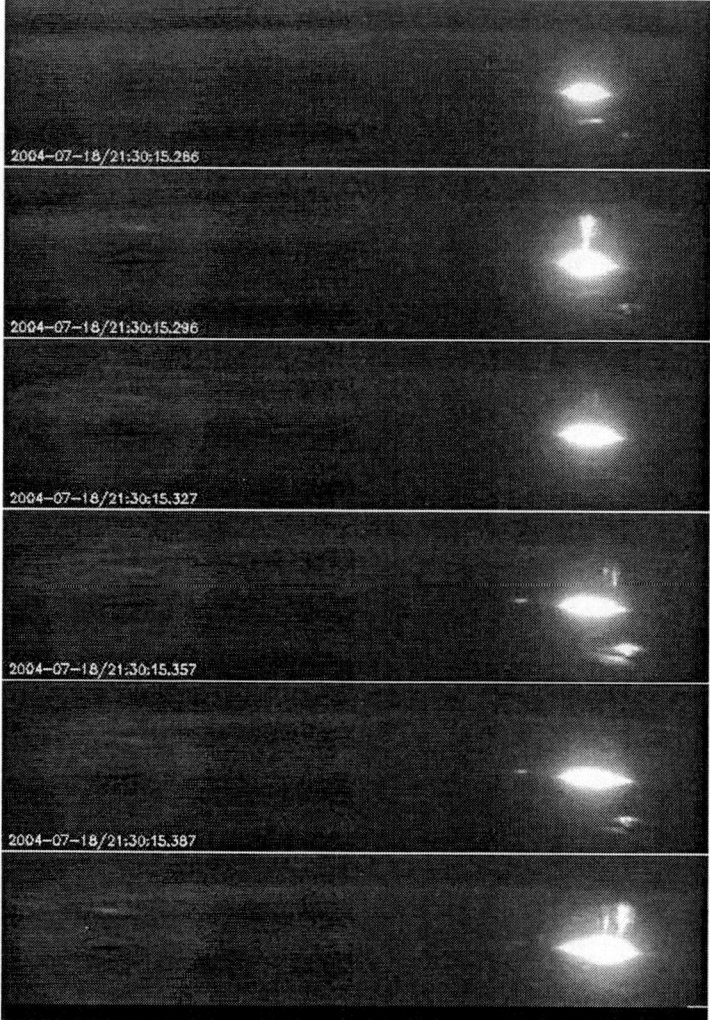

Figure 10. Sprite sequence that occurred on July 18, 2004 at 21:30:15.266 to 21:30:15.446

We see a strong contribution in channel 1 and 2 by the second feature which we believe is the sprite. Channel 3 is mostly sensitive to the N_2^+ 1st negative band and shows the presence of ion production in the sprite or the lightning. It appears that the first sprite had the strongest contribution to locally produced ions. Channel 4 is mainly responsive to the 1st positive and to some degree to the N_2^+ Meinel bands. The strong signal in channels 3 and 5 are showing the presence of energetic electrons that are capable of ionizing the atmosphere and they are characteristic of lightning. The set of peaks at 433 and 488 are consistent with image 6 showing that another lightning strike occurred at 433 msec, which was closely followed by a new sprite that can be seen in channels 1 and 2 at 438 msec. In Figures 12 and 13 we show an elve with minimal lightning

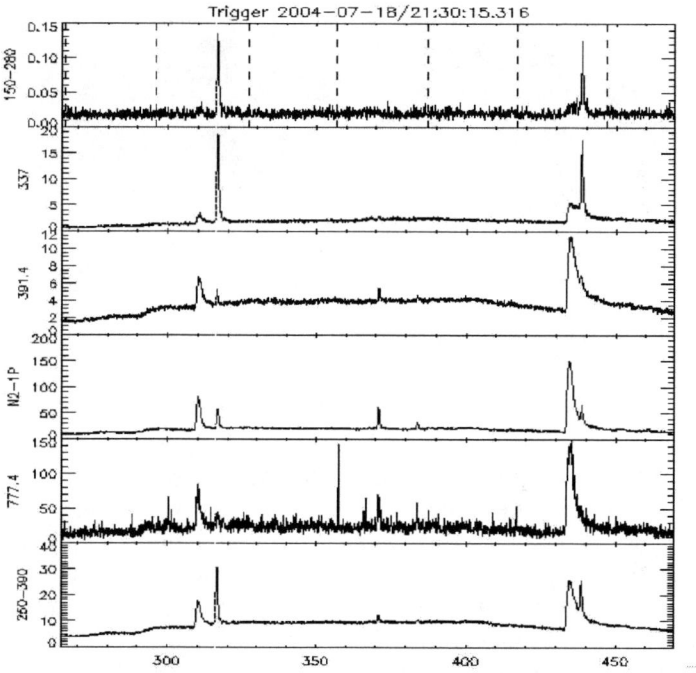

Figure 11. SP data for the sprite event of Figure 10. Vertical lines in channel 1 indicate the six (29 ms duration) exposures of the imager. The signal is shown calibrated in MR-s referenced to the wavelength of the peak filter transmission profile and to the solid angle subtended by the sprite on the second image (Figure 10). The x axis is time in ms. Channel 4 $N_2$1P filter wavelength profile is shown in Figure 8.

contamination. In Figure 13 the 777.4 nm channel shows that the lightning was a very sharp and sudden pulse at ∼456 msec. We assume that the elve caused the apparent longer lifetime of the luminosity in the other channels. There is a distinct amount of 391.4 nm emission showing that there is ionization during this event. Channel 6 the combined middle UV channel is actually saturated by this elve. The y axis was calibrated using the preflight calibrations and the angular size of the elve is derived from the camera image of Figure 12. The peak intensities reach very high values for example the $N_2 1^{st}$ positive (channel 4) reaches over 20 MR. In spite of this high intensity the elve appears to be relatively dim in the image because the phenomena is short- lived approximately 1 msec. Thus an auroral pulse of 1 sec duration would only need to be 20 kR for equivalent energy output.

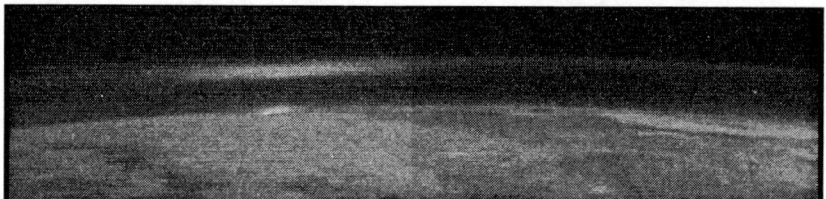

Figure 12. Elve observed on 04/07/05 with trigger time at 14:41:57.456

6.4 Summary

The new instrument the Imager of Sprites/Upper Atmospheric Lightning (ISUAL) has been in orbit since May 20, 2004 making new observations of TLEs from space. The ISUAL payload includes a visible wavelength intensified CCD imager, a boresighted six wavelength channel spectrophotometer, and a two channel Array Photometer (AP) with 16 vertically spaced horizon ally wide sensitive regions. The imager is equipped with 5 selectable filters on a filter wheel, for most observations a filter passing the N_2 1st positive band was used. The two channel AP is fitted with broad band blue and red filters. In this chapter we have presented examples of ISUAL imager and photometer data. One of these is a sequence of sprites with strong accompanying lightning activity. The other is an elve with only minimal lightning intensity. The photometer and the imager were calibrated pre-flight and using these numbers it was possible to compare the various responses by the photometer to the various phenomena. We have tabulated the results in Table 4 where we present intensity ratios to SP channel 4 which is a measurement of the N_2 first positive band. This band has been found to be the most intense in ground based observations of high altitude TLEs. The 1st SP channel with the UV pass band from 150-300 nm is mostly blind to lightning contamination because the atmosphere is opaque in the region well above the thunderclouds. In this channel the signals mainly represent N_2 LBH band emissions. Only the higher wavelength components of LBH can get through because of the strong O_2 absorptions for wavelength below 200 nm. This emission would also be quenched at low latitudes. Although there are strong signals in these bands due to sprites or elves, the intensity is less than a percent of the N_2 1st positive in terms of absolute

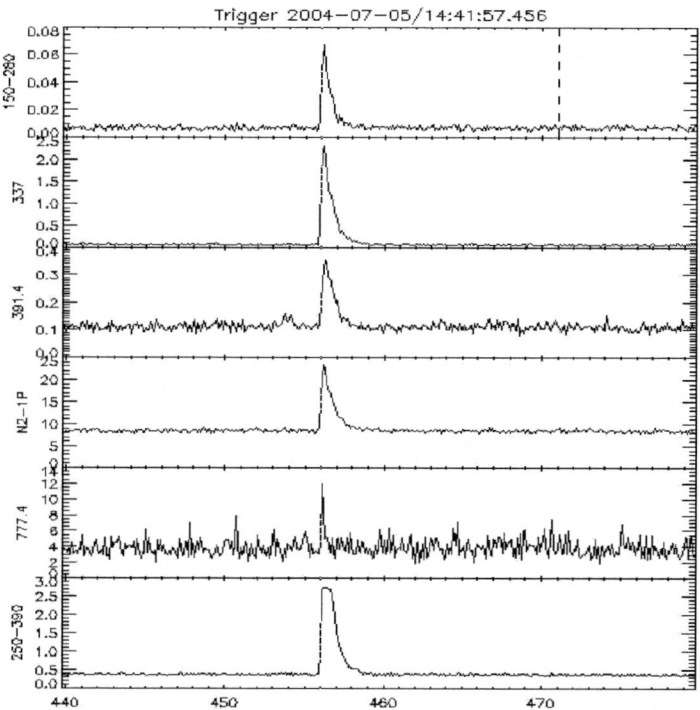

Figure 13. Photometer traces of the elve (shown in Figure 12).

quantities. Channel 2 is optimized for the detection of the N_2 2nd positive (0,0) band at 337 nm. This is very bright in TLEs. This was relatively bright in our sprite observation, 50% of the N_2 1st positive in the sprite and it was only about 12% in the elve. Channel 3, the 391.4 nm channel measures the $N_2^+ 1^{st}$ negative emission which is a direct indication of the presence of electrons with sufficient energy to ionize N_2. The lightning has the strongest relative signature in this emission (6%), the sprite is second with 4.3 and the elve produces the least, 1.6%. All these results are qualitatively consistent with the fact that lightning produces high energy electrons that can consistently ionize the gas whereas sprites produce a softer electron energy distribution while elves have an electron energy distribution with even softer characteristics.

The main focus of the ISUAL data analysis is to identify the temporal and spatial distribution of the emissions and compare them to the prediction of theoretically predicted processes leading to the various excitations seen. Using the satellite platform we expect to widen our case studies to larger samples and we hope to eventually produce statistical surveys of the global properties of TLEs.

Channel	Lightning	Sprite	Elve
1	0.00%	0.37%	0.35%
2	6.00%	53.33%	11.76%
3	6.00%	4.33%	1.65%
4	100.00%	100.00%	100.00%
5	120.00%	0.00%	0.00%
6	20.00%	66.67%	12.94%

Table 4. Approximate intensity ratios to channel 4 for the three types of events as derived from the measurements.

Acknowledgments

The ISUAL imager, spectrophotometer and the associated electronic package instrument was built by the Space Science Laboratory, University of California, Berkeley under a contract from National Space Program Office (NSPO) of Taiwan. The authors wish to express their gratitude to the technical staff of the Space Science Laboratory for their support in producing the ISUAL instrument. Thanks are also due to the staff of NSPO for their assistance and continued support.

The NCKU Team is supported in part by NSPO of Taiwan under Grant Numbers NSC90-NSPO (B)-ISUAL-FA09-01, NSC90-NSPO (B)-ISUAL-FA09-02, and NSC91-NSPO (B)-ISUAL-FA09-01.

Bibliography

Barrington-Leigh, C. P., Inan, U. S., and Stanley, M. (2001). Identification of sprites and elves with intensified video and broadband array photometry. *J. Geophys. Res.*, 106:1741–1750.

Boeck, W. L., O. H. Vaughan, Jr., Blakeslee, R. J., Vonnegut, B., Brook, M., and McKune, J. (1995). Observations of lightning in the stratosphere. *J. Geophys. Res.*, 100:1465.

Borst, W. L. and Zipf, E. C. (1970). Cross section for electron-impact excitation of the (0,0) first negative band of N_2^+ from threshold to 3 keV. *Phys. Rev. A*, 1(3):834–840.

Boys, C. V. (1926). Progressive lightning. *Nature*, 118:749.

Chamberlain, J. W. (1961). *Physics of Aurora and Airglow*. Academic Press.

Christian, Hugh J., Blakeslee, R. J., Boccippio, D. J., Boeck, W. L., Buechler, D. E., Driscoll, K. T., Goodman, S. J., Hall, J. M., Koshak, W. J., Mach, D. M., and Stewart, M. F. (2003). Global frequency and distribution of lightning as observed from space by the Optical Transient Detector. *J. Geophys. Res.*, 108(D1):ACL 4–1, doi:10.1029/2002JD002347.

Dick, K. A (1978). The auroral 2150 a feature - a contribution from lines of singly ionized atomic nitrogen. *Geophys. Res. Lett*, 5:273–274.

Everett, W. H. (1903). Rocket lightning. *Nature*, 68:599.

Franz, R. C., Nemzek, R. J., and Winckler, J. R. (1990). Television image of a large upward electrical discharge above a thunderstorm system. *Science*, 249:48–51.

Fukunishi, H., Takahashi, Y., Kubota, M., Sakanoi, K., Inan, U. S., and Lyons, W. A. (1996). Elves: lightning-induced transient luminous events in the lower ionosphere. *Geophys. Res. Lett.*, 23:2157.

Fukunishi, H., Takahashi, Y., Uchida, A., Sera, M., Adachi, K., and Miyasato, R. (1999). Occurrences of sprites and elves above the Sea of Japan near Hokuriku in Winter. *EOS*, 80(46):F217.

Gerken, E. A., Inan, U. S., and Barrington-Leigh, C. P. (2000). Telescopic imaging of sprites. *Geophys. Res. Lett.*, 27:2637–2640.

Hampton, D. L., Heavner, M. J., Wescott, E. M., and Sentman, D. D. (1996). Optical spectral characteristics of sprites. *Geophys. Res. Lett.*, 23:89–92.

Inan, U. S., Sampson, W. A., and Taranenko, Y. N. (1996). Space-time structure of optical flashes and ionization changes produced by lightning-EMP. *Geophys. Res. Lett.*, 23:133.

Lyons, W. A. (1996). Sprite observations above the U.S. high plains in relation to their parent thunderstorm systems. *J. Geophys. Res.*, 101:29,641.

Lyons, W. A., Nelson, T. E., and Faires, A. (2001). Electrical discharges into the stratosphere from the tops of intense thunderstorms. *EOS Trans. AGU - Fall Meet. Suppl.*, 82:47. Abstract AE22A-02.

Marinelli, W. J., Kessler, W. J., Green, B. D., and Blumberg, W. A. M. (1989). Quenching of $N_2(a^1\Pi_g, v = 0)$ by N_2, O_2, CO, CO_2, CH_4, H_2, and Ar. *J. Chem. Phys*, 90:2167–2173.

Mende, S. B., Rairden, R. L., Swenson, G. R., and Lyons, W. A. (1995). Sprite spectra; N_2 1PG band identification. *Geophys. Res. Lett.*, 22:2633.

Minschwaner, K., Anderson, G. P., Hall, L. A., and Yoshino, K. (1992). Polynomial coefficients for calculating O_2 Schumann-Runge cross sections at 0.5/cm resolution. *J. Geophys. Res.*, 97(D9):10103–10108.

Pasko, V. P. and George, J. J. (2001). Three-dimensional modeling of blue jets and blue starters. *EOS Trans. AGU - Fall Meet. Suppl.*, 82(47).

Pilling, M. J. A., Bass, A. M., and Braun, W. (1971). A curve of growth determination of the f values for the fourth positive band of CO and the Lyman Birge Hopfield band system of N_2. *J. Quant. Spectrosc. Rad. Trans.*, 11:1593.

Rairden, R. L. and Mende, S. B. (1995). Time resolved sprite imagery. *Geophys. Res. Lett.*, 22:3465.

Sentman, D. D., Wescott, E. M., Osborne, D. L., Hampton, D. L., and Heavner, M. J. (1995). Preliminary results from the Sprites94 aircraft campaign: 1. red sprites. *Geophys. Res. Lett.*, 22:1205.

Sera, M., Takahashi, Y., and H. Fukunishi: 2001, Takahashi, Y. (2001). Estimation of electron energy in sprites. In *Proc. of AP-RASC Meeting*, Tokyo.

Shemansky, D. E. (1969). N_2 Lyman-Birge-Hopfield band system. *J. Chem. Phys.*, 51:5487.

Stanley, M., Krehbiel, P., Brook, M., Moore, C., Rison, W., and Abrahams, B. (1999). High speed video of initial sprite development. *Geophys. Res. Lett.*, 26:3201–3204.

Strickland, D. J., Bishop, J., Evans, J. S., Majeed, T., Cox, R. J., Morrison, D., Romick, G. J., Carbary, J. F., and L. J. Paxton, C.-I. (2001). Meng Midcourse space experiment / ultraviolet and visible imaging and spectrographic imaging limb observations of combined proton / hydrogen / electron aurora. *J. Geophys. Res.*, 106:65–76.

Su, H. T., Hsu, R. R., Chen, A. B., Lee, Y. J., and Lee, L. C. (2002). Observation of sprites over the Asian continent and over oceans around Taiwan. *Geophys. Res. Lett.*, 29(4):10.1029/2001GL013737.

Takahashi, Y., Watanabe, Y., Uchida, A., Sera, M., Sato, M., and Fukunishi, H. (1998). Energy distributions of electrons exciting sprites and elves inferred from the fast array photometer observations. *EOS Trans. AGU*, 79:F175.

Vallance-Jones, A. (1974). *Aurora*. D. Reidel, Dordrecht, Holland. pp. 85-140.

Vaughan, O. H. Jr., Blakeslee, R., Boeck, W. L., Vonnegut, B., Brook, M., and J. McKune, Jr (1992). A cloud-to-space lightning as recorded by the space shuttle payload-bay TV cameras. *Mon. Wea. Rev.*, 120:1459.

Vaughan, O. H. Jr. and Vonnegut, B. (1989). Recent observations of lightning discharges from the top of a thundercloud into the clear air above. *J. Geophys. Res.*, 94:13179–1382.

Wescott, E. M., Sentman, D. D., Osborne, D. L., Hampton, D. L., and Heavner, M. J. (1995). Preliminary results from the Sprites94 aircraft campaign: 1. blue jets. *Geophys. Res. Lett.*, 22:1209.

Winckler, J. R., Lyons, W. A., Nelson, T. E., and Nemzek, R. J. (1996). New high-resolution ground-based studies of sprites. *J. Geophys. Res.*, 101:6997.

Yoshino, K., Cheung, A. S. C., Esmond, J. R., Parkinson, W. H., Freeman, D. E., Guberman, S. L., Jenouvrier, A., Coquart, B., and Merienne, M. F. (1988). Improved absorption cross-sections of oxygen in the wavelength region 205-240 nm of the Herzberg Continuum. *Planet. Space Sci.*, 36(12):1469–1475.

OBSERVATIONS OF SPRITES FROM SPACE AT THE NADIR: THE LSO (LIGHTNING AND SPRITE OBSERVATIONS) EXPERIMENT ON BOARD OF THE INTERNATIONAL SPACE STATION

Elisabeth Blanc [1], T. Farges [1], D. Brebion [1], A. Labarthe [2] and V. Melnikov [1]

[1] *Commissariat à l'Energie Atomique, Bruyères le Châtel, France.*

[2] *Centre National d'Etudes Spatiales, Toulouse, France.*

[3] *Rocket Space Corporation ENERGIA, Korolev, Russia.*

Abstract The experiment LSO (Lightning and Sprite Observations) on board of the International Space Station is the first experiment dedicated to sprite observations at the nadir. Such observations are difficult because the luminous emissions of sprites and lightning can be superimposed when they are observed from space at the nadir. Such observations are however needed for measuring simultaneously all possible emissions (radio, X-γ, high energy electrons) associated with sprites for a better understanding of the implied mechanisms. They are possible in specific spectral lines where sprites are differentiated from lightning. Absorption bands of the atmosphere are well adapted for this differentiation because the light emissions from sprites occurring in the middle and upper atmosphere are less absorbed in these bands than lightning emissions occurring more deeply in the atmosphere. The most intense spectral emission band of the sprites, corresponding to the $N_2 1P$ band at 761 nm, partly superimposed with the oxygen absorption A band of the atmosphere, is used by the LSO experiment. The experiment is composed of two micro-cameras, one in the visible and near infra red, the other equipped with an adapted filter. Only sprites, halos and superbolts, which correspond to a class of rare very intense lightning, are transmitted through the filter. Sprites, halos and superbolts are identified by the ratio of the intensities received through the filter and in the whole spectrum. This ratio is lower for superbolts than for sprites and halos. The response of the sprites is also more complex and variable than the response of superbolts which is very flat and comparable from an event to another. Finally, LSO observed 17 sprites, 3 halos and 9 superbolts. Several examples of differentiation of sprite and superbolts are given. The results of a first global statistical study are also presented.

7.1 Introduction

Most of the observations of sprites are performed from planes (Sentman et al., 1995) and from the ground in different parts of the world (Lyons et al., 2003; Hardman et al., 2000; Neubert et al., 2001; Su et al., 2002) at the horizon where sprites are spatially differentiated from the lightning flashes. Different types of emissions (jets, halos, elves) called TLE (Transient Luminous Events, Lyons et al., 2000) have been identified. Recent observations provide details of the space and time evolution of these phenomena (Gerken et al., 2000; Moudry et al., 2003). The first space observation of sprites was performed during thunderstorm observations (Boeck et al., 1998). Few experiments are now designed for sprite observations from space at the horizon: (i) MEIDEX onboard of the Space Shuttle performed 7 hours of sprite observations over thunderstorms (Yair et al., 2004), (ii) the ISUAL experiment is the first sprite experiment onboard a satellite (Chapter 6).

However sprites are complex phenomena and the emissions in the visible constitute only a part of the emissions related with sprites. Theoretical studies show that sprites can be produced by electrostatic electric fields above the altitude where the thunderstorm electric field exceeds the air breakdown electric field threshold (Pasko et al., 1997). This process predicts ELF electrostatic emissions which can be observed at the ground (Cummer et al., 1998; Füllekrug et al., 2001). Electromagnetic pulses are involved in the elve formation (Barrington-Leigh et al., 2001). Sprites have also been explained by relativistic runaway electrons triggered by cosmic radiation (Roussel-Dupré and A., 1996; Roussel-Dupré et al., 1998). The resulting high energy electron beam interacts with the atmosphere producing intense electromagnetic radio emissions in a large frequency range in the HF-VHF part of the spectrum and X-gamma emissions by bremsstrahlung process. Both conventional and runaway processes could occur in parallel (Roussel-Dupré et al., 2002). The runaway electron theory is supported by the observations of X and γ ray emissions from the Earth's atmosphere (Fishman et al., 1994; Feldman et al., 1995; Lopez et al., 2004). Ground based observations of energetic radiation up to many tens of MeV, produced during rocket triggered lightning shows that this process is more frequent than expected (Dwyer et al., 2004).

For a better understanding of these mechanisms, simultaneous measurements of all these emissions from space are needed. However, these observations are difficult to realize, because the light emissions of sprites are then superimposed on the intense light emissions of the lightning diffused by clouds.

The LSO (Lightning and Sprite Observations) experiment on board of the International Space Station (ISS) has been designed to perform sprite observations at the nadir using an original method of spectral differentiation between sprites and lightning by an adapted filter. The first sprite observations obtained in the frame of the Andromède and Odissea missions were shown by Blanc et al. (2004). The results of new observations performed in the frame of Cervantes and Delta missions are presented in this chapter. A first statistic of the global LSO observations obtained up to now is also presented.

7.2 Spectral Differentiation of Sprite and Lightning Emissions

The frequency band, used for the selective spectral measurements of sprites, corresponds to the most intense emission line of sprites $N_2 1P$ at 762.7 nm, or $N_2 \left(B^3 \Pi_g - A^3 \Sigma_u^+ \right)$ $(3-1)$. The interest of this spectral band is that it includes a significant part of the absorption band of the O_2 $\left(b^1 \Sigma_g^+ - X^3 \Sigma_g^+ \right)$ $(0-0)$ absorption band near 761.9 nm. For this reason, the sprite emission line $N_2 1P$ does not appear on the sprite spectra measured at the ground (Hampton et al., 1996; Morrill et al., 1998) where the dioxygen density is important. On the contrary, this emission will be observed from space because of a weaker dioxygen density above the sprite. The light emissions, within this band, from lightning, produced deeper in the atmosphere, will be absorbed, as well as all man made emissions from the ground surface. The spectral band selected for LSO corresponds to the most intense emission band of the sprites determined by using the theoretical spectrum of sprites provided by Milikh et al. (1998), see Figure 1. Figure 2 shows the solar light transmission from 754 to 770 nm (Solar Survey Archive-2000, 2005) superimposed with the filter response and the sprite spectrum $N_2 1P$ peak. The filter width of the LSO filter is 10 nm, it has been optimized for receiving the maximum of energy from the sprites and filtering the lightning emission through the atmospheric absorption. A narrower width selected where the absorption band is the most intense would optimize the spectral differentiation of sprites and lightning but a more sensitive camera is necessary to measure very low sprite intensities.

The possible emission of lightning in the same spectral band has been estimated by using the lightning spectrum measured by Orville and Henderson (1984) and the camera and filter responses. The intensity received through the filter is expected to be about 1% of the total spectrum.

7.3 Experiment

The first LSO measurements were performed in the frame of the flight of the French astronaut Claudie Haigneré (Andromède mission) on the International Space Station in October 2001. The experiment was developed by the Commissariat à l'Energie Atomique with the participation of the Centre National d'Etudes Spatiales. The measurements were realized with the collaboration of RKK Energia (Russia). Additional observations were performed in 2002

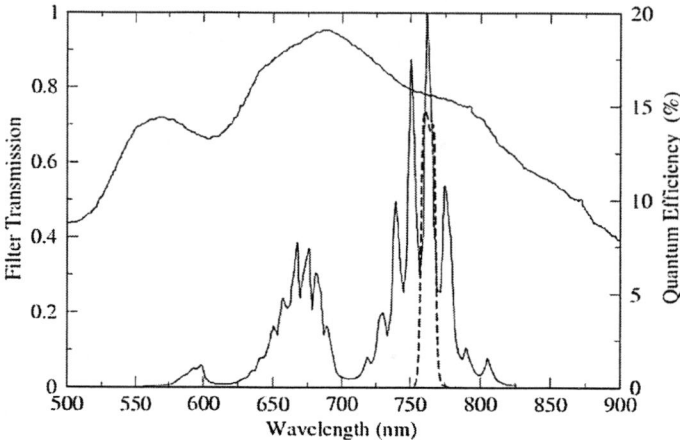

Figure 1. LSO filter response (in dashed bold line) is adapted to the observation of the most intense sprite emission band (Milikh et al., 1998). The camera response is also shown in dotted line.

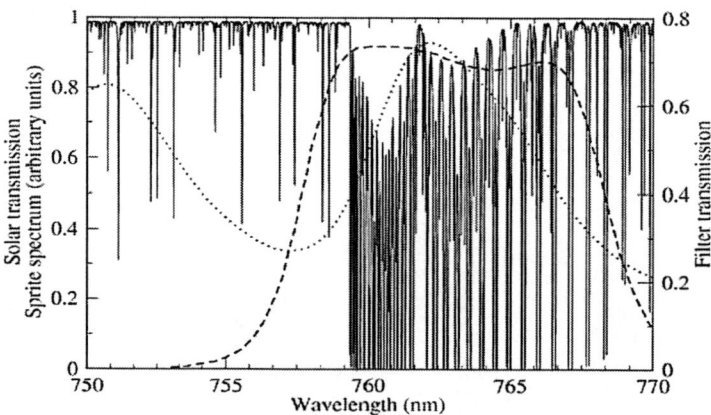

Figure 2. Solar light transmission (solid line) in the vicinity of the LSO filter (dashed line). The sprite spectrum in this band is indicated in the dotted line.

during the Odissea mission of the flight of the Belgium astronaut with the participation of the European Space Agency. The first results obtained during these missions are described by (Blanc et al., 2004). New results were also obtained from September 1st to 5th, 2003 and from 1st to 5th October 2003 in the frame of the ESA Cervantes mission (flight of the Spanish astronaut) and from April 24th to 26th, 2004 in the frame of the ESA Delta mission (flight of the Netherlands astronaut). However one camera failed during the Delta experiment and only the Cervantes data were used for sprite observations. In total, 19 h of effective observations are available up to now.

The ISS orbit is at about 400 km altitude, the inclination is 51.6°. This orbit is well adapted to the observation of sprites and lightning flashes which mainly occur at low and middle latitudes. The observations are performed when the station is stabilized during several days per month. The two-line norad orbital elements are used by LSO programs to predict the ISS orbit characteristics.

The experiment (Figure 3) is composed of two microcameras, one equipped with the filter adapted to the observations of sprites and the other in the total camera spectral range which extends from 400 up to 1000 nm (Figure 1). The camera response is maximum at 690 nm. The cameras are connected to an electronic box and to one Experiment Processing Computer which is only dedicated to the camera programs and data archiving. The digital space micro-cameras have been developed by CSEM (Centre Suisse d'Electronique et de Microelectronique, now Space-X). The objectives have a focal length of 14 mm, an aperture of f/3.5 and a field of view of 70°. The images are taken on 1024×1024 pixels with 10 bits dynamic range but only the central part of the CCD (512×512 pixels) is used (the effective field of view is then 39.8°). The pixel length is 14 μm with a pixel aperture ratio of 0.71 due to the anti-blooming system. The images of both cameras are taken simultaneously. Because of the very rapid development time of the experiment (3.5 months), it has not been possible to lower the data transmission time of both images time below 5.5 s. The time exposure is 1 s. The precision in time is one second. The cameras are fixed on an ISS window. Both cameras were calibrated for quanti-

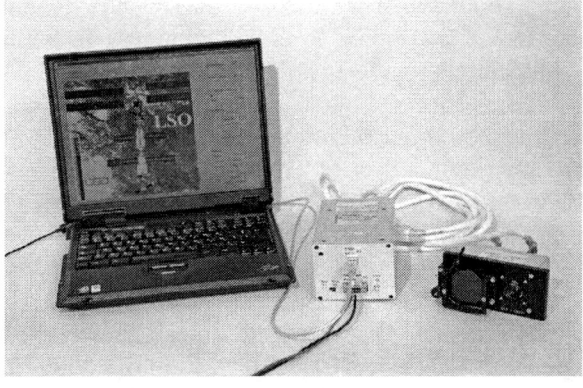

Figure 3. LSO experiment

tative measurements of brightness (Blanc et al., 2004). The spectral sensitivity of both cameras is 45 $\mu J/m^2/sr$ at 765 nm for 1 LSB (least significant bit) or about $2 \cdot 10^{-4}$ ft candles (~2 mLux), which is comparable with the sensitivity of the camera used for the first sprite measurements from space (Vaughan, 1994).

The measurements are automatic. The astronaut enters the dates of the beginning and of the experiment and the two lines norad elements needed by the on board computation program. Data are archived on a removable disk which is brought back to the ground by the astronaut.

Observations are performed during the night and mainly over continents, were most of the storms are expected according to the TRMM (Tropical Rainfall Measurement Mission) satellite observations (Christian et al., 2003). One observation area has been selected over the Pacific Ocean to represent ocean conditions. The observation areas covered during the Cervantes mission are shown in Figure 4.

7.4 Observations

At the end of the Odissea mission, LSO observed 60 transitory events with the camera in the visible and 13 events with both cameras. The first class of events observed by both cameras corresponds to sprites. It is defined by a ratio of both cameras' intensities higher than 3% (Figure 5, left and middle). These events are characterized by a complex spatial response and the ratio is variable inside a same event. Also the response is variable from one event to another. Ten events belonging to this class were identified in this first experiment set (Blanc et al., 2004). The second class of events, observed by both cameras, is characterized by a lower intensity ratio of about 1% (Figure 5, right), the ratio is quite stable over the event spatial coverage, in addition the response is comparable from one event to another. They have been identified as very intense lightning called superbolts with a power from $3 \cdot 10^{11}$ to $7 \cdot 10^{12}$ W (Turman, 1977). Only about 1% of lightning belongs to this lightning class. The LSO filter suppresses the response of most lightning, but the intensity of the superbolts is sufficient to provide a response in the filter.

The analysis of the new data obtained in the frame of the Cervantes mission confirms the first results about these two classes of events. An example of a sprite event observed in this data set is shown in the Figure 6 with the same color scale as in Figure 5. The complexity of this event arises from the presence of three different flashes, observed in the same image because of the 1 second of integration time of the observations. The three events are differentiated by the ratio of both cameras. The sprite appears in red in the picture of the ratio of the cameras' intensities, indicating a ratio value higher than 6%. Differently the two other emissions appear in blue with a ratio of about 1% or in yellow and green with a ratio of 2 or 3%. As a superbolt is characterized by a ratio of

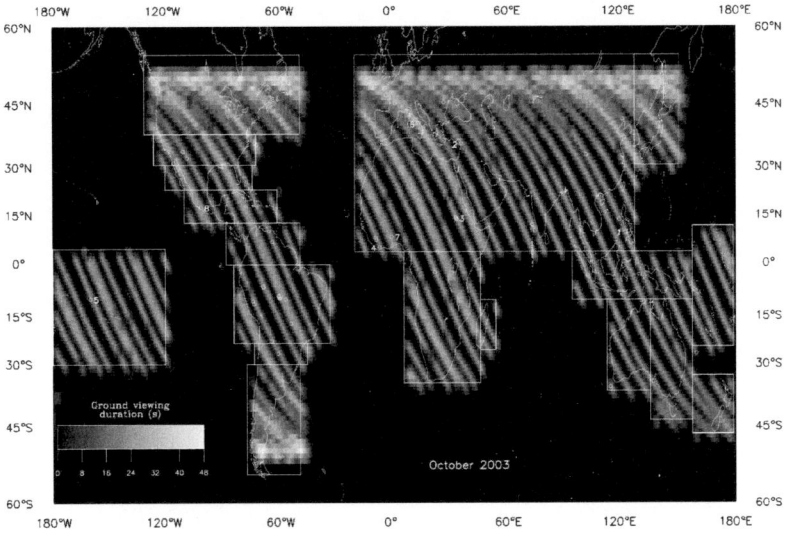

(a)

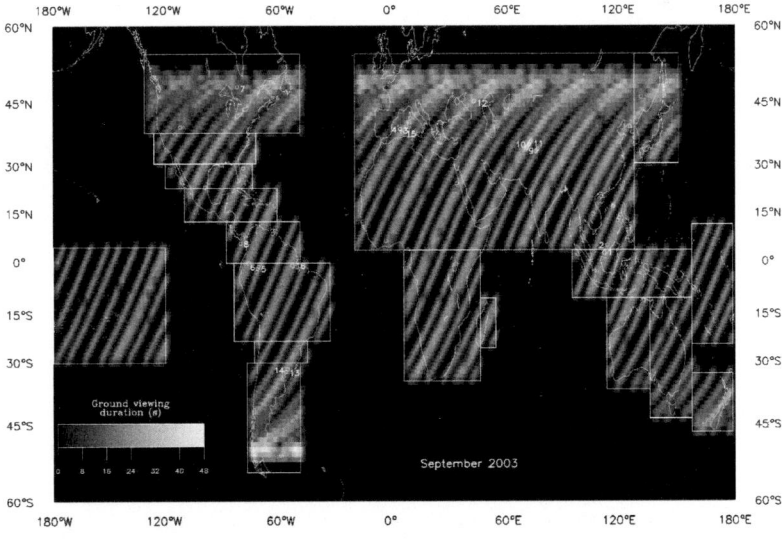

(b)

Figure 4. Duration and areas observed by LSO for the Cervantes mission.

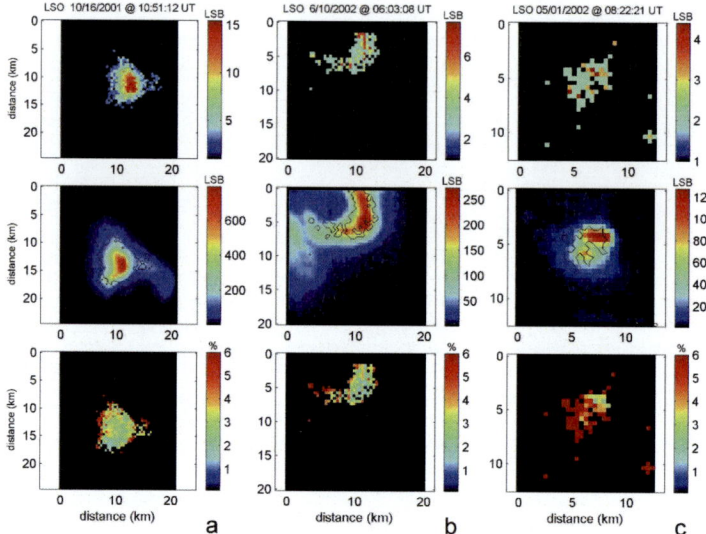

Figure 5. First observation of sprite from the nadir by LSO. 1. Top filtered images 2. Middle: images in the visible, 3. Bottom: ratio of both intensities. The event at the right is a superbolt while both events on the left are sprites. They are differentiated by the ratio of the intensities measured by both cameras, most intense for sprites than for lightning.

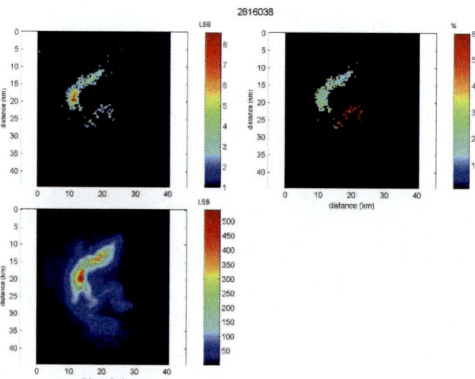

Figure 6. Simultaneous observation of a sprite and a superbolt performed on October 3rd, 2003, at 19:19:31 UT at $35.58°$N $36.18°$E. The ratio of both cameras is higher than 6 for the sprite. It is about 1% for the superbolt. The most intense event in the visible and in the filter corresponds to a ratio of 2 to 3% and could be explained by the presence of a halo.

1%, the ratio increase could indicate the presence of an halo superposed with the lightning. When isolated, halos have been identified by a ratio comparable or slightly lower than the sprite ratio, but with a larger geometric extension and a response more flat.

During the 19 hours of observations, 40 events were observed with both cameras. The first 13 hours correspond to the first data set analyzed by Blanc et al. (2004). Table 1 provides the characteristics of the additional events observed during the Cervantes mission. In total, 17 sprites were identified with 3 isolated halos. Three sprites, not clearly identified are not taken into account in this number. Also 9 superbolts were identified. The three, indicated in bold, are among the most intense events observed by LSO. The sprite observation frequency is then about 1.7 per hour. The number of sprites observed by MEIDEX onboard of the Space shuttle is 2.6 sprites per hour, and 4 TLE per hour (Yair et al., 2004). This number is higher but the cameras were oriented to observe the regions of interest according to the probability of lightning activity and the astronaut adjusted the camera pointing angle for lightning observations, increasing the probability to observe TLEs.

Figure 7 shows the local time distribution of the 40 events observed by both cameras of LSO. The local time is deduced from the universal time taking into account the position of the ISS at the measurement time. The maximum of activity is observed near midnight. The distribution is not related to the lightning nighttime distribution which is decreasing regularly from 18-7 LT (local time), the maximum being at about 14 LT (MacGorman and Rust, 1998).

LSO observed 180 flashes corresponding to the most intense lightning. The number of lightning flashes which effectively occurred during the 19 h of the LSO measurements can be estimated by using the LIS data measured on board of the TRMM satellite in the same period of observation (LIS/OTD webpage, 2002; Christian et al., 2003). This number of lightning flashes is estimated about 1100. This estimation takes into account the respective observation areas of LIS and LSO and the fact that LSO measures only over the continents, where 88% of the lightning occur. The occurrence of sprites is then 1.4 sprites for 100 lightning flashes. Taking into account that the global number of lightning determined from LIS observations is 44 flashes per second (Christian et al., 2003), the global number of sprites per minute could be about 37. This number is more important than the global number of sprites estimated by (Yair et al., 2004) which is 7.5 sprites per minute. The reason is that the LSO observations are performed over continents where most of the lightning flashes occur. This increases the probability to observe sprites.

Figure 8 shows the intensity distribution of the 203 events observed by LSO in the visible. This include sprites, lightning and superbolts. The superbolts correspond to the most intense events observed by LSO. Among the 5 most intense events, 4 are superbolts and 1 is a sprite.

7.5 Perspectives

The results obtained by LSO show that sprites and lightning flashes observed from space at the nadir can be differentiated by using adapted filters. This differentiation has been performed by the comparison of the response of two cameras, one in the visible and the other in the $N_2 1P$ sprite emission band, which partly coincides with the dioxygen absorption band. Sprites are also separated spatially from lightning. It is expected that the mean distance between sprites and lightning is about 40 km in agreement with ground based observa-

Table 1. Characteristics of the events observed by both cameras during the Cervantes mission.

Date	Universal Time	Longitude	Latitude	Event identification
01/09/2003	17:41:43	111.0	3.3	Lightning
01/09/2003	17:42:25	112.6	5.5	Sprite (?)
02/09/2003	03:06:58	4.0	38.8	Sprite
02/09/2003	03:07:03	4.3	39.0	Lightning
03/09/2003	04:59:07	-71.2	-1.7	Sprite (?)
03/09/2003	04:59:18	-70.8	-1.1	Lightning
03/09/2003	09:54:35	-82.0	0.5	Superbolt
04/09/2003	05:35:45	-80.3	6.2	**Superbolt**
04/09/2003	21:06:26	69.7	33.5	Lightning
04/09/2003	21:07:05	71.8	35.2	Sprite
04/09/2003	21:07:10	72.1	35.4	Sprite
05/09/2003	00:15:51	42.8	45.3	Halo
05/09/2003	01:20:23	-56.4	-32.5	Superbolt
05/09/2003	01:20:39	-55.5	-31.8	Sprite
05/09/2003	01:44:37	5.6	38.0	Sprite (?)
05/09/2003	03:03:18	-53.0	-0.7	**Superbolt**
01/10/2003	22:58:38	23.6	4.0	Lightning
02/10/2003	15:50:06	121.7	11.0	Lightning
03/10/2003	19:19:31	36.2	35.6	Sprite
03/10/2003	20:59:08	33.5	14.4	**Superbolt**
03/10/2003	22:36:45	22.2	-2.4	Lightning
04/10/2003	10:56:27	-159.2	-10.7	Sprite
04/10/2003	17:02:53	102.6	-4.3	Lightning
04/10/2003	19:51:26	13.8	40.6	Halo
04/10/2003	23:07:10	-0.3	8.7	Superbolt
05/10/2003	02:02:09	-71.8	35.3	Lightning
05/10/2003	05:12:38	-101.0	17.8	Sprite + Halo

tions (São Sabbas et al., 2003). In total, 17 sprites, 3 halos and 9 superbolts were identified in the LSO dataset available up to now.

This differentiation could be improved by increasing the time resolution of the measurements. The mean time difference between sprites and lightning

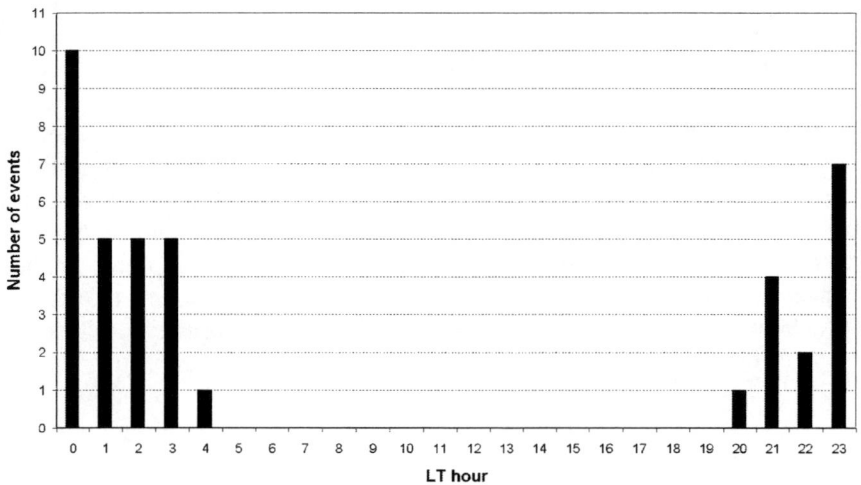

Figure 7. Local time distribution of the events observed by LSO.

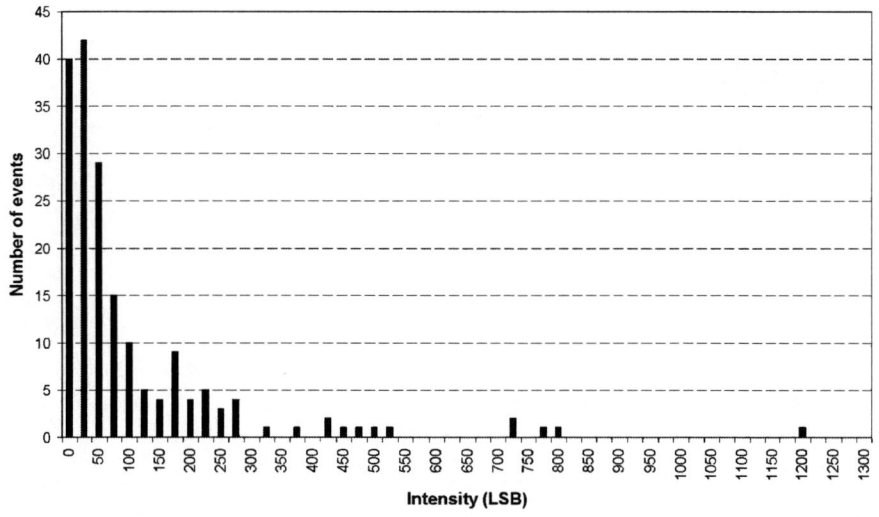

Figure 8. Distribution of all the LSO events.

flashes measured by the National Lightning detection network is about 20 to 30 ms (São Sabbas et al., 2003). Other observations using ELF-ULF measurements found that the difference can be smaller or larger, varying from an event to another (Bell et al., 1988). Cummer and Füllekrug (2001) showed that the time delay depends on the charge moment.

In addition to the spectral differentiation realized by LSO in the red part of the spectrum, measurements can be performed in the violet and ultraviolet part of the spectrum. The first results of ISUAL showed that the lightning lights completely disappeared in the 150-280 nm wavelength range and in the $N_2 2P$ sprite emission band at 337 nm (Chapter 6).

The measurement concept studied by the LSO experiment will be used by the microsatellite TARANIS (Tool for the Analysis of RAdiation from lightNIng and Sprites) dedicated to measure simultaneously the optical, X and γ emissions, the electric and magnetic field from ELF up to HF, and the high energy electrons associated with sprites for a better understanding of the implied mechanisms. The final goal is the study of the coupling between atmosphere, ionosphere and magnetosphere associated with thunderstorm activity and the effects of these phenomena on the Earth's environment (Blanc and Lefeuvre, 2003).

Bibliography

Barrington-Leigh, C. P., Inan, U. S., and Stanley, M. (2001). Identification of sprites and elves with identified video and broad-band array photometry. *J. Geophys. Res.*, 106(A2):1741–1750.

Bell, T. F., Reising, S. C., and Uman, U. S. (1988). Intense continuing currents following positive cloud-to-cloud lightning associated with red sprites. *Geophys. Res. Lett.*, 25(8):1285–1288.

Blanc, E., Farges, T., Roche, R., Brebion, D., Hua, T., Labarthe, A., , and Melnikov, V. (2004). Nadir observations of sprites from the International Space Station. *J. Geophys. Res.*, 109(A02306):1–8doi:10.1029/2003JA009972.

Blanc, E., Lefeuvre, F., et al. (2003). TARANIS: a project of microsatellite for the study of sprites and associated emissions. In *Paper presented at EGU*.

Boeck, W. L., Jr., O. H. Vaughan, Blakeslee, R. J., Vonnegut, B., and Brook, M. (1998). The role of the space shuttle videotapes in the discovery of sprites, jets and elves. *J. Atmos. Sol.-Terr. Phys.*, 60(7-9):669–677.

Christian, H. J., Blakeslee, R. J., Boccippio, D. J., Boeck, W. L., Buechler, D. E., Driscoll, K. T., Goodman, S. J., Hall, J. M., Koshak, W. J., Mach, D. M., and Stewart, M. F. (2003). Global frequency and distribution of lightning as observed from space by the Optical Transient Detector. *J. Geophys. Res.*, 108(D1):4005.

Cummer, S. A., Inan, U. S., Bell, T. F., and Barrington-Leigh, C. P. (1998). ELF radiation produced by electrical currents in sprites. *Geophys. Res. Lett.*, 25(8):1281–1284.

Cummer, S. C. and Füllekrug, M. (2001). Unusually intense continuing current in lightning produces delayed mesospheric breakdown. *Geophys. Res. Lett.*, 28(3):495–498.

Dwyer, J. R., Rassoul, H. K., Al-Dayeh, M., Caraway, L., Wright, B., Chrest, A., Uman, M. A., Rakov, V. A., Rambo, K. J., Jordan, D. M.,

Ferauld, J., and Smyth, C. (2004). A ground level gamma-ray burst observed in association with rocket triggered lightning. *Geophys. Res. Lett.*, 31(L05119):doi:10.1029/2003GL018771.

Feldman, W. C., Symbalisty, E. M. D., and Roussel Dupré, R. A. (1995). Association of discrete hard X ray enhancements with the eruption of Mount Pinatubo. *J. Geophys. Res.*, 100:23829.

Fishman, G. J., Bhat, P. N., Mallozzi, R., Horack, J. M., Koshut, T., Kouveliotou, C., Pendleton, G. N., Meegan, C. A., Wilson, R. B., Paciesas, W. S., Goodman, S. J., and Christian, H. J. (1994). Discovery of intense gamma ray flashes of atmospheric origin. *Science*, 264:1313–1316.

Füllekrug, M., Moudry, D. R., Dawes, G., and Sentman, D. D. (2001). Mesospheric sprite current triangulation. *J. Geophys. Res.*, 106(17):20,189–20,194.

Gerken, E. A., Inan, U. S., and Barrington-Leigh, C. P. (2000). Telescoping imaging of sprites. *Geophys. Res. Lett.*, 27(17):2637–2640.

Hampton, D. L., Heavner, M. J., Wescott, E. M., and Sentman, D. D. (1996). Optical spectral characteristics of sprites. *Geophys. Res. Lett.*, 23(1):89–92.

Hardman, S. F., Dowden, R. L., J. B., Brundell, and Bahr, J. L. (2000). Sprite observations in the Northern territory of Australia. *J. Geophys. Res.*, 105(D4):4689–4697.

LIS/OTD webpage, (PI H.J. Christian) (2002). LIS data are produced by the NASA. NASA website http://ghrc.msfc.nasa.gov. Available from the Global Hydrology Resource Center.

Lopez, L. I., Lin, R. P., Smith, D. M., and Barrington-Leigh, C. P. (2004). Detection of terrestrial gamma-ray flashes with the RHESSI spacecraft. In Fuellekrug, M., editor, *Sprites, Elves and Intense Lightning Discharges*, Corte, Corsica. NATO, Kluwer. Poster presentation.

Lyons, W. A., Nelson, T. E., Armstrong, R. A., Pasko, V. P., and Stanley, M. A. (2003). Upward electrical discharges from thunderstorm tops. *Bull. Am. Met. Soc.*, 84(4):445–454.

Lyons, W. A., Russell, A. R., Bering, E. A., and Williams, E. R. (2000). The hundred year hunt for the sprite. *EOS Trans. AGU*, 81:33.

MacGorman, D. R. and Rust, W. D. (1998). *The electrical nature of storms*. Oxford University Press.

Milikh, G., Valdivia, J. A., and Papadopoulos, K. (1998). Spectrum of red sprites. *J. Atmos. Sol.-Terr. Phys.*, 60:907–915.

Morrill, J. S., Bucsela, E. J., Pasko, V. P., Berg, S. L., Heavner, M. J., Moudry, D. R., Benesch, W. M., Wescott, E. M., and Sentman, D. D. (1998). Time resolved N_2 triplet state vibrational populations and emissions associated with red sprites. *J. Atmos. Sol.-Terr. Phys.*, 60:811–829.

Moudry, D., Stenbaek-Nielsen, H., Sentman, D., and Wescott, E. (2003). Imaging of elves, halos and sprite initiation at 1ms time resolution. *J. Atmos. Sol.-Terr. Phys.*, 65:509–518.

Neubert, T., Allin, T. H., Stenbaek-Nielsen, H., and Blanc, E. (2001). Sprites over Europe. *Geophys. Res. Lett.*, 28(18):3585–3588.

Orville, R. E. and Henderson, R. W. (1984). Absolute spectral irradiance measurements of lightning from 375 to 880 nm. *J. Atmos. Sci.*, 41:3180–3187.

Pasko, V. P., Inan, U. S., Bell, T. F., and Taranenko, Y. N. (1997). Sprites produced by quasi-electrostatic heating and ionization in the lower ionosphere. *J. Geophys. Res.*, 102(A3):4529–4561.

Roussel-Dupré, R. and A., Gurevich (1996). On runaway breakdown and upward propagating discharges. *J. Geophys. Res.*, 101:2297–2311.

Roussel-Dupré, R., Symbalisty, E., Taranenko, Y., and Yukhimuk, V. (1998). Simulations of high altitude discharges initiated by runaway breakdown. *J. Atmos. Sol.-Terr. Phys.*, 60(7-9):917–940.

Roussel-Dupré, R. A., Symbalisty, E. M. D., Tierny, H. E., and Triplett, L. (2002). New fully electromagnetic simulations of sprites initiated by runaway air breakdown. In *URSI XXVIIth General Assembly*, Maastricht.

São Sabbas, F. T., Sentman, D. D., Wescott, E. M., Jr., O. P. Pinto, Jr., O. Mendes, and Taylor, M. J. (2003). Statistical analysis of space time relation ships between sprites and lightning. *J. Atmos. Sol.-Terr. Phys.*, 65:525–535.

Sentman, D. D., Wescott, E. M., Osborne, D. L., Hampton, D. L., , and Heavner, M. J. (1995). Preliminary results from the sprites94 aircraft campaign: 1. red sprites. *Geophys. Res. Lett.*, 22(10):1205–1208.

Solar Survey Archive-2000 (2005). http://www.bass2000.obspm.fr.

Su, H. T., Hsu, R. R., Chen, A. B., and Lee, Y. J. (2002). Observation of sprites over the Asian continent and over oceans around Taiwan. *Geophys. Res. Lett.*, 29(4):10.1029/2001GL013737.

Turman, B. N. (1977). Detection of the lightning superbolts. *J. Geophys. Res.*, 82:2566–2568.

Vaughan, O. H. (1994). NASA Shuttle lightning research: observations of nocturnal thunderstorms and lightning displays as seen during recent space shuttle missions. In *Proc. SPIE*, volume 2266, pages 395–403.

Yair, Y., Israelevich, P., Devir, A. D., Moalem, M., Price, C., Joseph, J. H., Levin, Z., Ziv, B., A., Sternlieb, and A., Teller (2004). New observations of sprites from the space shuttle. *J. Geophys. Res.*, 109(D15201):1–10.

REMOTE SENSING OF THE UPPER ATMOSPHERE BY VLF

Craig J. Rodger and R. J. McCormick
Department of Physics, University of Otago, New Zealand.

Abstract The part of the electromagnetic spectrum described as VLF (Very Low Frequency) generally spans $3 - 30$ kHz (Kraus, 1984). In this chapter we will concentrate on observations made using strong VLF radiation to better understand the natural sources, such as lightning, and as long-range probes into the electrical properties of the upper atmosphere. However, in doing this we are restricting ourselves to a small subsection of VLF studies; the interested reader is directed to a recent and broad review article (Barr et al., 2000) for a wider description of studies involving VLF radio waves, and to the material of other contributors in this book.

8.1 Ionospheric Conductivity

The electrical conductivity of the middle atmosphere ranges over many orders of magnitude from troposphere to mesosphere. The electrical properties of the atmosphere depend on many parameters: the density of the neutral atmosphere, the number density of ionised particles and the presence of the geomagnetic field. In general the atmospheric electrical conductivity increases with height. However, in order to describe the electrical properties of regions stretching from the troposphere to thermosphere, this space must be divided into sections with different properties. At low altitudes where the electrical properties are collision dominated the effects of the geomagnetic field are unimportant, and the electrical conductivity is isotropic. At higher altitudes, say ~ 70 km, the collision frequency no longer dominates over the gyrofrequency and the medium is anisotropic. Thus the conductivity description depends greatly on the coordinate system used and the angle the field makes with this coordinate system. Above ~ 150 km, the conductivities along the geomagnetic field lines become very large compared with the transverse conductivities and the field lines can be treated as equipotentials (Maynard et al., 1984).

An example of representative conductivity profiles based on international standard models for the ionosphere and neutral atmosphere has been presented (McCormick et al., 2002).

8.2 Sources of VLF Electromagnetic (EM) Waves

VLF and Extremely Low Frequency (ELF) transient signals and noise are generated by various natural and man-made processes. On a global basis by far the most significant source of natural noise at ELF and VLF is that radiated by lightning discharges into the Earth-ionosphere waveguide. A VLF spectrum is shown in Figure 1. VLF communication and navigation transmitter are visible as the horizontal signals above 10 kHz, while local electrical noise is present below about 5 kHz. Strong sferics are common throughout the record, present in the spectra as vertical pulses.

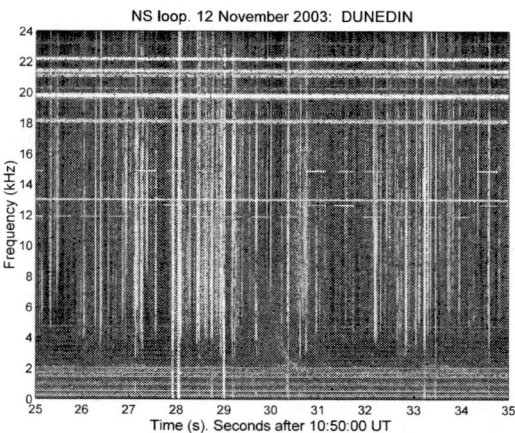

Figure 1. A VLF/ELF spectrum showing data received by a north-south aligned magnetic loop antenna located near Dunedin, New Zealand during the local nighttime. Strong VLF transmitters and lightning signals are present, with the strongest man-made transmitter being NWC at 19.8 kHz. A weak and "swishy" whistler is also present at ∼30 s.

8.2.1 Thunderstorms and Lightning

A lightning discharge is an electrical breakdown in which currents flow from the thundercloud to the ground, termed a cloud-to-ground or CG discharge, or more commonly within the thundercloud itself, termed an intra-cloud or IC discharge. The bulk of the power spectral density of lightning radiation is found in the Very Low Frequency (VLF, 3-30 kHz) and Extremely Low Frequency (ELF, 3-3000 Hz) bands (Pierce, 1977). Propagation of the ELF/VLF radiation in the waveguide formed by the ground and the lower boundary of the ionosphere disperses the initial sharp pulse of the lightning stroke into a wave train (termed a "sferic" or "atmospheric") of typically 1 to 10 ms duration. The sferic VLF radiation that propagates efficiently inside the Earth-ionosphere waveguide is dominated by the vertical component of the lightning return stroke current (Lee, 1989), such that most observed sferics are probably produced by cloud to ground lightning flashes. The vertical electric field can reach values as large as 1 V/m even at ranges of over 1000 km from the discharge (Taylor, 1960), and sferics can be detected from many thousands of kilometres away.

8.2.2 Man-Made VLF Radiation

A number of nations currently operate large VLF transmitters, primarily for communication with military submarines. While such transmitters originally served some commercial and governmental purpose, particularly for those nations involved in the building and maintenance of far-flung colonial empires (Byron, 1996), permanent operation is now usually only undertaken for military communications. To radiate electromagnetic waves efficiently one needs an antenna with dimensions on the order of the wavelength of the radiation. VLF waves, with frequencies from 3 to 30 kHz, have wavelengths from 100 to 10 km and this suggests that VLF antennas must be extremely large to be efficient. A ground based vertical electric monopole antenna[1] can operate with reasonable efficiency at VLF, especially at frequencies > 10 kHz, and were the mainstay of VLF communications systems for most of the 20th century. Such antennas are very large, typically many hundreds of metres long and are usually strung between high towers. However, VLF antennas often make use of natural geographic features, and have been strung across Fjords in Norway and over extinct volcanoes in Hawaii.

8.3 VLF Propagation in the Earth-Ionosphere Waveguide

Most of the energy radiated by lightning discharges and man-made VLF transmitters is trapped between the conducting ground (or sea) and the lower part of the ionosphere, forming the Earth-ionosphere waveguide. Such radiation is said to be propagating "subionospherically", i.e., beneath the ionosphere. While the creation and operation of man-made VLF transmitters is generally due to military requirements, the scientific use of the transmissions from these stations is well recognised (see the discussion in Barr et al., 2000).

---

[1] An antenna on which the current distribution forms a standing wave and which acts as one part of a dipole, the other part is formed by its electrical reflection in the ground or in an effective ground plane (Geller, 2003).

Because of the frequencies at which these transmitters broadcast, their high radiated power, and their nearly continuous operation, they are extremely well suited to long-range remote sensing of the lower ionosphere. Subionospheric VLF signals reflect from the D-region of the ionosphere, probably the least studied region of the Earth's atmosphere. These altitudes ($\sim 70 - 90$ km) are far too high for balloons and too low for most satellites, making in situ measurements extremely rare. Rocket lofted experiments have taken place in the D-region, but can only provide limited coverage due to their short-lived nature. Radio soundings made at frequencies >1 MHz (e.g., ionosondes), while successful for observing the upper ionosphere, generally fail in the D-region. The low electron number densities at D-region altitudes produce weak reflections, and hence measurement difficulties, particularly at night.

One of the few experimental techniques which can probe these altitudes uses very low frequency electromagnetic radiation trapped between the lower ionosphere and the Earth (Wait, 1996). Observations of the amplitude and/or phase of VLF transmissions have provided information on the variation of the D-region, both spatially and temporally. The nature of the received radio waves is largely determined by propagation between these boundaries e.g., (Cummer, 2000). Very long range remote sensing is possible; these signals can be received thousands of kilometres from the source (Crombie, 1964). In contrast, incoherent scatter radar techniques can make measurements in the D-region and above e.g., (Turunen, 1996), but are limited to essentially overhead measurements (i.e., a very narrow solid angle overhead). By using multiple VLF communication transmitters some understanding has been gained of the daytime lower ionosphere (McRae and Thomson, 2000). This method has not worked as well at nighttime, where the limited number of fixed paths (transmitter-receiver) has been insufficient to understand the complex nature of the lower ionosphere. It appears likely that a much larger number of transmitter-receiver paths is required to understand the nighttime D-region.

8.3.1 Variations in Subionospheric Propagation

Variations in the ionospheric D-region lead to changes in the propagation conditions for VLF waves propagating subionospherically, and hence changes in the observed amplitude and phase of VLF transmissions. Variations in space in the ground plate of the waveguide can also lead to changes in VLF propagation conditions (e.g. reflections from mountain ranges (Barr and Armstrong, 1996)). For daytime propagation conditions the D-region is particularly stable, with reflection heights occurring at about 70-75 km, the variation being strongly dominated by the change in Lyman-α flux with solar zenith angle (McRae and Thomson, 2000). Due to the dominant influence of the Sun, by day the propagation paths are largely stable and the received phases are reproducible, in quiet conditions away from dawn and dusk, to a very few microseconds or better than ~10° of phase e.g. (McRae and Thomson, 2000). Because of this the amplitude and phase of fixed frequency VLF transmissions varies in a consistent way during undisturbed conditions, as shown in Figure 2. Note that the received phase and amplitude vary in a fairly consistent manner when the transmitter-receiver path is in day time conditions (smooth variation). Night-time propagation at VLF frequencies is less stable and predictable than for day-time paths, although sufficient for communications purposes. The difference in stability reflects short-term variation in the night-time D-region and the lack of a dominant energy source (c.f., the Sun in daytime). In this case reflection occurs at about 80-90 km altitude. In addition, propagation across the day/night "terminator" leads to significant coupling of propagation modes and hence additional variability. The variations in phase and especially in amplitude can be clearly seen in the disturbed sections of Figure 2.

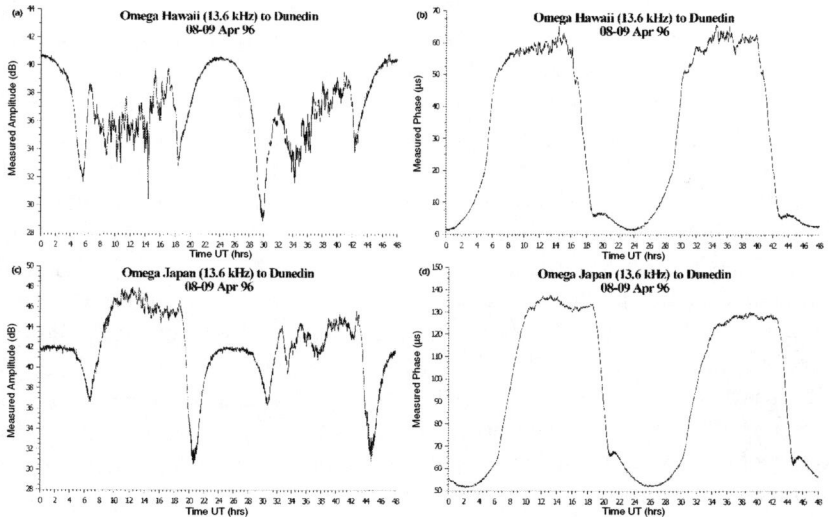

Figure 2. The amplitude and phase of fixed frequency VLF transmissions from navigation beacons received at Dunedin, New Zealand (local time (LT) is universal time (UT or GMT) + 12 hours) (McRae, 2000). Dawn is about 18 UT and dusk 30 UT in the upper two sections and about 20 UT and 32 UT respectively in the lower two sections.

8.3.2 TLE Associated Perturbations on VLF Transmissions

Much scientific attention has focused on short time-scale (~ 100 s) modifications of the D-region leading to VLF phase and amplitude perturbations, particularly those associated with thunderstorms (Rodger, 2003). It has become rather common practise to describe such short perturbations on subionospheric transmissions in terms of a phasor diagram (Dowden and Adams, 1988). Phasor subtraction of the unperturbed signal (immediately before onset), from the perturbed signal gives the phase and amplitude of the scattered signal relative to the phase and amplitude of the direct (unperturbed) VLF signal. Prior to the introduction of this phasor approach it was common to discuss the changes observed in the perturbed signal itself, which are rarely more than 1 dB in amplitude (Inan et al., 1993). These are the directly observed changes, rather than the "scatter phase" and "scatter amplitude" found through the phasor approach (also known as the "echo phase" and "echo amplitude").

Classic Trimpi Perturbations Classic Trimpi appear as rapid changes (<1 s) in the phase and amplitude of VLF transmissions, followed by a gradual (~ 30 s) relaxation to undisturbed levels. An example of a classic Trimpi perturbation is shown in Figure 3. Classic Trimpi are not caused by transient luminous events (TLE), but have a very similar appearance to perturbations which are associated to TLE. For this reason the TLE-associated perturbations were confused with classic Trimpi, and it took some time before it became clear they had their own distinct properties. Classic Trimpi perturbations are caused by whistler-induced electron precipitation (WEP) from the Van Allen radiation belts increasing the electron number density in the night-time D-region and hence altering the properties of the Earth-ionosphere waveguide (Helliwell et al., 1973). The energetic electron precipitation arises from lightning produced whistlers (Storey, 1953) interacting with cyclotron resonant radiation belt electrons in the equatorial zone e.g., (Tsurutani and Lakhina, 1997). The beginning of the VLF perturbation occurs ~ 0.6 s after the associated sferic (Armstrong, 1983), depending on the geomagnetic latitude; this delay is consistent with the time required for a whistler to propagate through the ionosphere and magnetosphere to the geomagnetic equator, interact with energetic electrons in the radiation belts, and for these electrons to arrive at the D-region as WEP. Such perturbations are known as Trimpi as they were first recognised by M. L. Trimpi during his time as a field scientist in Antarctica. Due to their production by WEP, classic Trimpi are also sometimes referred to as "WEP Trimpi". Classic Trimpi are caused by "patches" of modified ionosphere which are large, i.e., at least 600 km × 1500 km, with the longest axis orientated east-west (Clilverd et al., 2002), located on the great circle path between the transmitter and receiver. These sizes are considerably larger than the observed dimensions of the foot-print of the field-aligned ionization irregularities, termed "whistler ducts", which trap whistlers and extend between the conjugate hemispheres. In-situ satellite observations suggest that ducts have latitudinal extents of 25 km at 300 km altitude (Angerami, 1970), and thus one might have expected the ionospheric patch altered by WEP to have similar dimensions. However, large D-region patch dimensions have been suggested through a quasi-trapped whistler propagation theory in which ducted energy spreads at the magnetic equator (Strangeways, 1999), resulting in whistler-mode signals which have leaked outside their whistler duct still contributing to the horizontal lateral extent of WEP. Such leakage produces a significantly larger precipitation footprint than the actual dimensions of the whistler duct. A different mechanism also leading to large WEP patch dimensions comes through the precipitation caused by obliquely (nonducted) propagating whistlers, creating an ionospheric disturbance of 1000 km spatial extent (Lauben et al., 1999). As yet experimental studies have been unable to draw a definitive conclusion as to whether the whistlers leading to most WEP (and hence Trimpi) are ducted or nonducted (Clilverd et al., 2002). In contrast to their large horizontal dimensions, these patches are reasonably thin, e.g. spanning ~20 km in altitude (Lev-Tov et al., 1995; Dowden et al., 2001a). Examples of ionospheric modifications leading to a typical classic Trimpi have been presented by (Rodger et al., 2004) showing how the additional electron density relaxes back to undisturbed conditions.

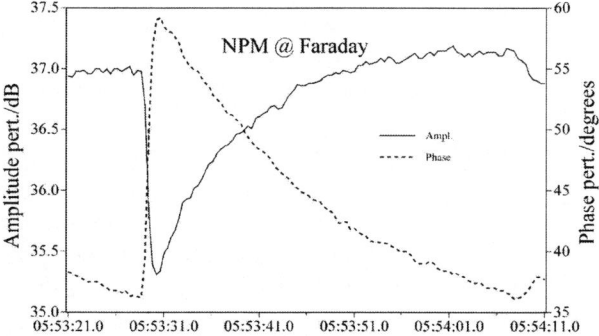

Figure 3. Example of a classic Trimpi perturbation observed on transmissions from the US Navy transmitter NPM (Hawaii; then 23.4 kHz) at Faraday (Antarctic Peninsula) on 23 April 1994. In this case the amplitude (sold line) decreases, while the phase (dotted line) increases (reproduced by permission of American Geophysical Union ©2001; Dowden et al., 2001b).

VLF Sprite Perturbations The changes in atmospheric electrical conductivity associated with luminous red sprites have also been shown to produce non-WEP Trimpi perturbations on VLF transmissions. The VLF perturbations associated with red sprites have been termed "VLF sprites" (Inan et al., 1995; Dowden et al., 1996), or "CID Trimpi" (Dowden et al., 2001a), where CID refers to "Cloud-Ionosphere Discharges", one of the early terms used to describe red sprites e.g., (Winckler, 1995) before the term "sprite" had been introduced. VLF sprites are almost certainly a subset of perturbations that were termed "early" Trimpi (Inan et al., 1988) as they may occur before one would expect the onset of a classic Trimpi, i.e., <0.1 s after the associated lightning discharge. It seems likely that VLF perturbations produced by sprite halos, sprites, and perhaps lightning EMP were initially described in the literature as early Trimpi perturbations. The majority of "early" Trimpi have the same visual decay form as a classic Trimpi, the significant difference being in onset times, relative to the associated lightning discharge. VLF sprites share some similarities with the strict definition of early Trimpi, in that they are transient VLF perturbations lasting of order minutes, with extremely small delays between the lightning discharge and the beginning of the VLF perturbations (a very small "onset"). However, VLF sprites do not share many important properties with early Trimpi. For example, the time delay between the lightning discharge and the beginning of the VLF perturbation is generally greater than 20 ms (Dowden et al., 2001a). This is to be expected, as the cooling time for the hot electrons in red sprites (Green et al., 1996) is of this order (or more) as shown in Section 8.4.1, such that the sprite-associated conductivity change will be transparent to VLF signals until significant cooling has taken place. At the same time, it is well known that some red sprites occur after significant delays from the associated CG lightning, such that the sprite producing the VLF perturbation does not appear until >20 ms after the discharge.

An example of a VLF sprite is shown in Figure 5, which was observed near Darwin, Australia.

The scattering pattern due to red sprite associated conductivity changes has allowed a greater understanding of the upper atmosphere electrical changes caused by red sprites. While the largest VLF sprite perturbations are observed for red sprites which occur on the transmitter-receiver great circle path (Hardman et al., 1998), detectable perturbations are also observed for sprite-events located well off the transmitter-receiver path, associated with sprites located within $\sim 500 - 1000$ km from the receiver (Rodger et al., 1999). It should be noted that the wide-angle scattering from red sprite conductivity changes produce much smaller perturbations than the "forward-scattered" signals along the transmitter-receiver path. The experimental evidence suggests that essentially all such nearby red sprite events lead to VLF sprite perturbations, irrespective of the sprites displacement from the transmitter-receiver path (Dowden et al., 1996; Corcuff, 1998; Hardman et al., 1998). In contrast, red sprites located near the transmitter-receiver path can produce VLF sprite perturbations when the sprite is located many megameters from the receiver. A schematic of scattering of VLF by red sprites is shown in Figure 4. This finding implies red sprite associated ionospheric conductivity changes that are relatively small with a high intensity. It has been shown that the magnitude of a VLF sprite perturbation decays in time with a strongly logarithmic dependence (Dowden et al., 1997, 1998), recovering to pre-perturbation conditions over $\sim 30 - 100$ s. An example of a large VLF perturbation which was measured to determine this dependence is shown in Figure 5. The logarithmic decay-signature is expected for ionization spread over a wide range of altitudes; the ionization relaxes back to ambient conditions faster at lower altitude

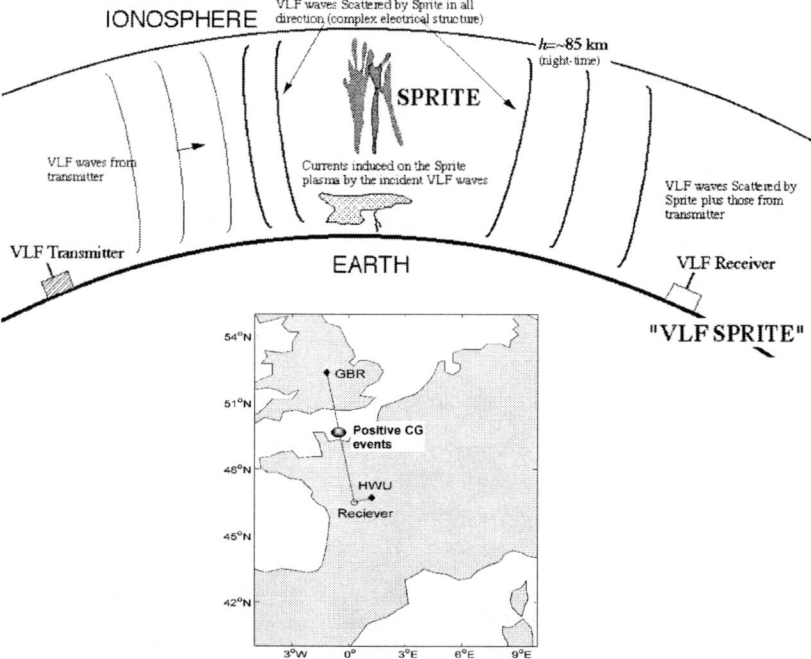

Figure 4. Schematic of VLF scattering from the electrical conductivity to produce a VLF perturbation at the receiver. As the sprites have a complex electrical structure, there can be significant scattering off the great circle path (GCP), right up to 180° (reproduced by permission of American Geophysical Union ©1999; Rodger, 1999). The lower panel shows the experimental setup used by (Corcuff, 1998). The shaded region shows the location of the positive CG discharges which caused simultaneous perturbations on the VLF transmitters GBR and HWU, consistent with highly structured ionization changes from red sprites.

than at upper, leading to a logarithmic decay pattern. Modelling of VLF sprite perturbations imply that the scattering comes from upper atmospheric electrical conductivity modifications which are structured enough (on $\sim$1 km scales) to cause significant wide-angle scatter e.g., (Rodger et al., 1998b; Rodger and Nunn, 1999), observed in the experimental setup shown in Figure 4. This is suggestive that the luminous structure seen in red sprites (in some cases, quasi-vertical columns) is present in the electrical structure, and that red sprites produce electrical conductivities significantly different from their surroundings. Due to the time over which the changes last, the conductivity changes are almost certainly determined by significant ionization present in red sprite events. Significant fine structure down to $\sim$30 m has been observed in red sprites (Gerken et al., 2000), although the dense fine structure seen in some red sprites is unlikely to be resolved using VLF wavelengths ($\sim$10 km). The observed VLF sprite perturbations are best explained by red sprite conductivity changes that involve high ionization changes (4-6 orders of magnitude at some heights) in comparison with the ambient night-time ionosphere. For example it has been found that a uniform conductivity of at least 30 μS/m is required to produce the observed VLF scattered signal strengths. This corresponds to an electron density at 70 km altitude of $\sim 10^4$ electrons per cm^3 ($\sim 10^{10}$ m^{-3}) and so about 10^5cm^{-3} at 55 km, over 6 orders of magnitude above the ambient density at 55 km (Dowden et al., 2001b). While these ionization levels are clearly huge, photometric data now appear to confirms that significant ionization in at least some red sprite events (Armstrong et al., 2000), with energetic "carrot" sprites appearing to exhibit ionization's between 10^4 and 10^5 electrons per cm^3

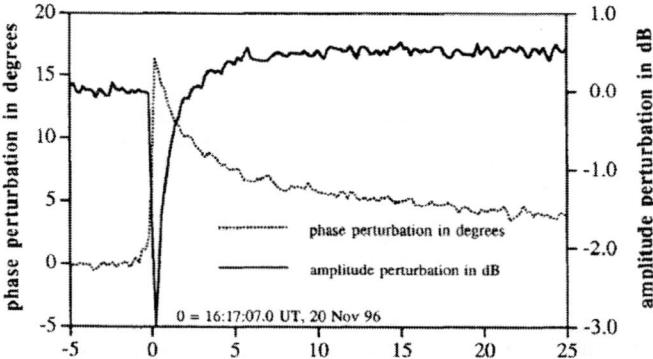

Figure 5. Relaxation of a large VLF perturbation observed on transmissions from NWC at Darwin, Australia. This is believed to be a VLF sprite perturbation. The x-axis in seconds after the origin time (reproduced with permission of American Geophysical Union ©1997; Dowden et al., 1997).

(10^{10} m^{-3} and 10^{11} m^{-3}). There appears to be growing convergence between the properties of red sprites determined through photometric measurements, VLF perturbations, and theoretical studies (e.g., streamer mechanisms, Pasko et al., 1998).

The time decay of VLF sprite perturbations are used to examine the vertical distribution of ionization in red sprites. Simple models, treating the ionization as a continuous column which decreases from the bottom up as the logarithm of time have shown that there is significant red sprite ionization at altitudes extending from 50 km to ~80 km altitude (Dowden and Rodger, 1997). It should be noted that VLF perturbation studies are poorly suited to determining the lower altitudes limits of red sprite associated changes, as in the first ~100 ms of a red sprite event (near the time resolution of many VLF studies), almost all ionization located below ~50 km altitude will disappear through chemical recombination and attachment processes (Nunn and Rodger, 1999, Figure 2). The later modelling also indicates that it may be difficult to detect the long-lived (>60 s) upper-most portions of the red sprite produced plasma (>80 km), as the scattered VLF signals will be both small in comparison with the typical noise levels, and changing very slowly, and thus difficult to discriminate from other variations.

Perturbations from QE Thunderstorm Fields Strictly speaking, this is the only class of non-WEP VLF perturbations which should be referred to as "early Trimpi" events. Early Trimpi perturbations exhibit a rapid onset (<20 ms between causative sferic and the beginning of the perturbation), are associated with amplitude changes of magnitude $\geqslant$ 0.2 dB, and last for longer than 10 s e.g., (Inan et al., 1996b). Examples of early Trimpi are shown in Figure 6. The minimum amplitude threshold is determined by the background VLF radio atmospheric noise levels, while the onset time parameter (<20 ms) comes from the time resolution of the narrow band monitors. Early Trimpi are believed to be caused by quasi-electrostatic (QE) fields with large intensities that exist at high altitudes above thunderstorms, leading to conductivity changes over large areas through the heating of ionospheric electrons (and in some cases the production of additional ionization). A schematic of this situation is shown in Figure 7. For particularly intense discharges the QE fields appear to be intense enough to cause break-down and produce the highly structured ionization observed in red sprites (Pasko et al., 2000). In the case of QE fields strong enough to produce additional high-altitude ionization, the early VLF perturbations will be associated with sprite halos (Moore et al., 2003). Indicative electron density profiles for this case have been shown by the authors of this chapter. It has been suggested that early Trimpi perturbations linked to much weaker lightning may be due to heating and/or cooling of the ionospheric electrons without significant ionization. Immediately after a

178   SPRITES, ELVES AND INTENSE LIGHTNING DISCHARGES

lightning discharge the transient QE fields produce large high altitude conductivity changes through ionospheric electron heating. After these transients, however, relatively stable thunderstorm charge distributions are believed to produce ionospheric QE fields that maintain the ionospheric electrons at a persistently heated level well above their ambient thermal energy. Changes in the thundercloud charge (those involved with lightning discharges) lead to ionospheric electron heating or cooling above or below this level, observed as sudden subionospheric VLF signal changes, occurring simultaneously with lightning discharges, and thus producing early Trimpi VLF perturbations. The recharging of the thunderstorm returns the ionospheric electrons to their previous levels and thus produces the decay of the observed perturbation. As the mechanism relies upon relative changes in the QE fields, discharges of either polarity or direction (CG or cloud-to-cloud) might produce an ionospheric

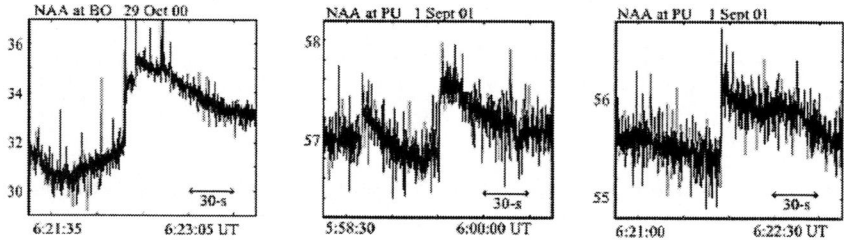

Figure 6. Examples of early Trimpi perturbations observed by the Stanford HAIL array on transmissions from NAA. The onsets of these events are coincident with the causative sferic within the 20 ms sampling resolution (reproduced by permission of American Geophysical Union ©2003; Moore et al., 2003).

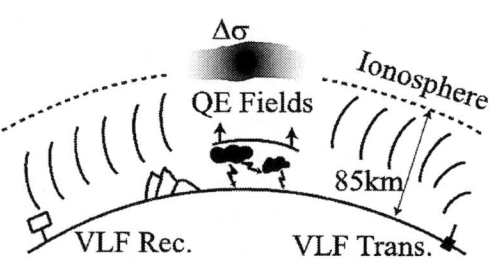

Figure 7. Schematic of a proposed mechanism leading to early Trimpi perturbations. Changes in the quasi-electrostatic fields produced by the charging and discharging of thunderstorms produce conductivity changes in the upper atmosphere above the storm (reproduced by permission of American Geophysical Union ©2003; Moore et al., 2003).

conductivity change sufficient to generate an observable VLF perturbation. In addition, the creation of a VLF perturbation also depends on the background level, thereby explaining how small lightning discharges (<20 kA) are sometimes (but not always) observed to produce an early Trimpi.

The QE thunderstorm changes which lead to early Trimpi perturbations are associated with large, "smooth" ionospheric conductivity changes. This is consistent with the scattering pattern for such events, which are strongly forward directed (forward along the transmitter-receiver path). Early Trimpi have forward scattering patterns exhibiting 15 dB beam widths of less than 30°, consistent with a disturbed region in the lower ionosphere with horizontal extents of 90 ± 30 km (Johnson et al., 1999). These early Trimpi events do not show significant wide-angle scattering, as would be expected from a small, dense, ionospheric modification, as is the case for VLF sprites. This is consistent with earlier findings that the ionospheric disturbances which produce early Trimpi must lie within 50 km of the transmitter-receiver great circle path, with the associated lightning discharge located under the disturbance (Inan et al., 1996b,a). The recovery signatures of early Trimpi and classic (WEP) Trimpi indicates that the majority of early Trimpi recover more rapidly towards pre-event levels during the first 20 s of the perturbation than do WEP Trimpi (Sampath et al., 2000). It appears that early Trimpi events have a wider range of recovery times than classic Trimpi ($\sim 60 - 240$ s compared to $\sim 120 - 180$ s). As yet it is unclear if the decay signature of early Trimpi fit the strong logarithmic pattern reported for VLF sprites – however, one might expect that changes driven by the cooling of ionospheric electrons driven by QE field changes might generally not show a logarithmic signature as this is due to altitude dependent ionization relaxation times (see Section 8.4.2).

Lightning EMP Perturbations When elves were first reported, their name included the observation of simultaneous VLF perturbations, e.g., Emissions of Light and VLF perturbations due to Electromagnetic pulse Sources (elves) (Fukunishi et al., 1996). While it was originally reported that elves are always accompanied by "large amplitude VLF perturbations" (Fukunishi et al., 1996), later reports have suggested that the mean scattered amplitude for elve events are smaller (~ -11 dB) than those associated with red sprites (~ -2.5 dB) (Hobara et al., 2001). The term "VLF elves" has been put forward to describe such events (Dowden, 1996). The strong lightning produced electromagnetic pulse (EMP) which creates elves does so through the heating of high-altitude electrons. Such a heating effect would last only a few ms, as discussed below. Theoretical modelling indicates that sufficiently strong lightning EMP also leads to changes in ionization (Cho and Rycroft, 1998), which at elve-altitudes (up to 105 km) will be relatively long lasting (Rodger et al., 2001).

Although lightning EMP and elve production appears to be fairly well understood, the secondary issue of modifications to long-wave propagation inside the Earth-ionosphere waveguide has received little attention. Lightning EMP produces a large ($\sim$ 500 km), relatively smoothly varying ionospheric disturbance at high altitudes ($\sim$ 85 km), near the nighttime VLF reflection height. Subionospheric VLF perturbations associated with these disturbances would display narrow forward scattering along the transmitter-receiver path (Wait, 1964). For the case of ionization increases below the VLF reflection height, the relaxation of such perturbations to pre-event levels would be expected to be extremely slow due to the long lifetimes of electrons at elve-altitudes. Such perturbations would appear as sudden step-like perturbations in amplitude and phase without a clear relaxation, this being masked by other variations in the subionospheric signal occurring over hundreds of seconds. Perturbations of this type have been occasionally reported e.g., (Inan et al., 1988), although they have not been linked to elves. Those perturbations which have been linked to elves in the literature are short-lived (Hobara et al., 2001), Figure 4. However, such a perturbation is not consistent with high altitude increases in ionization-levels. However, it may be that ionization changes occur above the VLF reflection height, while the conductivity modification "seen" by the subionospheric VLF transmissions is entirely due to EMP-heating. Intense heating, occurring near the VLF reflection height ($\sim$ 85 $-$ 90 km) without ionization production might lead to VLF perturbations lasting $\sim$ 0.5 s.

The event shown by (Hobara et al., 2001) is rather similar to RORD perturbations (Rapid Onset, Rapid Decay; Dowden et al., 1994) which are characterised by a very small delay ($<$ 20 ms (Inan et al., 1996b)) between the associated lightning and the beginning of the perturbation, but also show a rapid decay ($\sim$ 1 s) compared with early Trimpi perturbations. At one stage RORD events were thought to be part of a VLF sprite perturbation (a RORD combined with a classic Trimpi), but this idea has been withdrawn (Dowden et al., 1997). Modelling has shown that RORDs are consistent with the VLF response to electron temperature changes (without significant ionization changes) due to heating by QE fields (Pasko et al., 1995), although it has been speculated that reported daytime RORDs might perhaps be caused by red sprites occurring during the day, when only the lowest, shortest-lived parts of the sprite plasma would exist below the day-time VLF reflection height ($\sim$75 km) (Rodger, 1999). It seems most likely that RORDs are due to non-ionizing QE fields or perhaps to sprite ionization located only at lower altitudes. As such, the VLF perturbation shown by (Hobara et al., 2001) may actually be due to a thunderstorm QE field, and not from lightning EMP.

8.4 Relaxation of High-Altitude Ionospheric Modifications

VLF propagation in the Earth-ionosphere waveguide is used as a remote sensing technique to measure changes in ionospheric conductivity. Changes in the electrical conductivity of the Earth-ionosphere waveguide will be due to changes in the effective collision frequency, $v_{eff}(z)$, and the ionization number density. Changes in the collision frequency (e.g., heating) or ionization density (e.g., addition of extra ionization) will tend to relax back towards ambient levels over time. The relaxation time-scales involved provide additional information about the conductivity changes and the processes involved, as these parameters relax over very different times. Because ions are very much heavier we will tend to focus on electrons in our discussion, which will generally dominate the conductivity for VLF waves. The cooling time of electrons in the lower ionosphere has been found in experimental measurements to be very fast, characteristically ~ 1 ms at 90 km (Stubbe et al., 1981). In contrast, the relaxation in electron number density takes place over tens of seconds, if not more.

8.4.1 Temperature Relaxation

Using the theory and discussion previously presented by the authors (Rodger et al., 1998a), we can estimate the relaxation time of heated electrons in the lower ionosphere. In the altitude range we consider (50-90 km) the ambient temperature (T_0) is about 200 K (± 50 K). We shall take a fairly extreme example, where the electrons are heated to 300,000 K. Note, however, that spectral measurements of red sprites suggest that the maximum electron temperatures (T_e) present could be this high (Green et al., 1996).

In order to estimate the cooling time, we consider two cases: where the electrons are taken to have cooled once their temperature is within 20% of ambient and the situation where "cooling" is reached at 100% of ambient (i.e., a factor of 2). This is shown in Figure 8, estimated assuming $T_e \gg T_0$ using collision frequencies and ambient temperatures from CIRA-1986 for Dunedin in late July.

8.4.2 Ionization Relaxation

The time-evolution of electron density in the lower ionosphere is a fundamental part of understanding this altitude range, and indeed in using VLF measurements as a remote sensing technique. The change with time in scattered signals due to ionization modifications has allowed estimates to be made of ionization levels – as well as increasing our understanding of the chemistry of these high altitude regions.

Using previously developed theory (Rodger et al., 1998a), we can estimate the relaxation time for ionization modifications of the lower ionosphere. In the simplest model the ionospheric electron density is described using a Wait ionosphere (Wait and Spies, 1964) where the electron number density (i.e., electrons per m^{-3}), N_e, increases exponentially with altitude z,

$$N_e(z) = 1.4265 \times 10^{13} \exp(-0.15 h') \times \exp((\beta - 0.15)(z - h')) \qquad (8.1)$$

where β is given per kilometre and reflects the sharpness of the profile, and h' is a reference height (in kilometres). Both β and h' (but especially the latter) vary depending on the time of day. We take the ambient electron number density over the altitude range we consider (50-90 km) to be given by a nighttime Wait ionosphere with $h' = 86$ km and $\beta = 0.5$ km^{-1}. In order to demonstrate the lifetimes we again consider two cases in which the ionization modification involves a doubling in electron number density, and the situation where the ionization levels are one hundred times larger than ambient. These relaxation times are shown in Figure 8, where the time given is that for the electron number density to reach 5% of ambient conditions. The ambient neutral densities and temperatures are taken from CIRA-1986 for Dunedin in late July, while the attachment and recombination rates are those previously developed by the authors (Rodger et al., 1998a).

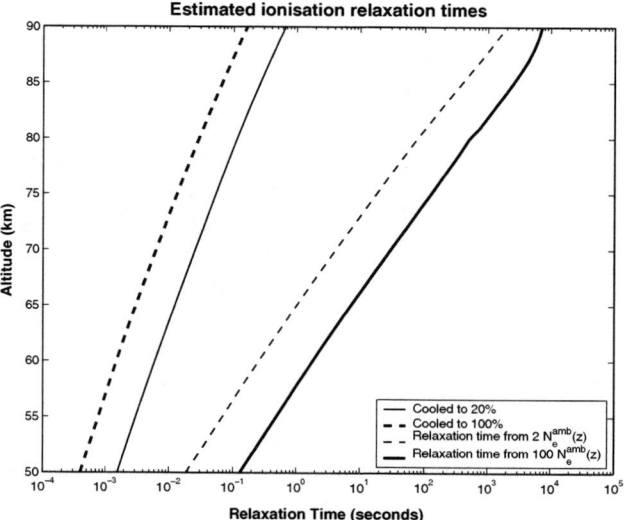

Figure 8. Estimated cooling and ionization relaxation times for electrons in the lower ionosphere, using published expressions (Rodger et al., 1998a) under the assumption of $T_e \gg T_0$.

By combining the expressions used to model the time dependant collision frequency and charged particle density one can determine the time varying electrical conductivity.

8.5 Summary

In this text we have attempted to provide an overview on the use of VLF techniques to probe the electrical nature of the upper atmosphere, with a particular focus on lightning and the parameterisation of high altitude transient luminous events (TLE's). We have described how VLF techniques have been applied to TLE-linked electrical conductivity changes, and our current understanding. The reader will have noticed that there are still significant areas in which our understanding is not complete. The story is not over, and there is still much work to do, particularly by linking the observations from multiple techniques. For example, there are clearly multiple ways in which thunderstorm associated phenomena can directly impact upon the conductivity of the lower-ionosphere. While experimentally observed as VLF perturbations much like early Trimpi, the conditions leading to these events are poorly understood. Some events show wide-angle scattering, others do not. In some cases the experimental data suggest that very small lightning can produce large modifications to the ionosphere. Clarifying the processes by which such changes occur is very important for determining their wider significance. While the initial (primarily optical) experiments significantly advanced our understanding, questions remain which require measurements made over the entire electromagnetic spectrum.

Bibliography

Angerami, J. J. (1970). Whistler duct properties deduced from VLF observations made with the OGO 3 satellite near the magnetic equator. *J. Geophys. Res.*, 75:6115–6135.

Armstrong, R. A., Suszcynsky, D. M., Lyons, W. A., and Nelson, T. (2000). Multi-color photometric measurements of ionization and energies in sprites. *Geophys. Res. Lett.*, 27:653–656.

Armstrong, W. C. (1983). Recent advances from studies of the Trimpi effect. *Antarctic J. of USA*, 18:281–283.

Barr, R. and Armstrong, T. R. (1996). Reflection of VLF radio waves from the canadian rockies. *Radio Science*, 31:533–546.

Barr, R., Jones, D. L., and Rodger, C. J. (2000). ELF and VLF radio waves. *J. Atmos. Sol.-Terr. Phys.*, 62:1689–1718.

Byron, W. J. (1996). The monster antennas. *Communications Quarterly*, 6(2):5–24.

Cho, M. and Rycroft, M. J. (1998). Computer simulation of the electric field structure and optical emission from cloud-top to the ionosphere. *J. Atmos. Terr. Phys.*, 60:871–888.

Clilverd, M. A., Nunn, D., Lev-Tov, S. J., Inan, U. S., Dowden, R. L., Rodger, C. J., and Smith, A. J. (2002). Determining the size of lightning-induced electron precipitation patches. *Geophys. Res.*, 107(A8):10.1029/2001JA000301.

Corcuff, Y. (1998). VLF signatures of ionospheric perturbations caused by lightning discharges in an underlying and moving thunderstorm. *Geophys. Res. Lett.*, 25(13):2385–2388.

Crombie, D. D. (1964). Periodic fading of VLF signals received over long paths during sunrise and sunset. *Journal of Research National Bureau of Standards, Radio Science*, 68D:27–34.

Cummer, S. A. (2000). Modeling electromagnetic propagation in the earth-ionosphere waveguide. *IEEE Transactions on Antennas and Propagation*, 48(9):1420–1429.

Dowden, R., Brundell, J., Lyons, W., and Nelson, T. (1996). Detection and location of red sprites by VLF scattering of subionospheric transmissions. *Geophys. Res. Lett.*, 23:1737–1740.

Dowden, R., Rodger, C., Brundell, J., and Clilverd, M. (2001a). Decay of whistler-induced electron precipitation and cloud-ionosphere electrical discharge Trimpis: Observations and analysis. *Radio Science*, 36:151–169.

Dowden, R. L. (1996). Comment on "VLF signatures of ionospheric disturbances associated with sprites" by inan et. al. *Geophys. Res. Lett.*, 22(23):3421.

Dowden, R. L. and Adams, C. D. D. (1988). Phase and amplitude perturbations on sub-ionospheric signals explained as echoes from lightning induced electron precipitation ionisation patches. *J. Geophys. Res.*, 93:11543–11550.

Dowden, R. L., Adams, C. D. D., Brundell, J. B., and Dowden, P. E. (1994). Rapid onset, rapid decay (RORD), phase and amplitude perturbations of VLF subionospheric transmissions. *J. Atmos. Terr. Phys.*, 56:1513–1527.

Dowden, R. L., Hardman, S. F., Rodger, C. J., and Brundell, J. B. (1998). Logarithmic decay and doppler shift of plasma associated with sprites. *J. Atmos. Sol.-Terr. Phys.*, 60:741–75.

Dowden, R. L. and Rodger, C. J. (1997). Decay of a vertical plasma column: A model to explain VLF sprites. *Geophys. Res. Lett.*, 24:2765–2768.

Dowden, R. L., Rodger, C. J., and Brundell, J. B. (1997). Temporal evolution of very strong Trimpis observed at Darwin, Australia. *Geophys. Res. Lett.*, 24:2419–2422.

Dowden, R. L., Rodger, C. J., and Nunn, D. (2001b). Minimum sprite plasma density as determined by VLF scattering. *IEEE Antennas and Propagation Magazine*, 43(2):12–24.

Fukunishi, H., Takahashi, Y., Kubota, M., Sakanoi, K., Inan, U. S., and Lyons, W. A. (1996). Elves: Lightning-induced transient luminous events in the lower ionosphere. *Geophys. Res. Lett.*, 23:2157–2160.

Geller, E., editor (2003). *McGraw-Hill Dictionary of Scientific and Technical Terms*. McGraw-Hill, 6th edition.

Gerken, E. A, Inan, U. S., and Barrington-Leigh, C. P. (2000). Telescopic imaging of sprites. *Geophys. Res. Lett.*, 27:2637–2640.

Green, B. D., Fraser, M. E., Rawlins, W. T., L. Jeong, W. A. M. Blumberg, Mende, S. B., Swenson, G. R., Hampton, D. L., Wescott, E. M., and Sentman, D. D. (1996). Molecular excitation in sprites. *Geophys. Res. Lett.*, 23(16):2161–2164.

Hardman, S., Rodger, C. J., Dowden, R. L., and Brundell, J. B. (1998). Measurements of the VLF scatter pattern of the structured plasma of red sprites. *IEEE Antennas and Propagation Magazine*, 40(2):29–38.

Helliwell, R. A., Katsufrakis, J. P., and Trimpi, M. L. (1973). Whistler-induced amplitude perturbation in VLF propagation. *J. Geophys. Res.*, 78:4679–4688.

Hobara, Y., Iwasaki, N., Hayashida, T., Hayakawa, M., Ohta, K., and Fukunishi, H. (2001). Interrelation between ELF transients and ionospheric disturbances in association with sprites and elves. *Geophys. Res. Lett.*, 28(5):935–938.

Inan, U. S., Bell, T. F., Pasko, V. P., Sentman, D. D., Wescott, E. M., and Lyons, W. A. (1995). VLF signatures of ionospheric disturbances associated with sprites. *Geophys. Res. Lett.*, 22:3461–3464.

Inan, U. S., Pasko, V. P., and Bell, T. F. (1996a). Sustained heating of the ionosphere above thunderstorms as evidenced in "early/fast" events. *Geophys. Res. Lett.*, 23(10):1067–1070.

Inan, U. S., Rodriguez, J. V., and Idone, V. P. (1993). VLF signatures of lightning-induced heating and ionisation of the nighttime D-region. *Geophys. Res. Lett.*, 20(21):2355–2358.

Inan, U. S., Shafer, D. C., Yip, W. Y., and Orville, R. E. (1988). Subionospheric VLF signatures of nighttime D-region perturbations in the vicinity of lighting discharges. *J. Geophys. Res.*, 93:11455–11472.

Inan, U. S., Slingeland, A., Pasko, V. P., and Rodriguez, J. V. (1996b). VLF and LF signatures of mesospheric/lower ionospheric response to lightning discharges. *J. Geophys. Res.*, 101:5219–5238.

Johnson, M. P., Inan, U. S., Lev-Tov, S. J., and Bell, T. F. (1999). Scattering pattern of lightning-induced ionospheric disturbances associated with early/fast VLF events. *Geophys. Res. Lett.*, 26:2363–2366.

Kraus, J. D. (1984). *Electromagnetics*. McGraw-Hill, 3rd edition.

Lauben, D. S., Inan, U. S., and Bell, T. F. (1999). Subionospheric VLF signatures of oblique (nonducted) whistler-induced precipitation. *Geophys. Res. Lett.*, 26:2633–2636.

Lee, A. C. L. (1989). Ground truth confirmation and theoretical limits of an experimental VLF arrival time difference lightning flash locating system. *Quart. J. Roy. Met. Soc.*, 115:1147–1166.

Lev-Tov, S. J., Inan, U. S., and Bell, T. F. (1995). Altitude profiles of localized D-region density disturbances produced in lightning-induced electron-precipitation events. *J. Geophys. Res.*, 100:21375–21383.

Maynard, N. C., Hale, L. H., Mitchell, J. D., Schmidlin, F. J., Goldberg, R. A., Barcus, J. R., Søraas, F., and Croskey, C. L. (1984). Electrical structure in the high-altitude middle atmosphere. *J. Atmos. Terr. Phys.*, 46:807–817.

McCormick, R. J., Rodger, C. J., and Thomson, N. R. (2002). Reconsidering the effectiveness of quasi-static thunderstorm electric fields for whistler duct formation. *J. Geophys. Res.*, 107(A11):1396, doi:10.1029/2001JA009219.

McRae, W. M. (2000). *Modelling the D-Region Electron Density using VLF Phase and Amplitude Recordings*. Phd thesis, University of Otago.

McRae, W. M. and Thomson, N. R. (2000). VLF phase and amplitude: Daytime ionospheric parameters. *J. Atmos. Terr. Phys.*, 62(7):609–618.

Moore, R. C., Barrington-Leigh, C. P., Inan, U. S., and Bell, T. F. (2003). Early/fast VLF events produced by electron density changes associated with sprite halos. *J. Geophys. Res.*, 108:doi:10.1029/2022JA009816.

Nunn, D. and Rodger, C. J. (1999). Modeling the relaxation of red sprite plasma. *Geophys. Res. Lett.*, 26:3293–3296.

Pasko, V. P., Inan, U. S., and Bell, T. F. (1998). Spatial structure of sprites. *Geophys. Res. Lett.*, 25:2123–2126.

Pasko, V. P., Inan, U. S., and Bell, T. F. (2000). Fractal structure of sprites. *Geophys. Res. Lett.*, 27:497–500.

Pasko, V. P., Inan, U. S., Taranenko, Y. N., and Bell, T. F. (1995). Heating, ionization and upward discharges in the mesosphere due to intense quasi electrostatic thundercloud fields. *Geophys. Res. Lett.*, 22(4):365–368.

Pierce, E. T. (1977). Atmospherics and radio noise. In Golde, R. H., editor, *Lightning 1: Physics of Lightning*, pages 351–384. Academic Press, London.

Rodger, C. J. (1999). Red sprites, upward lightning, and VLF perturbations. *Rev. Geophys.*, 37(3):317–336.

Rodger, C. J. (2003). Subionospheric VLF perturbations associated with lightning discharges. *J. Atmos. Sol.-Terr. Phys.*, 65(5):591–606.

Rodger, C. J., Cho, M., Clilverd, M. A., and Rycroft, M. J. (2001). Lower ionospheric modification by lightning-EMP: Simulation of the nighttime ionosphere over the United States. *Geophys. Res. Lett.*, 28:199–202.

Rodger, C. J., Molchanov, O. A., and Thomson, N. R. (1998a). Relaxation of transient ionization in the lower ionosphere. *J. Geophys. Res.*, 103(4):6969–6975.

Rodger, C. J. and Nunn, D. (1999). VLF scattering from red sprites: Application of numerical modelling. *Radio Science*, 34:923–932.

Rodger, C. J., Nunn, D., and Clilverd, M. A. (2004). Investigating radiation belt losses though numerical modelling of precipitating fluxes. *Ann. Geophys.* (in press).

Rodger, C. J., Thomson, N. R., and Wait, J. R. (1999). VLF scattering from red sprites: Vertical columns of ionisation in the Earth-ionosphere waveguide. *Radio Science*, 34(4):913–921.

Rodger, C. J., Wait, J. R., Dowden, R. L., and Thomson, N. R. (1998b). Radiating conducting columns inside the Earth-ionosphere waveguide: Application to red sprites. *J. Atmos. Sol.-Terr. Phys.*, 60:1193–1204.

Sampath, H. T., Inan, U. S., and Johnson, M. P. (2000). Recovery signatures and occurrence properties of lightning-associated subionospheric VLF perturbations. *J. Geophys. Res.*, 105:183–191.

Storey, L. R. O. (1953). An investigation of whistling atmospherics. *Phil. Trans. Roy. Soc. (London)*, 246:113–117.

Strangeways, H. J. (1999). Lightning induced enhancements of D-region ionization and whistler ducts. *J. Atmos. Sol.-Terr. Phys.*, 61:1067–1080.

Stubbe, P., Kopka, H., and Dowden, R. L. (1981). Generation of ELF and VLF waves by polar electrojet modulation: Experimental results. *J. Geophys. Res.*, 86:9073–9078.

Taylor, W. L. (1960). Daytime attenuation rates in the very low frequency band using atmospherics. *Journal of Research of the National Bureau of Standards*, 64D:349–355.

Tsurutani, B. T. and Lakhina, G. S. (1997). Some basic concepts of wave-particle interactions in collisionless plasmas. *Rev. Geophys.*, 35(4):491–501.

Turunen, E. (1996). Incoherent scatter radar contributions to high latitude D-region aeronomy. *J. Atmos. Sol.-Terr. Phys.*, 58(6):707–725.

Wait, J. R. (1964). Influence of a circular depression on VLF propagation. *Journal of Research National Bureau of Standards, Radio Science*, 68D(6):709–715.

Wait, J. R. (1996). *Electromagnetic waves in stratified media.* IEEE Press.

Wait, J. R. and Spies, K. P. (1964). Characteristics of the Earth-ionosphere waveguide for VLF radio waves. Tech. note 300, National Bureau of Standards.

Winckler, J. R. (1995). Further observations of cloud-ionosphere electrical discharges above thunderstorms. *J. Geophys. Res.*, 100:14335–14345.

MEASUREMENTS OF LIGHTNING PARAMETERS FROM REMOTE ELECTROMAGNETIC FIELDS

Steven A. Cummer
Electrical and Computer Engineering Department, Duke University, Durham, NC 27708, USA

Abstract For obvious reasons, optical measurements have played a crucial role in advancing our understanding of lightning-generated transient luminous events (TLEs) and the lightning-mesosphere-ionosphere coupling processes behind them. But while not as visually spectacular, measurements of the electromagnetic radiation produced by TLE-producing lightning and, special cases, the TLE itself have also revealed important information and insight into lightning-ionosphere coupling processes. Optical observations give a direct view into the processes that happen inside the phenomenon, and lightning-related electromagnetic measurements give a quantitative view into the processes that drive the phenomenon from below. Each by itself is valuable, but the combination of the two is probably the most powerful tool for TLE research.

We describe here the techniques involved in recording andanalyzing lightning-related electromagnetic fields, and briefly summarize the main scientific findings that have come from such analysis. We focus primarily on lightning measurements related to sprites because sprites are the best documented transient luminous event (TLE) and have been the most-studied from a lightning perspective.

9.1 Background and Motivation

After the discovery of sprites (Franz et al., 1990), it soon became clear that the phenomenon was linked to lightning and was probably driven from below by individual lightning strokes (Winckler et al., 1996). But what specific lightning parameters are important in generating sprites? The first attempts to answer this question were mostly qualitative. Boccippio et al. (1995) and Reising et al. (1996) measured the distant ($>$2000 km), low frequency ($<$20 kHz) electromagnetic impulses (often called radio atmospherics or sferics) radiated by lightning strokes. They showed unequivocally that sprite-producing lightning strokes are very strong radiators at the lowest frequencies detected by each system ($\lesssim$ 100 Hz). Figure 1 demonstrates this by comparing sferics recorded on a system with 50 Hz to 6 kHz bandwidth from one lightning stroke that did produce a sprite and one that did not, both originating in the same storm. Both

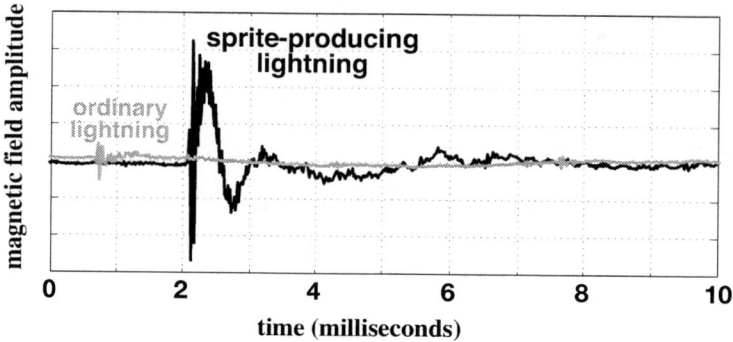

Figure 1. Two sferics that demonstrate the typical difference between signals radiated by sprite-producing and non-sprite-producing lightning strokes.

signals contain comparably large fast variations at the beginning of the waveform, but the sprite-producing stroke radiates a much larger slowly-varying component and thus contains significantly more low frequency energy. This qualitative observation has very specific consequences for the source parameters of sprite-producing lightning that are sketched in Figure 2. First we note that the effective source of distant, low frequency radiation is not lightning current but rather current moment (the product of current and the length of the current channel). This will be shown more explicitly in later sections of this chapter. If the lightning current moment waveform is strictly positive or strictly negative and non-oscillatory, it is easy to show that its Fourier transform (the source spectrum) must be monotonically decreasing as shown in the left panel of Figure 2. (Note that even when the source current moment is non-oscillatory, the radiated field waveforms are bipolar and sometimes oscillatory because of the high-pass filtering nature of the propagation.) Consequently the source spectra must look as they do in the left panel. These spectra imply that the time domain source current moment waveforms must look as they do in the right panel. The higher frequency source amplitude is largely controlled by the current moment risetime, thus the risetimes in sprite-producing and non-sprite-producing lightning are often similar. But the stronger low frequency radiation in sprite-producing strokes necessarily implies greater total charge moment change (the time integral of current moment) and thus longer duration current, as shown in the right panel of Figure 2. In short, Boccippio et al. (1995) and Reising et al. (1996) showed indirectly that the primary difference between sprite-producing and non-sprite-producing lightning is that sprite-producing lightning strokes contain larger charge moment changes and thus transfer more charge from the cloud to the ground. Around the same time, the first theoretical

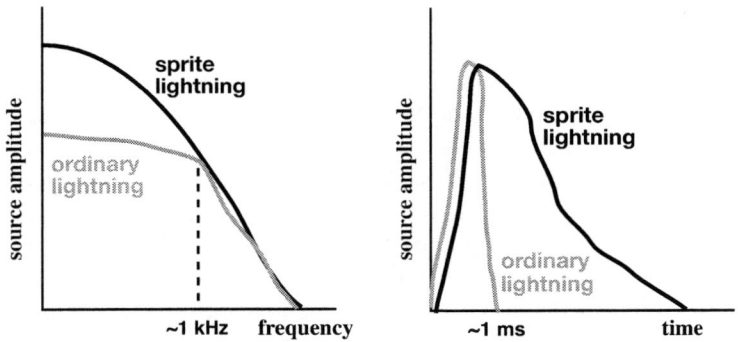

Figure 2. Comparison of source spectra and waveforms in sprite-producing and non-sprite-producing lightning. Sprite-producing lightning radiates more strongly at lower frequencies than more typical lightning, as shown in the left panel. This implies directly that sprite-producing lightning must have a longer duration and thus transfers more charge from the cloud to the ground than non-sprite-producing lightning, as shown in the right panel.

models of sprites were being developed. The conventional breakdown model (Pasko et al., 1995, 1997) suggested that post-lightning electric field transients produced by the lightning charge transfer exceeded the dielectric breakdown field of the atmosphere at high altitudes where the background pressure (and thus breakdown field strength) is low. This idea was inspired in part by numerical simulations of these fields by Baginski et al. (1988), and it is interesting to note that C. T. R. Wilson realized that the electric field amplitude would decay more slowly with altitude than the breakdown field strength, thereby essentially predicting sprites 70 years before their documented discovery (Huang et al., 1999).

A different sprite mechanism was proposed around the time, when conventional breakdown models were developed. The runaway breakdown model (Roussel-Dupré and Gurevich, 1996) suggested that these same post-lightning quasi-electrostatic fields were accelerating $\sim$1 MeV electrons into an avalanching upward beam of high energy particles that generated the observed optical emissions. Although the runaway and conventional breakdown mechanisms contain entirely different internal physics, the driver is the same in each. Substantial lightning charge moment change is required in order to generate electric fields high enough to initiate each process. These models were thus in good qualitative agreement with the qualitative observations of sprite-producing lightning at the time. But the models made specific predictions about the charge moment change magnitudes needed to initiate the conventional breakdown and runaway breakdown processes. Consequently there was a real need for quantitative measurements of charge moment changes in sprite-producing lightning.

The first attempt to measure lightning current moment and charge moment change in sprite-producing lightning was made by Cummer and Inan (1997), who analyzed the distant low frequency ($\lesssim 1$ kHz) magnetic fields radiated by sprite-producing lightning strokes. A similar approach was employed by Bell et al. (1998). Other measurements of sprite-producing lightning have been reported that also used the same or related techniques (Cummer et al., 1998; Huang et al., 1999; Cummer and Stanley, 1999; Füllekrug and Constable, 2000; Cummer and Füllekrug, 2001; Hobara et al., 2001; Hu et al., 2002; Sato and Fukunishi, 2003). It is worth noting that at the time of earlier reported measurements, some sprite details were not known and thus the interpretation of the measurements in the context of theory is not always complete. In general, improved modeling has lowered the expected charge moment change required to create sprites with the observed altitude extents, and lightning measurements continue to show that large lightning strokes are more common than previously thought. Hu et al. (2002) analyzed many sprite-producing strokes after carefully accounting for confirmed delay times between the lightning stroke and sprite initiation. Figure 3 contains their basic conclusions. By using an assumed distribution of charge moment changes in all positive lightning strokes, they derived a statistical distribution of sprite initiation probability based on charge moment change. The threshold they found ($\sim$600–1000 C·km) is consistent with conventional breakdown theory. Despite this consistency, open questions remain concerning the relationship between lightning charge moment changes and sprite initiation. In some cases, unexpectedly small light-

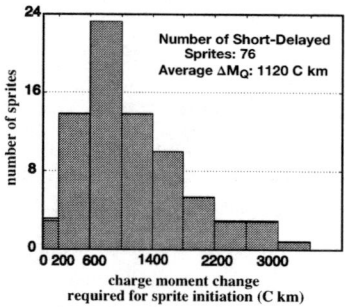
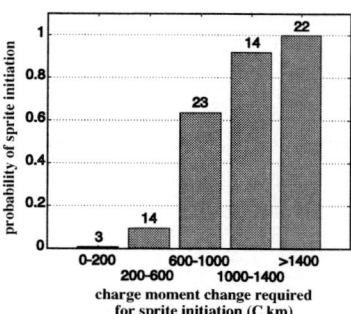

Figure 3. Measured charge moment changes responsible for sprite initiation. Left panel: The distribution of lightning charge moment changes at sprite initiation measured from 76 sprites recorded on 17 different nights. Right panel: Assuming a particular distribution of charge moment change in all +CG lightning, the fraction of lightning strokes that generate sprites versus charge moment change.

ning charge moment changes generated sprites (Hu et al., 2002). Sprites are generated essentially exclusively by positive cloud-to-ground lightning strokes (+CGs) that transfer positive charge from cloud to ground (Barrington-Leigh et al. (1999) reported the only published observations of sprites generated by negative CGs). But conventional breakdown theory does not predict a strong dependence of sprite initiation on lightning polarity. Electric current inside sprites themselves (Cummer et al., 1998) has been documented in approximately 10% of sprites in the U.S. High Plains (Cummer, 2003). Although conductivity changes inside sprites would produce sprite currents like those observed (Pasko et al., 1998), it is not known why this only occurs in a small fraction of sprites. And, lastly, little is known about the global distribution (or for that matter, the global characteristics) of sprites because the overwhelming majority of those observed have occurred in North America.

In the sections below we describe in detail the techniques behind measuring lightning current moment and charge moment change from distant low frequency electromagnetic fields.

9.2 Remote Lightning Parameter Measurements

The science of measuring lightning parameters from remote electromagnetic fields goes back decades (e.g., Norinder and Knudsen, 1956). In most cases it is difficult to measure lightning parameters directly. Triggered lightning measurements (Rakov et al., 1998) and instrumented towers (Berger et al., 1975) are exceptions, but one has little control over the specific lightning strokes measured by these techniques. Lightning parameter measurements from the analysis of remote electromagnetic fields are, in practice, the only measurements capable of probing specific, sprite-producing lightning.

Information about different lightning parameters is embedded in different ways in the radiated electromagnetic fields. Fast (>1 MHz) and nearby measurements contain much information about the fine details of the lightning stroke currents and are routinely used to test and validate lightning return stroke models (Thottappillil and Uman, 1993). Relatively high frequency (a few hundred kHz) and close (a few hundred km) measurements are used by the U.S. National Lightning Detection Network (NLDN) to routinely measure the peak current in most lightning return strokes (Cummins et al., 1998).

As described above, in sprite research, the most important lightning parameters are current moment and charge moment change. This is because these parameters are most closely linked to the high altitude electric field above a thunderstorm. The relevant time scale is on the order of milliseconds; sprites can initiate as fast as 1 ms (Cummer and Stanley, 1999) or as long as hundreds of milliseconds after a return stroke (Füllekrug and Reising, 1998; Cummer and Füllekrug, 2001). Thus the desired information is contained in sferics at frequencies below roughly $(1\ \text{ms})^{-1} = 1$ kHz. This is an extremely important constraint because energy at Very Low Frequency (VLF, 3–30 kHz) and below can propagate for very long distances in the waveguide formed by the ionosphere and the surface of the Earth – the so-called Earth-ionosphere waveguide. Thus the information needed for sprite studies is the very information that can be detected the greatest distance from the source lightning. Consequently, sprite-producing lightning can be studied with sensors operating a substantial distance (thousands of km) from the lightning itself.

There are a number of ways to measure the Extremely Low Frequency (ELF, 3–3000 Hz) and lower frequency fields that contain the needed information about lightning charge moment changes. The technique that has the longest history in measuring lightning charge transfer and charge moment changes is based on measuring the electrostatic field change created by a lightning stroke (Krehbiel et al., 1979, e.g,). Immediately following a lightning stroke, there is a dipole electrostatic field created by the annihilated charge and its image in the ground. See Krehbiel et al. (1979) for numerous examples of lighting-generated electrostatic field changes. This electrostatic field is directly proportional to the lightning charge moment change. The difficulty is that this electrostatic field decays with distance as r^{-3}; practical considerations regarding instrument sensitivity and signal strength require that the sensor must usually be closer than a few tens of km to the lightning stroke for the electrostatic field to be detected. Sprites, while not uncommon, are sporadic and sufficiently distributed geographically that it has proven difficult to measure electrostatic field changes from sprite-producing lightning strokes.

Another technique depends on measuring the electromagnetic fields in the Schumann Resonance (SR) band, roughly 5-50 Hz. Each lightning stroke on Earth excites the spherical shell cavity formed by the ionosphere and ground. Exceptionally large lightning strokes produce Schumann band fields that are detectable everywhere on the surface of the planet. Figure 4 shows an example Schumann-band magnetic field waveform produced by a large lightning stroke. Magnetic field sensors in this frequency range tend to perform better than electric field sensors (magnetic sensors typically have greater long term stability and are less sensitive to highly local noise sources like space charge), but either type can be used. Schumann resonances have a long research history with applications in a variety of fields from lightning detection to global change; (see Nickolaenko and Hayakawa, 2002, for a review). Burke and Jones (1996) showed how the source lightning charge moment change can be inferred from these distant Schumann resonance fields, and related techniques have since been applied by others to the sprite problem (Füllekrug and Reising, 1998; Huang et al., 1999; Hobara et al., 2001; Sato and Fukunishi, 2003). An obvious strength of the Schumann resonance technique is its global range; sufficiently large sprite-producing lightning all over the planet can be analyzed with just

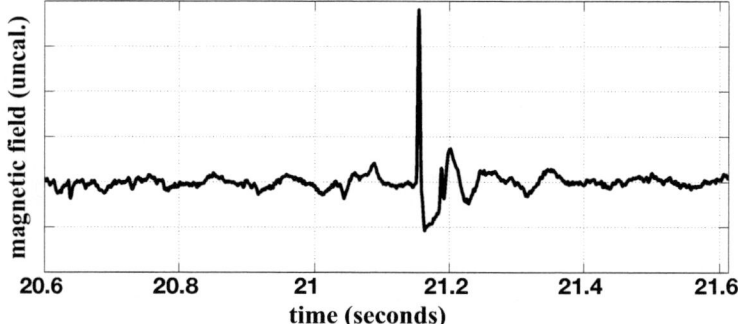

Figure 4. A sample measured Schumann-band magnetic field waveform from a distant lightning stroke. The significant background noise is produced by a continuum of lightning strokes all over the planet and is thus inescapable (data courtesy of M. Sato, Tohoku University).

a few sensors. But a significant shortcoming is the lack of time resolution or bandwidth in the extracted source waveform that results from a maximum signal frequency of roughly 50 Hz. Lightning charge moment changes can only be resolved with approximately 20 ms time resolution, which is substantially longer than the typical lightning-to-sprite delay time. Consequently questions regarding the charge moment change required to initiate a sprite are difficult to address with the Schumann resonance technique. It has unquestioned value to the field due to its global reach, but conclusions must be carefully drawn and its limitations understood. Falling somewhere between these techniques is one that uses the ELF and VLF radiation from lightning strokes to measure lightning current moment and charge moment change. Lightning is a strong radiator of electromagnetic energy across the entire Schumann resonance-ELF-VLF bands. All of these are guided by the Earth-ionosphere waveguide, but only the lowest frequencies are sufficiently low loss to generate a global resonance. Nevertheless, Extremely Low Frequency (ELF) and VLF sferics are easily detectable many thousands of km away from the lightning stroke and thus can provide substantial lightning remote sensing range. Figure 5 shows an example ELF-VLF magnetic field waveform radiated by distant lightning. Electric field measurements are also possible and are fairly similar in form. The chief advantage of using this band for lightning remote sensing is that the wider bandwidth and thus sub-millisecond time resolution obtainable in inferred current moment waveforms. Most applications of this technique have intentionally filtered the higher frequencies (>1 kHz) out of the signals to significantly simplify the analysis. Doing so still yields roughly 1 ms time resolution in current moment waveforms, which is sufficient for detailed analysis of

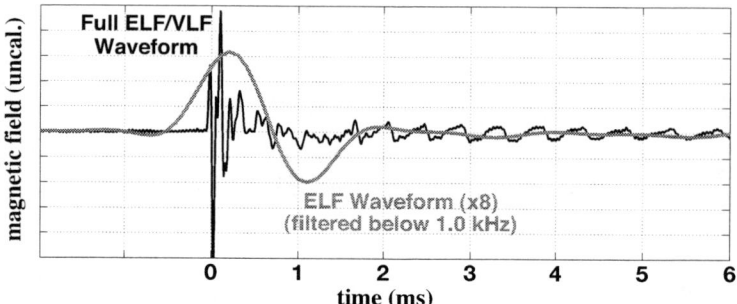

Figure 5. A sample measured ELF-VLF magnetic field waveform from a distant lightning stroke. The full waveform (up to ~25 kHz) and the filtered ELF (<1 kHz) waveforms are shown for comparison.

sprite initiation and other fast processes that are beyond the reach of Schumann resonance techniques.

9.3 Data Analysis Techniques

The same general technique for inferring the lightning current moment and charge moment change can be used for all three types of remote electromagnetic field measurements described above. This can be seen from the mathematical expressions that describe how each of these fields are generated by a lightning stroke. The vertical electrostatic field change following a lightning stroke at distance r from the sensor that transfers charge Q over channel length l is, from basic electrostatics,

$$E_{\text{vert}}(r) = \frac{2Ql}{4\pi\epsilon_0 (l^2 + r^2)^{3/2}}. \tag{9.1}$$

The azimuthal horizontal magnetic field component in the Schumann band produced by a lightning stroke with current spectrum (i.e., the Fourier transform of the current waveform) $I(\omega)l$ is

$$B_\phi(\theta) = \frac{I(\omega)l}{4\pi\epsilon_0 a^3 h_1} \sum_{n=1}^{\infty} \frac{(2n+1)P_n^1(\cos\theta)}{(\omega - \omega_n)(\omega + \omega_n^*)}, \tag{9.2}$$

where a is the radius of the Earth, θ is the polar angle describing the lightning to sensor separation, $P_n^1(x)$ is the associated Legendre function of order 1, and h_1 and ω_n are empirical parameters that describe the ionospheric upper boundary of the spherical cavity (Sentman, 1990; Füllekrug and Constable, 2000). The superscript * denotes the complex conjugation operation.

The azimuthal ELF-VLF magnetic fields produced by a lightning stroke as a function of lightning-sensor distance r are given by

$$B_\phi(r) = \frac{j\mu_0 k I(\omega) l}{2h} \sum_{n=1}^{\infty} \sin\theta_n H_1^{(2)}(kr\sin\theta_n), \tag{9.3}$$

where $H_1^{(2)}$ is the Hankel function of the second kind and order 1, h is an effective Earth-ionosphere waveguide reference height, and θ_n is the eigenangle of a specific waveguide mode that depends on the specific ionospheric profile assumed (Wait, 1970).

The critical similarity between all of these expressions is that the remote fields depend linearly on either the source current moment or charge moment change. This linearity means that scaling the source amounts to equivalent scaling of the output fields. In other words, if by some method one can estimate or simulate the fields produced by a *known* source, for example an impulsive 1 C·km charge moment change, then one can simply scale the simulated fields to match the measured fields to estimate the actual source strength. There are, of course, complications in this scaling due to noise in the signal and uncertainty in how well the simulated fields represent the true fields (in other words, how precisely the ionospheric transfer function is known). But in general this approach is a powerful technique for remotely measuring lightning current moment and charge moment change. Because the quantity of interest for sprite studies is current moment or charge moment change, it is actually a benefit that channel length and current or charge alone do not appear in any of these expressions for low frequency remote electromagnetic fields.

Figure 6 shows a block diagram of the components involved in such a measurement. The first component is the measurement of the remote fields. We do not discuss in detail the sensors and hardware involved in such a measurement. Because the target frequencies are quite low ($\lesssim$30 kHz), the sensor and electronics are fairly simple, usually involving an air core or high permeability core wire loops (for a magnetic field sensor) or a charge collecting surface such as a wire, a plate, or a sphere (for an electric field sensor). Absolute system timing at the millisecond level or better is usually required in order to unambiguously compare data from different sources; this is easily done using Global Positioning System (GPS) receivers. One must also know in detail the frequency response (amplitude and phase) of the overall system because the simulated signals needed for the quantitative analysis must represent what the sensor would have seen from a known source. Consequently this frequency response must be included in the simulations. The remaining components are field modeling and data inversion. These are discussed in some detail below using the ELF technique as an example because it is probably best suited for

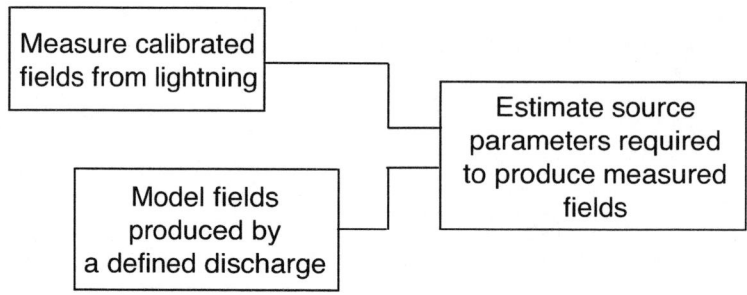

Figure 6. Block diagram of the general technique of measuring lightning current moment and charge moment change from measured remote fields of some variety and modelled fields produced by a defined source.

sprite studies. The general techniques we describe are easily adaptable to the analysis of other electromagnetic field measurements.

9.3.1 Electromagnetic Field Modeling

Equation 9.3 above describes ELF and VLF propagation in the Earth-ionosphere waveguide with deceptive simplicity. The complication is how the eigenangles θ_n that describe how each waveguide mode propagates depend on the specific ionospheric electron density and collision frequency profiles. Except in few cases, this is a problem that must be solved numerically. Decades of effort have been put into numerical codes for precisely this problem (Pappert and Ferguson, 1986). See Budden (1962); Wait (1970); Cummer (2000) for more details concerning the precise mathematical formulation of this problem and techniques for its numerical solution.

As noted above, this problem can be simplified significantly by focusing only on the <1 kHz energy that propagates in only one waveguide mode (the quasi-transverse electromagnetic or QTEM mode). In this case, a fairly accurate approximate formulation due to Greifinger and Greifinger (1979) can be employed, and we briefly describe that here. Greifinger and Greifinger (1979) showed that the vertical electric field at the ground in the QTEM mode as a function of frequency ω and distance from the source r satisfies the equation

$$E_z(r) = \frac{\mu_0 \omega I(\omega) l}{4(h_0 - i\pi\zeta_0/2)} S_0^2 H_0^{(2)}(kS_0 r), \qquad (9.4)$$

where

$$S_0 = \frac{h_1(h_1 + i\pi\zeta_1)}{(h_1 + i\pi\zeta_1/2)(h_0 - i\pi\zeta_0/2)}. \qquad (9.5)$$

The parameters h_0, h_1, ζ_0, and ζ_1 are the altitudes and scale heights at which particular conditions are met, and they are easily computed from arbitrary electron density and collision frequency profiles using expressions given by Greifinger and Greifinger (1979).

As an example and using the above approximate formulas and assuming a propagation distance of 1000 km, a source with $I(\omega)l$=1 C·km (an impulsive source), and a sensor with single order high- and low-pass cutoff frequencies of 100 Hz and 1 kHz, respectively, Figure 7 shows the calibrated signal spectrum and signal waveform that would be measured under these conditions. Typical midlatitude nighttime electron density and collision frequency profiles taken from Cummer (2000) were used for this calculation. The shape of the pulse agrees qualitatively with the measured data shown in Figure 5. Also plotted are the signal spectrum and waveform computed using the more complete LWPC mode theory propagation code (Pappert and Ferguson, 1986). They agree fairly closely, with the approximate form missing some of the finer details and overestimating signal amplitude above about 800 Hz. This is not surprising because the approximations made become less valid as frequency increases. Nevertheless the approximate form is a simple and reliable method of predicting the propagation impulse response (the expected signal for a known impulsive lightning stroke) under realistic conditions.

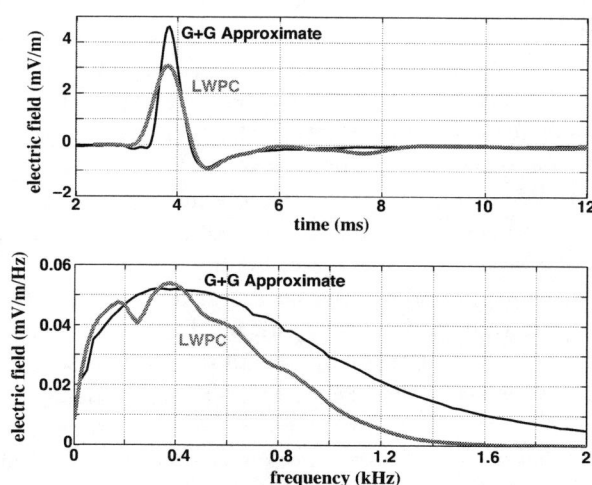

Figure 7. Demonstration of ELF propagation impulse response computation. The computed vertical electric field signals received 1000 km from a 1 C·km impulsive lightning stroke are shown in the time (top) and frequency (bottom) domains. The simple approximate form agrees fairly well with the numerical mode theory (LWPC) computation.

9.3.2 Data Inversion

Once the propagation impulse response is estimated, the goal is to find the proper combination of impulse responses that reproduces the measured signal with acceptable accuracy. The combination of source impulses that reproduces this signal is then the inferred source current moment, with the charge moment change easily computable from this by a simple time integration. This straightforward matching must be done with great care because this inverse problem is inherently ill-conditioned. In other words, there are usually many different possible combinations of impulsive sources that acceptably match the measured signal. Because of this it is particularly easy to overestimate the source current moment and charge moment change. For example, the impulse responses shown in Figure 7 have no zero frequency energy. Consequently a very slowly changing source current moment contributes very little to the output signal. Thus a quasi-DC signal that contains substantial charge moment change could be added to the source without perturbing the received signal. In light of this uncertainty, a rational way of posing the inverse problem is that we are searching for the source with the smallest charge moment change that is consistent with the noisy data.

A simple and robust technique that works well when the source signal of interest is fairly impulsive is simple scaling. Good agreement between the measured and simulated signals can sometimes be obtained by simply scaling the propagation impulse response. The left panels of Figure 8 demonstrate this with a real signal. By simply scaling the 1 C·km propagation impulse by a factor of 1000 (remember that linearity is assumed), the simulated and measured waveforms are in close quantitative agreement, indicating that a reasonable result has been obtained. The imperfect match is attributable to differences between the model ionosphere assumed in the simulation and the real ionosphere. We can thus conclude that the source current moment waveform contains a total charge moment change of ~ 1000 C·km and moves this charge on a time scale that is impulsive on ELF time scales $\lesssim 1$ ms.

In some cases, simple multiplicative scaling is not sufficient to give a good fit, as shown in the right panels of Figure 8. This occurs when the actual source current waveform is not impulsive. In this case a more sophisticated technique is needed to solve the inverse problem appropriately. A variety of approaches have been used in the literature. For Schumann resonance analysis, a two-parameter (amplitude and time constant) decaying exponential current moment waveform is often assumed, and the parameters are determined by a least squares fit to the measured data using a modelled propagation impulse response (Burke and Jones, 1996; Huang et al., 1999). The wider ELF bandwidth means that more information than just two parameters is contained in the waveform, and more general approaches have been used. One technique

ELF REMOTE SENSING

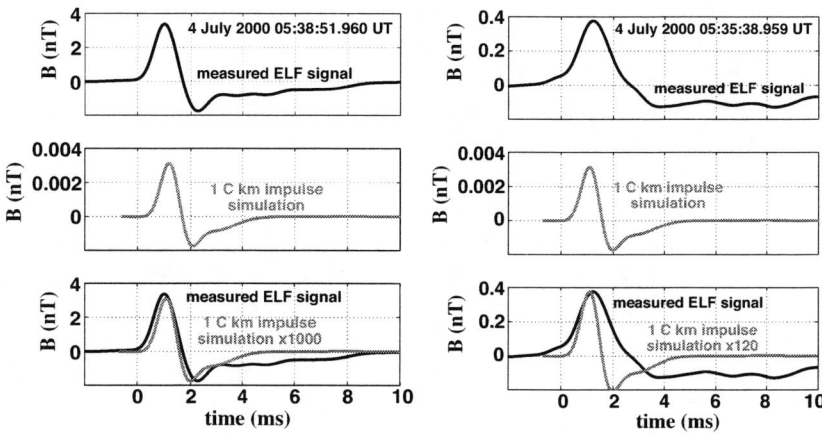

Figure 8. Demonstration of simple multiplicative scaling of the propagation impulse to match the measured data. In the left panels, after scaling by a factor of 1000, the agreement is good and consequently the source measurement is simple. In the right panels, a slow, non-impulsive discharge results in a broader measured waveform that cannot be matched with a scaled impulse.

called CLEAN, which originated in astronomical image restoration, has been applied to this problem with success (Cummer and Inan, 1997). Regularized least squares fitting has also been used (Cummer and Inan, 2000) (see Press et al. (1992) for a good practical description of this approach). In the end, no matter what technique is used, it is fairly simple to confirm that the results of such a current moment extraction are reasonable. The relationship between the current moment and the resulting field waveform is a simple convolution with the propagation impulse response, or

$$F(t) = \int_{-\infty}^{\infty} M_i(\tau) h_{\text{prop}}(t - \tau) d\tau \tag{9.6}$$

where $M_i(t)$ is the current moment waveform, $h_{\text{prop}}(t)$ is the propagation impulse response, and $F(t)$ is the resulting electromagnetic field waveform. If the inverse analysis has been done well, then the extracted $M_i(t)$ should have three features. First, the $M_i(t)$ waveform should be strictly positive or negative because lightning current does not normally change polarity in a single stroke. Second, the measured field waveform $F_{\text{meas}}(t)$ and the reconstructed field waveform defined by the simple convolution $F_{\text{recon}}(t) = \int_{-\infty}^{\infty} M_i(\tau) h_{\text{prop}}(t - \tau) d\tau$ should be in good agreement. Good agreement is dictated by the background noise and the degree of mismatch expected between the simulated and true propagation impulse responses. Third, the extracted current moment waveform should be the waveform that satisfies the previous condition with the minimum possible total charge moment change $M_q = \int_0^\infty M_i(t)dt$. In this way, the extracted source is the smallest one consistent with the noisy data. Because, as described above, there can be much larger charge moment changes also consistent with the data, this lower bound is the most scientifically meaningful.

The application of this technique in a specific form was described in detail by Cummer and Inan (2000). To demonstrate it, Figure 9 shows two examples. The post-filtered measured ELF magnetic field waveforms are shown in the top panels. After deconvolution with the regularized least squares technique, the extracted source current moment and charge moment change waveforms are shown in the lower panels. The top panels then also show the agreement between the reconstructed (as defined above) and the measured field waveforms. In the first case, a very strong signal is fit closely by a smoothly varying, strictly positive, and physically reasonable current moment waveform. In the second case, a smaller signal with poorer signal to noise ratio is also fit by a much smaller but still smooth and physically reasonable current moment waveform. Note that the fit in this second case is not as close. This is achieved on purpose because in light of the significant noise in the second waveform, the reconstructed sferic *should not* fit the original data too closely; if it does then the current moment waveform would have been extracted from the signal and the noise, not just the signal. In practice this often leads to current moments larger than justified by the data. These two examples attempt to demonstrate a good outcome from an application of any deconvolution or inverse technique to extract the lightning current moment waveform.

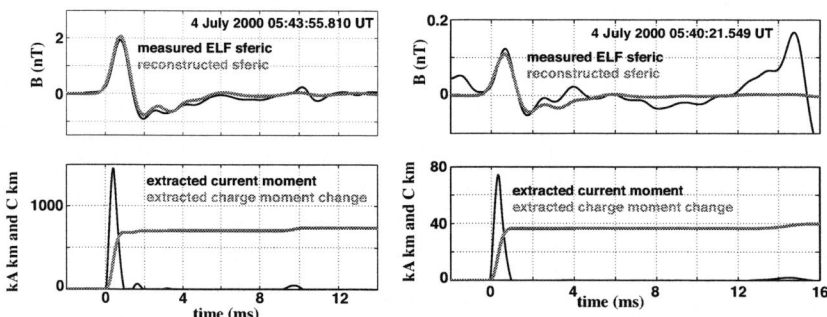

Figure 9. Demonstration of source parameter extraction. The source current moment and charge moment change waveforms (lower panels) were extracted from the measured waveforms in the upper panels. The good agreement between the measured and reconstructed waveforms in the upper panels indicates that the overall result is reasonable.

9.4 Summary

We have described a general approach for measuring lightning current moment and charge moment change from the distant electromagnetic fields produced by an individual lightning stroke. Such a measurement is possible with almost any variety of remote electromagnetic fields, from nearby electrostatic field changes to global Schumann resonance fields. The different fields have different strengths and weaknesses in terms of achievable time resolution and range from the lightning source. For studying sprites, the optimal signals are ELF sferics from lightning (essentially 1 kHz and below) because their propagation distance is large and they provide time resolution on the $\sim$1 ms time scales relevant for sprite initiation. Care is required in the application of this technique to ensure reliable results, but it is a powerful technique because lightning parameters can be measured over a huge geographic range with very few sensors. Past application of the techniques described here have already provided many important experimental results, but many important questions that could be addressed with the proper combination of electromagnetic and optical measurements remain unanswered.

Bibliography

Baginski, M. E., Hale, C., L., and Olivero, J. J. (1988). Lightning-related fields in the ionosphere. *Geophys. Res. Lett.*, 15(8):764–767.

Barrington-Leigh, C. P., Inan, U. S., Stanley, M., and Cummer, S. A. (1999). Sprites triggered by negative lightning discharges. *Geophys. Res. Lett.*, 26(24):3605–3608.

Bell, T. F., Reising, S. C., and Inan, U. S. (1998). Intense continuing currents following positive cloud-to-ground lightning associated with red sprites. *Geophys. Res. Lett.*, 25(8):1285–8.

Berger, K., Anderson, R. B., and Kroninger, H. (1975). Parameters of lightning flashes. *Electra*, 80:223–237.

Boccippio, D. J., Williams, E. R., Heckman, S. J., Lyons, W. A., Baker, I. T., and Boldi, R. (1995). Sprites, ELF transients, and positive ground strokes. *Science*, 269(5227):1088–91.

Budden, K. G. (1962). The influence of the Earth's magnetic field on radio propagation by wave-guide modes. *Proc. Roy. Soc. A*, 265:538.

Burke, C. P. and Jones, D. L. (1996). On the polarity and continuing currents in unusually large lightning flashes deduced from ELF events. *J. Atmos. Terr. Phys.*, 58(5):531–40.

Cummer, S. A. (2000). Modeling electromagnetic propagation in the Earth-ionosphere waveguide. *IEEE Transactions on Antennas and Propagation*, 48:1420.

Cummer, S. A. (2003). Current moment in sprite-producing lightning. *J. Atmos. Sol.-Terr. Phys.*, 65:499–508.

Cummer, S. A. and Füllekrug, M. (2001). Unusually intense continuing current in lightning causes delayed mesospheric breakdown. *Geophys. Res. Lett.*, 28:495.

Cummer, S. A. and Inan, U. S. (1997). Measurement of charge transfer in sprite-producing lightning using ELF radio atmospherics. *Geophys. Res. Lett.*, 24(14):1731–4.

Cummer, S. A. and Inan, U. S. (2000). Modeling ELF radio atmospheric propagation and extracting lightning currents from ELF observations. *Radio Science*, 35:385–394.

Cummer, S. A., Inan, U. S., Bell, T. F., and Barrington-Leigh, C. P. (1998). ELF radiation produced by electrical currents in sprites. *Geophys. Res. Lett.*, 25(8):1281–4.

Cummer, S. A. and Stanley, M. (1999). Submillisecond resolution lightning currents and sprite development: Observations and implications. *Geophys. Res. Lett.*, 26(20):3205–3208.

Cummins, K. L., Murphy, M. J., Bardo, E. A., Hiscox, W. L., Pyle, R. B., and Pifer, A. E. (1998). A combined TOA/MDF technology upgrade of the US National Lightning Detection Network. *J. Geophys. Res.*, 103(D8):9035–9044.

Franz, R. C., Nemzek, R. J., and Winckler, J. R. (1990). Television image of a large upward electrical discharge above a thunderstorm system. *Science*, 249(4964):48–51.

Füllekrug, M. and Constable, S. (2000). Global triangulation of intense lightning discharges. *Geophys. Res. Lett.*, 27(3):333–336.

Füllekrug, M. and Reising, S. C. (1998). Excitation of Earth-ionosphere cavity resonances by sprite-associated lightning flashes. *Geophys. Res. Lett.*, 25(22):4145–4148.

Greifinger, C. and Greifinger, P. (1979). On the ionospheric parameters which govern high-latitude ELF propagation in the earth-ionosphere waveguide. *Radio Science*, 14(5):889–895.

Hobara, Y., Iwasaki, N., Hayashida, T., Hayakawa, M., Ohta, K., and Fukunishi, H. (2001). Interrelation between ELF transients and ionospheric disturbances in association with sprites and elves. *Geophys. Res. Lett.*, 28:935–938.

Hu, W., A., Cummer S., Lyons, W. A., and Nelson, T. E. (2002). Lightning charge moment changes for the initiation of sprites. *Geophys. Res. Lett.*, 29(8):10.1029/2001GL014593.

Huang, E., Williams, E., Boldi, R., Heckman, S., Lyons, W., Taylor, M., Nelson, T., and Wong, C. (1999). Criteria for sprites and elves based on Schumann resonance observations. *J. Geophys. Res.*, 104(D14):16943–64.

Krehbiel, P. R., Brook, M., and McCrory, R. A. (1979). An analysis of the charge structure of lightning discharges to ground. *J. Geophys. Res.*, 84(C5):2432–2456.

Nickolaenko, A. P. and Hayakawa, M. (2002). *Resonances in the Earth-Ionosphere Cavity*. Kluwer Academic Publishers, Dordrecht.

Norinder, H. and Knudsen, E. (1956). Pre-discharges in relation to subsequent lightning strokes. *Arkiv För Geofysik*, 2(27):551–571.

Pappert, R. A. and Ferguson, J. A. (1986). VLF/LF mode conversion model calculations for air to air transmissions in the earth-ionosphere waveguide. *Radio Science*, 21:551–558.

Pasko, V. P., Inan, U. S., Bell, T. F., and Reising, S. C. (1998). Mechanism of ELF radiation from sprites. *Geophys. Res. Lett.*, 25(18):3493–6.

Pasko, V. P., Inan, U. S., Bell, T. F., and Taranenko, Y. N. (1997). Sprites produced by quasi-electrostatic heating and ionization in the lower ionosphere. *J. Geophys. Res.*, 102(A3):4529–61.

Pasko, V. P., Inan, U. S., Taranenko, Y. N., and Bell, T. F. (1995). Heating, ionization and upward discharges in the mesosphere due to intense quasi-electrostatic thundercloud fields. *Geophys. Res. Lett.*, 22(4):365–368.

Press, W. H., Teukolsky, S. A., Vetterling, W. T., and Flannery, B. P. (1992). *Numerical Recipes in FORTRAN - The Art of Scientific Computing*. Cambridge University Press, Cambridge.

Rakov, V. A., Uman, M. A., Rambo, K. J., Fernandez, M. I., Fisher, R. J., Schnetzer, G. H., Thottappillil, R., Eybert-Berard, A., Berlandis, J. P., Lalande, P., Bonamy, A., Laroche, P., and Bondiou-Clergerie, A. (1998). New insights into lightning processes gained from triggered-lightning experiments in Florida and Alabama. *J. Geophys. Res.*, 103(D12):14117–14130.

Reising, S. C., Inan, U. S., Bell, T. F., and Lyons, W. A. (1996). Evidence for continuing current in sprite-producing cloud-to-ground lightning. *Geophys. Res. Lett.*, 23(24):3639–3642.

Roussel-Dupré, R. A. and Gurevich, A. V. (1996). On runaway breakdown and upward propagating discharges. *J. Geophys. Res.*, 101:2297.

Sato, M. and Fukunishi, H. (2003). Global sprite occurrence locations and rates derived from triangulation of transient schumann resonance events. *Geophys. Res. Lett.*, 30(16):Art. No. 1859.

Sentman, D. D. (1990). Approximate schumann resonance parameters for a 2-scale-height ionosphere. *J. Atmos. Terr. Phys.*, 52(1):35–46.

Thottappillil, R. and Uman, M. A. (1993). Comparison of lightning return-stroke models. *J. Geophys. Res.*, 98(D12):22903–22914.

Wait, J. R. (1970). *Electromagnetic Waves in Stratified Media*. Pergamon, Oxford.

Winckler, J. R., Lyons, W. A., Nelson, T. E., and Nemzek, R. J. (1996). New high-resolution ground-based studies of sprites. *J. Geophys. Res.*, 101:6997–7004.

LOCATION AND ELECTRICAL PROPERTIES OF SPRITE-PRODUCING LIGHTNING FROM A SINGLE ELF SITE

Yasuhide Hobara [1*], M. Hayakawa [2], E. Williams [3], R. Boldi [4] and E. Downes [5]

[1*] *Swedish Institute of Space Physics, Kiruna, Sweden.*

[2] *The University of Electro-Communications, Department of Electronic Engineering, Tokyo, Japan.*

[3] *Parsons Laboratory, Massachusetts Institute of Technology, Cambridge, MA, USA.*

[4] *Lincoln Laboratory, Massachusetts Institute of Technology, Lexington, MA, USA.*

[5] *Physics Department, Massachusetts Institute of Technology, Cambridge, MA, USA.*

[*] *now at: The University of Sheffield, Department of Automatic Control and Systems Engineering, Sheffield, U.K.*

Abstract   Recently discovered TLEs (Transient Luminous Events) such as red sprites and elves provided a great opportunity to revisit the electromagnetic waves in the lower ELF (Extremely Low Frequency) region known as the Schumann resonances (SR). The resonance behavior is afforded by the low attenuation experienced by electromagnetic waves in this frequency range. Since TLEs are caused by energetic lightning with abundant energy in this range, these so-called ELF transients can be analyzed on a global basis from single measurement stations. In particular, the geographical location and the vertical charge moment of the lightning flash may be determined remotely. In this chapter, we aim at providing readers with an overview of electromagnetic waves from lightning in the SR frequency band. Then we introduce the technique to determine the location and demonstrate the global mapping of lightning for different thresholds of charge moment change based on the ELF transient observations in Rhode Island, USA. Meteorological interpretations of the global maps are also provided. Furthermore the sprite-producing winter lightning activity is characterized in Hokuriku by using the ELF field site in Moshiri, Japan. The generation condition for winter sprites and their coupling to the tropospheric lightning and to the ionosphere are also presented.

10.1 Introduction
10.1.1 Lightning Activity

The conventional thundercloud is generally characterized by a positive dipole. Negative charges are distributed mainly in the mid-region of the cloud and the positive charges are at higher altitude. Such clouds are typically about as wide as they are tall. In contrast, a Mesoscale Convective System (MCS) has a horizontal extent more than 10 times its depth. This system has a significant lateral extent with a large, positive charge layer near cloud base, which in these systems is often close to the 0°C isotherm. One of the important characteristics in MCSs is its inverted dipole structure in comparison with the conventional isolated thundercloud (Williams, 1998; Lyons et al., 2003).

A lightning flash that lowers positive (negative) charge to the ground is so-called positive (negative) cloud-ground lightning . The vast majority of lightning ground flashes worldwide are negative, although some exceptional cases such as lightning activity in wintertime over the Sea of Japan show a predominance of positive polarity (see Section 10.3 in this chapter).

Two different types of positive ground flashes are well known. Such discharges may be initiated from the upper region of the cloud, which leads to a long vertical extent to the ground in the case of the conventional thundercloud (e.g., Rust et al., 1981). But the large positive charge reservoir near the base of the MCS stratiform anvil (e.g. a few thousand coulombs) can also contribute a substantial amount of positive charge to the ground (Lyons et al., 2003). As a simplified model, current I flow along the lightning channel with length ds has a current moment $Ids(t)$. This current moment generates a vertical electric field pulse which can be received by ELF antennas, as described later by Equation (10.1).

10.1.2 Terrestrial ELF Electromagnetic Signals

Maxwell's equations govern the behavior of electromagnetic waves for all frequency ranges. For example, in the so-called radio wave (range from 300 kHz to 300 MHz), we have radio and TV broadcasting. Radar systems and microwave ovens use the frequency range (1 GHz to 30 GHz), the so-called microwave region. Furthermore the visible frequency range is situated in a rather narrow frequency band from 400 THz to 800 THz (7500 Å to 3750 Å, 1 Å = 10^{-10}m). Among the various frequency ranges, we focus here on the frequencies lower than 3 kHz (so-called ELF Extremely Low Frequency) where radiation from lightning discharges are rather strong and propagates within a natural waveguide.

The conductive Earth and overlying ionosphere can be recognized as a wave guide or transmission line (Madden and Thompson, 1965) for low frequency electromagnetic waves. The waves are nearly perfectly reflected from the ground while the reflection property of the ionosphere varies depending on the frequency ranges. In the case of the ELF frequency range (f<3 kHz), the imperfectly conducting ionosphere plays a major role on the loss of electromagnetic energy. For example, at a frequency of a few Hz, the reflection coefficient at the ionosphere is close to unity and waves are almost perfectly reflected so the propagation loss is 0.1 dB/Mm (1 Mm=1000 km), while rather large loss (about 60 dB/Mm) can be expected in the frequency range near 2 kHz, the so-called 'cutoff region' (Jones, 1999, pp. 171). The low frequency component of electromagnetic radiation from lightning discharges can propagate global distances (round the Earth) in the lower ELF frequency range, the so-called Schumann Resonance (SR) band.

In this favorable frequency range there are two different aspects of Schumann resonances (SR) driven by lightning – the 'background' and the 'transients'. The background resonances are set up by overlapping contributions from about 100 flashes per second of global lightning flash rate. Superimposed on the background signal are the transient events – the extraordinarily energetic flashes that dominate all the lightning on the planet for short periods ($\sim$100 ms) and set up their own global resonances. Expected resonance frequencies in a hypothetical lossless cavity are given by the formula ($f_n = c\sqrt{n(n+1)}/(2\pi a)$), where c is the velocity of the light, and a is the radius of the Earth). At $n=1$ one has the fundamental frequency f_1=10.6 Hz and higher harmonics are found at f_2=18.4 Hz, f_3=26.0 Hz, etc. However the observed resonance frequencies in the damped cavity with peak intensity at a fundamental $\sim$8 Hz and its harmonics are all lower than predictions for a lossless cavity. Variations of these resonance frequencies and peak intensities reflect the spatio-temporal dependence of the global thunderstorm activity (Nickolaenko et al., 1996; Füllekrug and Constable, 2000). Figure 1 shows the frequency spectrum of the Schumann resonances observed at the Moshiri sta-

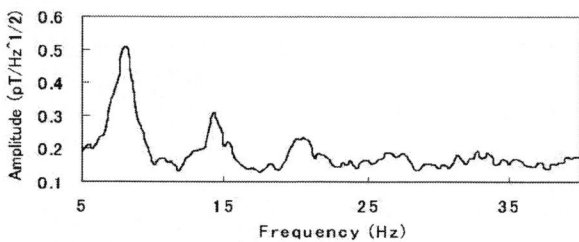

Figure 1. Schumann resonance spectrum (north-south magnetic field component) observed in Moshiri, Hokkaido, Japan (10-minute average from 16:39 UT on April 19, 1999). Spectral peaks near 8, 14, 20 and 26 Hz are evident.

tion in Hokkaido, Japan (Hobara et al., 2000a). Readers can find more detailed description of SR in Sentman (1995), Hayakawa and Nickolaenko (2001) and Nickolaenko and Hayakawa (2002).

10.1.3 Sprites and Elves and Causative Lightning Properties

The red sprite is a transient luminous event generally in the altitude range from 55-80 km (mesosphere) with lifetimes of milliseconds to tens of milliseconds. Numerous observations have shown that the above-mentioned positive ground flash is the predominant cause for the sprite (Boccippio et al., 1995; Hu et al., 2002; Lyons et al., 2003). Yet the majority of positive ground flashes do not produce sprites, including the ones with a peak current (>50 kA) in comparison to the mean peak current for ground flashes. Recent experimental results have shown that the large charge moment change (deduced remotely from the observation of ELF electromagnetic fields from the lightning, Section 10.2 in this chapter) is the key parameter for the initiation of a sprite (Huang et al., 1999; Hu et al., 2002). On the other hand, elves are short-lived (life time is shorter than 1 ms) optical events appearing at the lower edge of the ionosphere (80 to 90 km altitude) and are likely caused by ionization by the electromagnetic pulse (EMP) from ground flashes (e.g., Fukunishi et al., 1996; Inan et al., 1997). The corresponding ELF signature often shows an impulsive current waveform (Huang et al., 1999).

10.1.4 Contents of this Chapter

Two different topics are presented in this chapter in relation to ELF transient signals.

Firstly in Section 10.2 of this chapter, we describe the extraction of the location and electrical properties of lightning discharges from electromagnetic measurements made from a single ELF receiving station (Rhode Island, USA). We then demonstrate practical results from these procedures and their physical interpretation.

Secondly, in Section 10.3 of this chapter, we give attention to the deduced electrical properties of the lightning source from ELF transients by using the method mentioned in Section 10.2 of this chapter. The vertical charge moment Qds is extensively used as one of key parameters to characterize the winter lightning activity in Hokuriku, Japan, the source for winter TLEs. Furthermore, the coupling mechanism between tropospheric lightning, mesospheric TLEs, and the lower ionosphere has been studied.

10.2 Locating Distant ELF Sources and Quantifying their Electrical Properties

10.2.1 Theory

The localization and characterization of gigantic lightning flashes on a global basis from a single measurement station are feasible in the Schumann resonance region (3-50 Hz) of the ELF frequency range. By 'superlative lightning flashes' we mean flashes for which ELF amplitudes are ten times (or more) greater than the Schumann Resonance (SR) 'background', a quasi-continuous signal maintained by ordinary lightning flash activity and the 'noise' for the analysis of individual flashes. Sprite lightning flashes are in this superlative category, and represent something like 0.01% of all lightning flashes, a small minority.

The procedure is conceptually straightforward and involves the merging of the full 3-component field measurements (H_x, H_y, E_z) with analytic theory for a uniform Earth-ionosphere cavity. (The tangential electric field at the Earth's surface is close to zero, and the vertical component of magnetic field vanishes for this waveguide mode.) The theoretical equations (Wait, 1996) are

$$E_z = i \frac{I(f)\,ds}{4a^2 \varepsilon_o 2\pi f h} \frac{\nu(\nu+1) P_\nu^0(-\cos\theta)}{\sin(\pi\nu)} \left[\text{Vm}^{-1}\text{Hz}^{-1}\right] \quad (10.1)$$

$$H_\varphi = -\frac{I(f)\,ds}{4ah} \frac{P_\nu^1(-\cos\theta)}{\sin(\pi\nu)} \left[\text{Am}^{-1}\text{Hz}^{-1}\right] \quad (10.2)$$

$I(f)ds$ is the current moment of the lightning flash and characterizes the source, in which ds is the vertical distance over which the lightning current $I(f)$ flows. $P_\nu^{0,1}$ are associated Legendre functions with complex subscripts ν. The quantity h is the waveguide thickness, ε_o is the dielectric constant of free space, and a is the radius of the Earth. The angle θ is the great circle angular distance between source and receiver. The propagation effects of the ionosphere-topped waveguide are represented by the complex eigenvalue $\nu(f)$. Accurate values of ν for the uniform Earth-ionospheric cavity are considered by Mushtak and Williams (2002). Values given in Ishaq and Jones (1977) are as accurate as any, and have been used in the work in West Greenwich, Rhode Island, USA (Huang et al., 1999). The eigenvalue ν is calculated by the following equations

$$\nu(f)(\nu(f)+1) = (kaS)^2$$

$$S = \left(\frac{C}{V}\right) - i\left(5.49\frac{\alpha}{f}\right)$$

$$\frac{C}{V} = 1.725 - 0.275 \ln f + 0.052 (\ln f)^2 + 0.0036 (\ln f)^3$$

$$\alpha = 0.063 f^{0.64} \,[dB/Mm]$$

where k is the wave number, S is the sine of the mode eigenangle, C is the speed of light, V is the phase velocity and α is the ELF attenuation factor.

The geometry for global lightning location from the MIT station in Rhode Island, is shown in Figure 2a. Great circle paths emanate from Rhode Island and reconverge on the antipode on the opposite side of the Earth – off the southwest coast of Australia. Lines of equal distance θ along great circle paths are also shown at 2 Mm intervals.

The 3-component electromagnetic measurements involve two components of magnetic field, H_x and H_y aligned with geographical coordinates, to resolve H_φ in Equation (10.2) as

$$H_\varphi = \sqrt{H_x^2 + H_y^2}. \qquad (10.3)$$

The vertical electric field E_z is measured with an electrode on a 7-meter high tower of stacked insulator columns (Figure 2b). The large insulators provide for mechanical stability and minimize vibration in conditions of strong wind. Great circle paths are determined by comparing H_x and H_y in standard

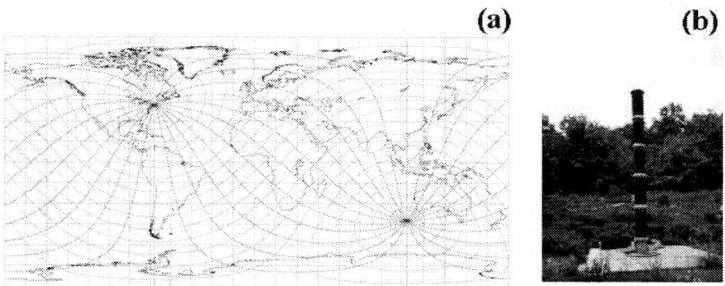

Figure 2. The geometry for global lightning location from the MIT station in Rhode Island (lines are great circle paths and isolines for distance from the receiver in Rhode Island), (b) The vertical electric field E_z antenna at the MIT station in Rhode Island.

crossed-loop methods (Huang et al., 1999). The ambiguity about direction of the lightning source along the great circle path is resolved by computing the direction of the Poynting flux, $E \times H$. Well-recognized (but poorly understood) site error corrections in magnetic bearing are implemented based on comparisons with a large number of ground flashes recorded by the National Lightning Detection Network (with $\sim$1 km location error) in North America, and by comparisons with events detected optically from space (Boccippio et al., 1998).

The distance θ to the source is determined by appeal to the wave impedance, a method suggested by J. R. Wait, implemented initially by D. L. Jones and his colleagues (Kemp, 1971; Jones and Kemp, 1971) and exploited more recently in numerous studies enabled by modern digital signal processing methods (Burke and Jones, 1995, 1996; Cummer and Inan, 1997; Huang et al., 1999; Hu et al., 2002; Hobara et al., 2001).

Equations (10.1) and (10.2) each involve a frequency-dependent current moment for the lightning source, but it is identical for the electric and magnetic field. The wave impedance is defined as the quotient of these fields

$$Z(f) = \frac{E_Z(f)}{H_\varphi(f)} = -i \frac{\nu(\nu+1) P_\nu^0(-\cos\theta)}{a\varepsilon_o 2\pi f P_\nu^1(-\cos\theta)} \text{ [Ohm]} \qquad (10.4)$$

and so the source property $I(f)ds$ is eliminated by this division. For a waveguide with globally uniform ionospheric properties, i.e., $\nu(f)$, $Z(f)$ is now a unique function of the great circle distance θ. This uniqueness enables a determination of θ. $Z(f)$ is an oscillatory function of frequency, with a frequency of oscillation that decreases monotonically with distance from 0 to 20 Mm, receiver to antipode (Bliokh et al., 1980). Figure 3 shows the amplitude of the wave impedance as a function of frequency for a source that is 8 Mm from the receiver. Numerous algorithms have been devised (Kemp, 1971; Bliokh et al., 1980; Nickolaenko and Kudintseva, 1994; Huang et al., 1999) for extracting distance θ from specific features of $Z(f)$ with Equation (10.4). In practice, the error in determining distance by these methods is in the range of 0.5 Mm (Boccippio et al., 1998).

Given the great circle bearing and the distance on that great circle path, one can easily compute the latitude and longitude of the lightning source using basic spherical trigonometry.

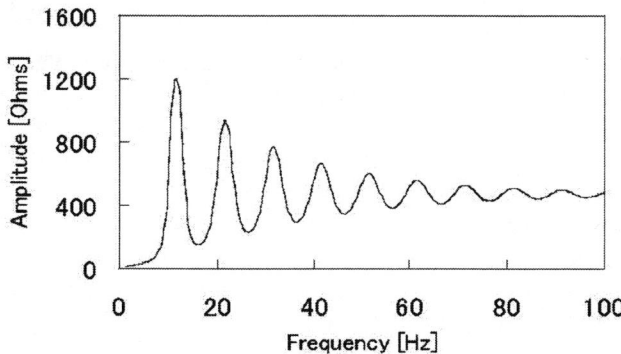

Figure 3. Amplitude of the wave impedance for a 8 Mm distant source.

Given the location of the flash and the measured (calibrated) spectra for the electric and magnetic field, $E_z(f)$ and $H_\varphi(f)$, one can use either Equation (10.1) or Equation (10.2) to extract the source current moment $I(f)ds$. A simplification of the source term arises when the characteristic duration of the lightning flash is small in comparison to the propagation of light around the world (120-140 ms) (Sentman, 1996). (This is the typical condition for lightning, although the long continuing current in extreme events may violate this condition). In this so-called "impulse approximation" of short-duration lightning, it is easy to show that the vertical current moment in the frequency domain reduces to a simple charge moment Qds(C·km). This quantity is particularly convenient because it is the direct measure of lightning's tendency to induce dielectric breakdown in the mesosphere (Pasko et al., 1995; Huang et al., 1999; Williams, 2001) and thereby initiate a sprite, following the ideas introduced by Wilson (1925).

Wilson's concept is rather easily illustrated in graphical form (Figure 4). The dielectric strength of the atmosphere is proportional to air density and declines exponentially with altitude. In contrast, the electric field from the vertical dipole change of a cloud-to-ground lightning discharge declines as the reciprocal cube of the altitude. This difference in altitude dependences guarantees that at some altitude the field increase due to the lightning discharge will exceed the dielectric strength of the atmosphere and thereby initiate a sprite. Ordinary cloud-to-ground discharges in convective scale (~10 km) summer thunderclouds have charge moments too small (<100 C·km) to initiate sprites in a range of altitude where the atmosphere can be considered a dielectric. A charge moment change of 750 C·km is sufficient to initiate breakdown at 75 km, a realistic height for sprite initiation. This simple electrostatic treatment does not fully take into account the electrodynamic process involved, but does provide a rough estimate for sprite initiation.

10.2.2   Global Map of Lightning by Rhode Island Station

In the case of the Rhode Island station, all procedures for global location and source characteristic of lightning flashes are automated, following extensive manual analysis of individual events. Calculations have been made for tens of thousands of flashes, thereby enabling the plotting of detailed global maps. Figure 5a shows a map of all flashes registering positive polarity for a five year period and with computed charge moments exceeding 500 C·km, ~33% less than the sprite threshold for initiation at 75 km altitude (~750 C·km; Williams, 2001). The correlation coefficient between the theoretical (Equation (10.4)) and measured wave impedance spectrum is greater than 0.8 for these events – evidence that they are reasonably well located. The three tropical continental 'chimney' regions are apparent with Africa clearly dominant. The African

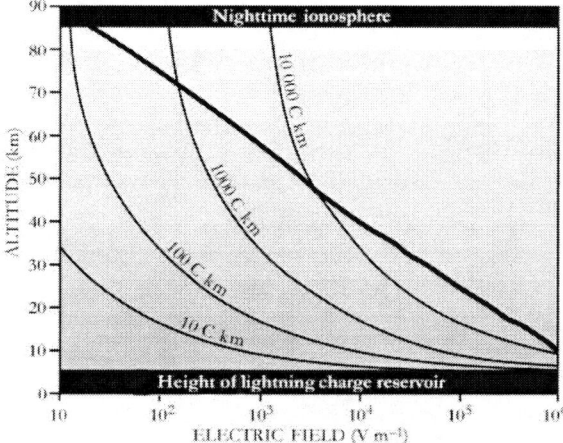

Figure 4. Physical conditions between conductive Earth and conductive ionosphere necessary for dielectric breakdown and the initiation of sprites, as predicted by Wilson (1925). The bold curve represents the density-dependent dielectric strength of air, while the curved lines represent the electric field strength imposed by lightning for various values of lightning charge moment (Williams, 2001). Reprinted with permission from Physics Today, 54(11), 2001, p. 41. Copyright 2001, American Institute of Physics.

dominance of global lightning has also been documented recently by Williams and Sátori (2004). Figure 5b shows the same map but for events with negative polarity. Overall, the negative events are reduced in number relative to the positives. The negative events also tend to spill out over the ocean to a greater extent. Indeed, the Maritime Continent appears to be the most abundant of the three 'chimney' regions in the population of negative flashes. The relative prevalence of large negative events over oceans is consistent with independent documentation by (Füllekrug et al., 2002). Figures 6a and 6b show similar maps for positive and negative polarity events, respectively, but now with charge moments greater than 1000 C·km, and therefore well above the value needed for sprites. This map is strong evidence that sprites, like lightning in general, are dominated by continental storms. Africa shows clear dominance over all other regions, where deep MCSs are also in greatest abundance (Toracinta and Zipser, 2001). The Great Plains of the U.S. and the Argentinian region of South America also stand out for similar meteorological reasons. The Amazon Basin of South America is relatively sparsely populated in comparison, again consistent with the relative prevalence of large and energetic MCSs in the two regions. Again the Maritime Continent appears to harbor the largest population of large negative flashes, with the South Pacific also well represented.

10.3 Winter TLEs and Associated Electromagnetic Phenomena in Japan

10.3.1 Winter Thunderstorm Activity and TLEs in the Hokuriku Region

Lightning characteristics in wintertime in the Hokuriku area exhibit some similarities and differences with those in summer continental storms (such as MCSs in the USA) and the summer storms in Japan (Brook et al., 1982; Takeuti and Nakano, 1983; Michimoto, 1993). Recent reports show some similarity to

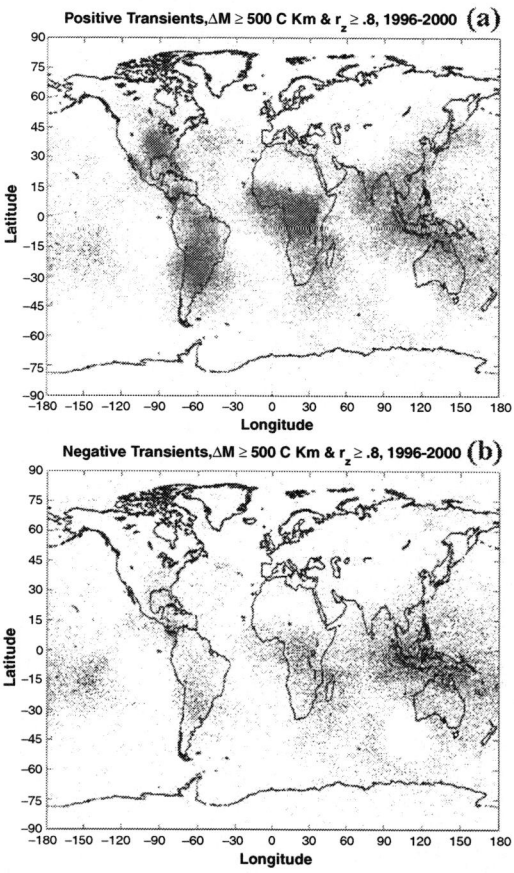

Figure 5. (a) A map of all flashes registering positive polarity for a five year period and with computed charge moment exceeding 500 C·km, (b) Same map as (a) but for the negative. The empty circular zone southwest of Australia is centered on the antipode for the Rhode Island receiver (see Figure 2a). The light ring of sources surrounding the antipode is an artifact.

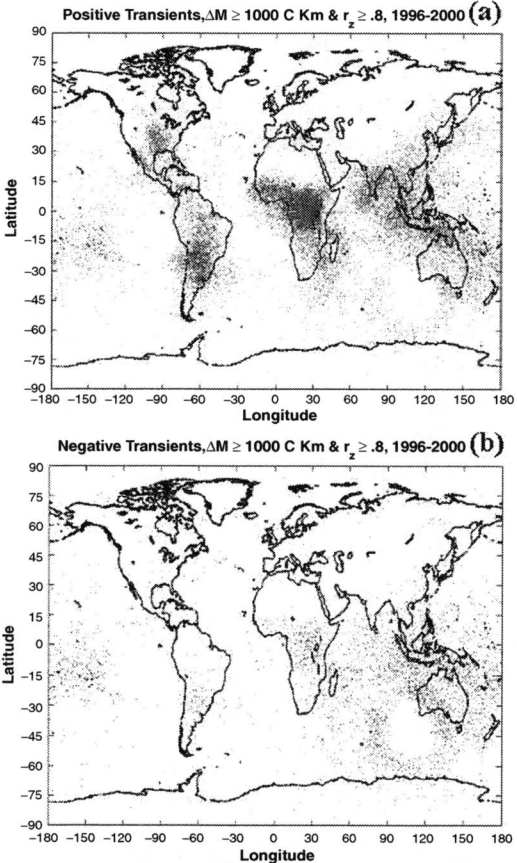

Figure 6. (a) A map of all flashes registering positive polarity for a five year period and with computed charge moments exceeding 1000 C·km, (b) Same map as (a) but for the negative polarity.

the summertime picture in Mesoscale Convective Systems (e.g., Lyons et al., 2003) – a large area of deep convection (Zipser, 1982) with attendant stratiform regions. For example, Saito et al. (2003) demonstrate the location of the positive charge region in winter thunderstorm clouds being close to the 0°C isotherm, a common feature of the summertime MCS stratiform regions. More detailed meteorological issues are discussed in the accompanying chapters by Lyons and by Williams and Yair, Chapters 2 and 3, respectively. Some of the typical electrical characteristics of the winter thunderstorm being different from ordinary summer thunderstorms are (1) a relatively high occurrence rate of positive cloud to ground discharges, (2) frequent upward flashes, (3) very large, slowly varying current leading to large charge transfer to ground. These characteristics may be attributed to the proximity of the cloud charge to the ground and the relatively large horizontal extent of the charge (e.g., Chapter 8 in Rakov and Uman, 2003). The distribution of peak current for the Hokuriku winter lightning for three days in Figure 7b clearly indicates the characteristics (1) because the distribution of lightning peak current (positive and negative) shows a rough balance of positive and negative flashes as was also shown in early results (Brook et al., 1982; Takeuti and Nakano, 1983). The implication of characteristic (3) for charge transfer is observed during our campaigns for two different years (Hobara et al., 2001; Hayakawa et al., 2004b). Only the vertical charge moment change Qds, and not the transferred charge, Q, is available from ELF observation since we do not know the exact height of the charge concentration lowered to ground. Although estimates of Qds range from $\sim$200 C·km to >500 C·km (Figure 12 and Figure 11b), which corresponds to rather large charge transfer Q from 40 C to 100 C even with a charge height ds=5 km.

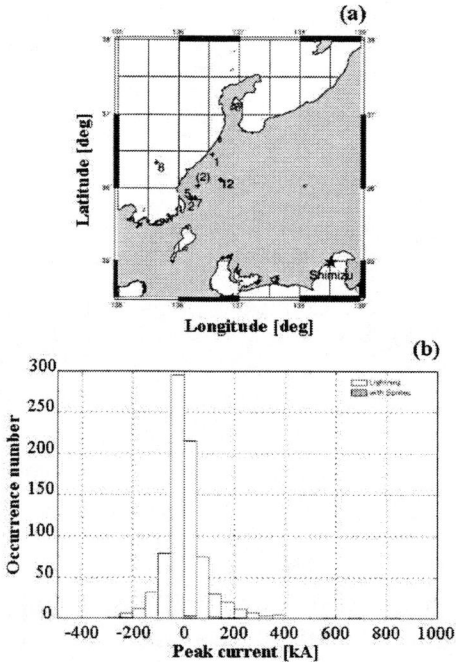

Figure 7. (a) Locations of sprite-producing cloud to ground discharges on two different days (Dec. 14, 2001 and Jan. 28, 2002), (b) Distribution of peak currents in Hokuriku winter lightning, shading refers to the lightning associated with sprites (Hayakawa et al., 2004b).

The locations of observed TLEs during winter thunderstorm activity on two different days (Dec 14, 2001 and Jan 28, 2002) are illustrated in Figure 7a. These locations are based on the onset place of the parent positive ground flashes for TLEs based on the Japan Lightning Detection Network (JLDN), by assuming that TLEs occurred directly above the causative ground flash. As seen from Figure 7, most TLE-producing ground flash locations are in the coastal area of the Sea of Japan where winter thunderstorm activity is significant (Kitagawa and Michimoto, 1994). A similar spatial distribution of ground flash locations was obtained in the 1998/1999 winter sprite campaign (Hobara et al., 2001). Figure 8 show two examples of TLE (sprite) images that were captured by the team from the University of Electro-Communications during the 2001/2002 sprite winter campaign. The group of column sprites in Figure 8a and a carrot-type sprite in Figure 8b are from our optical observation site in Shimizu (Figure 7a). Among 12 sprites observed during the campaign in 2001/2002, 11 sprites have the column shape. The difference in shape (column or carrot-type) may be related to the charge moment change of the causative lightning discharge. However an important statement here is that the morphological characteristics of sprites are not exceptional in comparison with sprites observed in U.S. campaigns (e.g., Sentman and Wescott, 1995; Lyons, 1996; Williams, 2001); instead they have simpler shapes. The altitude distribution of the sprites is found to be in the range from 50 km to 90 km.

Figure 8. Two examples of sprites caused by Hokuriku winter lightning observed during 2001/2002. (a) A group of column sprites, (b) Carrot-type sprite surrounded by a couple of columnar sprites; streaks are reflected light from power transmission lines near the observation site (Hayakawa et al., 2004b).

10.3.2 ELF Radiation Associated with the Hokuriku TLEs and the Generation Condition of Winter Sprites

The majority of lightning discharges associated with winter TLEs in Japan produce ELF radiation observed as ELF transients just as in the USA. Two typical examples of time series of observed TLE-associated ELF transients at the Moshiri field site (44.2°N, 142.2°E) during the 1998/1999 sprite winter campaign are shown in Figures 9a and b. The observation system for these transients (Hobara et al., 2000a) is similar to the one in Rhode Island USA (Section 10.2). The system records three field components (E_z, H_{ns}, H_{ew}) in the frequency range from 1 to 800 Hz with a sampling frequency of 2 kHz. An elve-producing ELF transient in Figure 9a has large peak amplitude with short decay time, which is more preferable for the EMP (electromagnetic pulse) heating of electrons and ionizing of the lower ionosphere (Inan et al., 1991, 1997), whereas the sprite-related ELF transient in Figure 9b has a smaller amplitude with long decay time, suggesting rather large charge transfer due to long-lasting continuing current. These tendencies are in good agreement with the previously reported results of Huang et al. (1999) for a summertime Mesoscale Convective Complex, with a white-noise-like spectrum for elves and a red-noise-like spectrum for sprites.

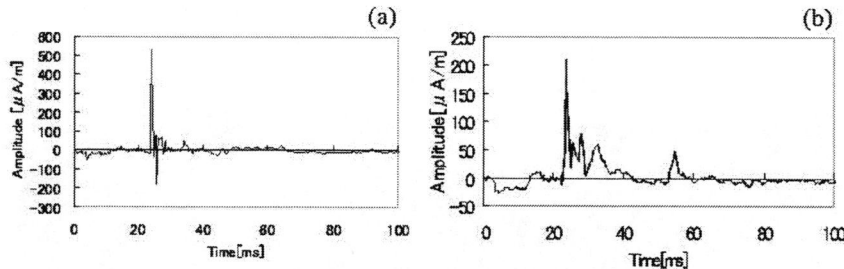

Figure 9. (a) Time series for the ELF waveform for one of the magnetic field components associated with an elve, (b) The same variation for a carrot-type sprite during 1998/1999 sprite winter campaign (Hobara et al., 2003).

The search for the essential conditions for sprite generation with winter lightning in the Hokuriku region is interesting because sprite-producing lightning discharges in summer continental conditions, and in Hokuriku have different characteristics (see Subsection 10.3.1 of this chapter). In this section, two different electrical and meteorological features are given attention, (1) spatial scale of the thundercloud, (2) the polarity of peak current and the charge moment Qds.

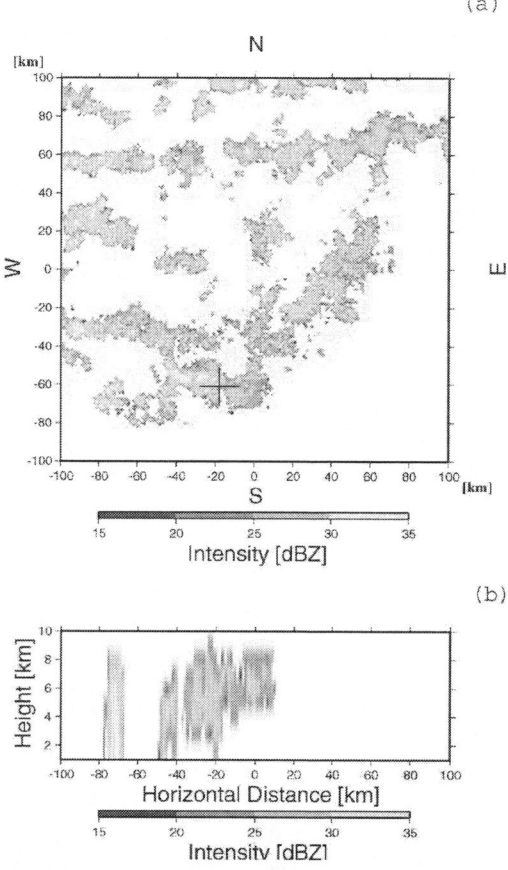

Figure 10. (a) Radar CAPPI at a height of 5 km at the time around the carrot-type sprite. (b) Vertical cross section of radar image along the East-West plane crossing the position (+) of the lightning discharge leading to that sprite (Hayakawa et al., 2004b).

Firstly, the typical Hokuriku winter thundercloud is not a large MCS (nor is it as compact as a conventional summer thundercloud). An example of the spatial scale of a sprite-producing winter thunderstorm morphology is shown in Figure 10 to examine the condition (1). In the radar CAPPI (Constant Altitude Plan Position Indicator) at the height of 5 km (Figure 10a), the plus sign indicates the position of a positive ground flash associated with a carrot sprite. The horizontal scale of the sprite-producing cloud from the CAPPI is 20-30 km, which is considerably smaller than the scale of MCS defined by Zipser (1982) and (Mohr and Zipser, 1996), but it is still greater in area by one order of that of a typical isolated thundercloud (MacGorman and Rust, 1998;

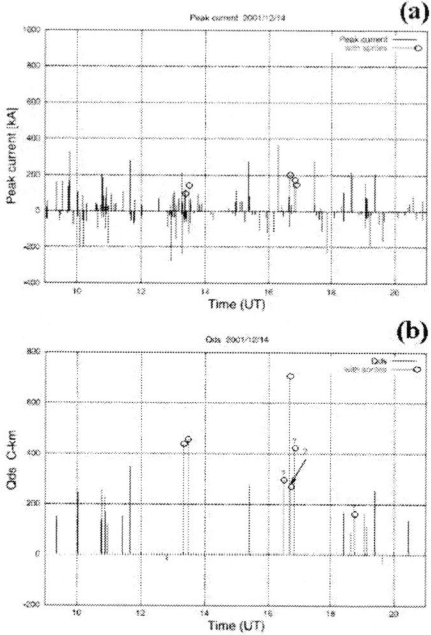

Figure 11. (a) Temporal dependence of the lightning information: (a) peak current and (b) charge moment (Qds) on 14 December 2001 (Hayakawa et al., 2004b).

Williams, 1998). The vertical extent of the sprite-producing cloud forms a flat-topped shape but only from 2 up to 6 km (Figure 10b). It is considerably lower than that of MCS (typically 10-14 km deep stratiform region). It is important to note however that the vertical extent of the storm in the mixed phase region (0°C to -40°C) is not very different from that in summertime MCSs.

Secondly, positive peak current (positive ground flashes) is observed for Hokuriku sprites (Figure 11a) and there is no significant difference in the threshold value of Q_{ds} for sprite-producing lightning between Hokuriku and U.S. sprites. The threshold value of Q_{ds} for Hokuriku sprite generation from our two different winter seasons is around 200 C·km (Hobara et al., 2001; Hayakawa et al., 2004b), which is in the range of values for the sprites in summer continental conditions in the USA, ranging from 200-400 C·km (Huang et al., 1999; Hu et al., 2002). One problem with making a detailed quantitative comparison is the lack of good quantitative estimates of the initiation heights for sprites in the Hokuriku winter storm context.

Other important considerations for winter sprite generation in addition to the Q_{ds} threshold and positive ground flash are important. They can be related to the contiguity of charge among the thundercloud cells by the self-organization of the winter thundercloud because the observed thundercloud cells form a group, connecting with each other, and extend horizontally as a spider-like structure. This structure may increase the macroscopic quantity Qds (Hayakawa et al., 2005). A generation model for the red sprite by the electromagnetic pulse from a horizontal fractal lightning discharge is proposed by Valdivia et al. (1997). It is also possible that the accumulation of positive charge at the altitude of the 0°C isotherm inferred from the observation by Saito et al. (2003) may be important for the sprite-producing positive flash.

10.3.3 Atmosphere-Mesosphere-Ionosphere Coupling

The electrodynamic coupling between tropospheric lightning and the mesosphere and overlying lower ionosphere has been studied by combined results from coordinated measurements in the sprite winter campaign. We found strong correlations between several pairs of physical parameters indicating that significant coupling exists between different regions. During the campaign in 1998/1999, optical observations were carried out by Tohoku University, in parallel with measurements of ELF and VLF scattering measurements to estimate the ionospheric perturbation (in electron temperature and density) by the VLF scattered amplitude. A more detailed explanation of VLF Trimpi measurements can be found in Chapter 8 by Rodger and McCormick. These measurements (VLF and ELF) were organized by the University of Electro-Communications (Hobara et al., 2000b, 2001; Hayakawa et al., 2004a). In addition, information on positive lightning discharges (peak current, polarity and locations) has been obtained from the Japan Lightning Detection Network.

Pairs of different parameters are examined for winter TLE-producing lightning to search for the coupling agents between tropospheric ground flashes, mesospheric TLEs, and the lower ionosphere (namely, peak current and polarity, type of TLEs (sprite or elves), ELF transient peak amplitude, VLF scattered amplitude, and vertical charge moment change from the ELF transient).

The most important pair with high correlation coefficient (r >0.9) is the VLF scattered amplitude as a function of the vertical charge moment change for the December 19, 1998, events (A events in Figure 12). The positive correlation (with $r=0.97$) for A events in Figure 12 clearly indicates that the QE (quasi electrostatic (QE)) stressing due to the sudden removal of positive charge characterized by large Qds by the TLE-producing ground flash significantly affects the ionosphere perturbation. Among observed TLEs, sprites have larger Qds than those for elves and have much larger ionospheric perturbations than those for elves because of larger-scale ionization. On the contrary, two elves in A events in Figure 12 have significantly larger peak current with positive polarity than the observed sprite-producing discharges, thereby satisfying the general condition of elve generation by EMP (Inan et al., 1997). Thus relatively large Qds in tropospheric lightning related to sprite generation in the mesosphere is the main determinant of the amplitude of the lower ionospheric perturbation. On the contrary, on January 27, 1999, (B events in Figure 12) sprite events do not show a clear linkage to the ionospheric perturbation ($r \sim 0.28$). The quantitative discrepancy in r for two different days is not very clear but presumably is due to the differences in the ambient conductivity of the D-region produced by QE thundercloud fields during the storm activity (e.g. changing thunderstorm locations and strength of activity). Moreover, one unsolved problem is how effective the sprite ionization columns in the mesosphere are as reflectors of the VLF energy.

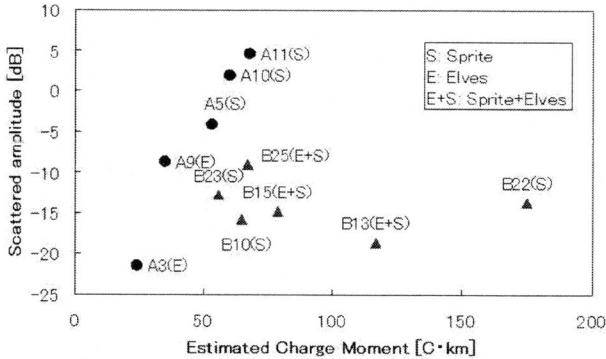

Figure 12. VLF scattered amplitude (indicating the ionospheric perturbation) as a function of the vertical charge moment change from ELF transient measurement. (Hobara et al., 2003) Dots (A events) for the events on December 19, 1998, and triangles (B events) for events on January 27, 1999.

10.4 Conclusion

ELF methods to locate the energetic TLE-producing lightning discharges by receiving ELF transients at a single field site have been presented. The associated electrical properties such as polarity, peak current and vertical charge moment change have been deduced for each lightning discharge.

Global maps of energetic lightning using the ELF method based on a five-year record at Rhode Island USA clearly indicate three tropical continental 'chimney' regions. Among the regions, Africa has the strongest activity on average. The map with charge moment change (>1000 C·km) indicates the fact that sprites are dominated by continental storms consistent with the distribution of large and energetic MCSs. In contrast, a large population of negative discharges appears over the world's oceans and the Maritime Continent.

The ELF method was also used to deduce local thunderstorm electrical properties during the winter lightning campaign in Japan. The campaigns in two different years provided the following results. (1) A rather strong coupling between tropospheric lightning, TLEs and the overlying ionosphere is observed. (2) The generation conditions for winter sprites are not very different from those in the continental U.S. ($Qds > 200 - 300$ C·km, +CG) in spite of the rather small cloud spatial scale. Hence another condition such as the clustering or self-organizing effect of the thunderclouds storing large amount of positive charges may be important.

Acknowledgments

We are indebted to E. Huang for key assistance in the processing and organization of transient events detected from Rhode Island. We are also thankful to T. Nagao of Tokai University for the optical field site and to K. Ohta of Chubu University for VLF measurements.

Bibliography

Bliokh, P. V., Nickolaenko, A. P., and Filippov, Yu. F. (1980). *Schumann Resonances in the Earth-ionosphere Cavity*. Peter Peregrinus, Oxford.

Boccippio, D. J., Williams, E. R., Heckman, S. J., Lyons, W. A., Baker, I. T., and Boldi, R. (1995). Sprites, ELF transients, and positive ground strokes. *Science*, 269:1088–1091.

Boccippio, D. J., Wong, C., Williams, E. R., Boldi, R., Christian, H. J., and Goodman, S. J. (1998). Global validation of single-station Schumann resonance lightning location. *J. Atmos. Sol.-Terr. Phys.*, 60:701–712.

Brook, M., Nakano, M., Krehbiel, P., and Takeuti, T. (1982). The electrical structure of the Hokuriku winter thunderstorms. *J. Geophys. Res.*, 87:1207–1215.

Burke, C. P. and Jones, D. L. (1995). Global radiolocation in the lower ELF frequency band. *J. Geophys. Res.*, 100:26263–26272.

Burke, C. P. and Jones, D. Llanwyn (1996). On the polarity and continuing currents in unusually large lightning flashes deduced from ELF events. *J. Atmos. Terr. Phys.*, 58:531–540.

Cummer, S. A. and Inan, U. S. (1997). Measurement of charge transfer in sprite-producing lightning using ELF radio atmospherics. *Geophys. Res. Lett.*, 24:1731–1734.

Fukunishi, H., Takahashi, Y., Kubota, M., Sakanoi, K., Inan, U. S., and Lyons, W. A. (1996). Elves: Lightning-induced transient luminous events in the lower ionosphere. *Geophys. Res. Lett.*, 23:2157–2160.

Füllekrug, M. and Constable, S. (2000). Global triangulation of lightning discharges. *Geophys. Res. Lett.*, 27:333–336.

Füllekrug, M., Price, C., Yair, Y., and Williams, E. R. (2002). Intense oceanic lightning. *Ann. Geophys.*, 20:133–137.

Hayakawa, M., Molchanov, O. A., and team, NASDA/UEC (2004a). Summary report of NASDA's earthquake remote sensing frontier project. *Phys. Chem. Earth*, 29:617–625.

Hayakawa, M., Nakamura, T., Hobara, Y., and Williams, E. (2004b). Observation of sprites over the Sea of Japan and conditions for lightning-induced sprites in winter. *J. Geophys. Res.*, 109(A0):doi:10.1029/2003JA009905.

Hayakawa, M., Nakamura, T., Iudin, D., Michimoto, K., Suzuki, T., Harada, T., and Shimura, T. (2005). On the fine structure of thunderstorms leading to the generation of sprites and elves: Fractal analysis. *J. Geophys. Res.*, 110(D6):doi:10.1029/2004JD004545.

Hayakawa, M. and Nickolaenko, A. P. (2001). Lightning effects in the mesosphere and associated ELF radio signals. *Proc. Indian Nat. Science Academy*, 67A(4-5):509–529.

Hobara, Y., Hayakawa, M., Ohta, K., and Fukunishi, H. (2003). Lightning discharges in association with mesospheric optical phenomena in Japan and their effect on the lower ionosphere. *Adv. Pol. Upp. Atmos. Res.*, 17:30–47.

Hobara, Y., Iwasaki, N., Hayashida, T., Hayakawa, M., Ohta, K., and Fukunishi, H. (2001). Interrelation between ELF transients and ionospheric disturbances in association with sprites and elves. *Geophys. Res. Lett.*, 28:935–938.

Hobara, Y., Iwasaki, N., Hayashida, T., Tsuchiya, N., Williams, E. R., Sera, M., Ikegami, Y., and Hayakawa, M (2000a). New ELF observation site in Moshiri, Hokkaido Japan and the results of preliminary data analysis. *J. Atmos. Electr.*, 20:99–109.

Hobara, Y., Watanabe, H., Yamaguchi, T., Akinaga, Y., Koons, H. C., Roeder, J. L., and Hayakawa, M. (2000b). Wide-band ULF/ELF magnetic field measurement in Seikoshi, Izu Japan and some results from preliminary data analysis in relation with seismic activity. *J. Atmos. Electr.*, 20:111–121.

Hu, W., Cummer, S. A., Lyons, W. A., and Nelson, T. E. (2002). Lightning charge moment changes for the initiation of sprites. *Geophys. Res. Lett.*, 29(8):doi:10.1029/2001GL014593.

Huang, E., Williams, E., Boldi, R., Heckman, S., Lyons, W., Taylor, M., Nelson, T., and Wong, C. (1999). Criteria for sprites and elves based on Schumann resonance observations. *J. Geophys. Res.*, 104:16943–16964.

Inan, U. S., Barrington-Leigh, C., Hansen, S., Glukhov, V. S., Bell, T. F., and Rairden, R. (1997). Rapid lateral expansion of optical luminosity in lightning-induced ionospheric flashes referred to as 'elves'. *Geophys. Res. Lett.*, 24:583–586.

Inan, U. S., Bell, T. F., and Rodriguez, J. V. (1991). Heating and ionization of the lower ionosphere by lightning. *Geophys. Res. Lett.*, 18:705–708.

Ishaq, M. and Jones, D. L. (1977). Methods of obtaining radiowave propagation parameters for the Earth-ionosphere duct at ELF. *Electron. Lett.*, 13:254–255.

Jones, D. L. (1999). ELF sferics and lightning effects on the middle and upper atmosphere. In Stuchly, M.A., editor, *Modern Radio Science*. Wiley.

Jones, D. L. and Kemp, D. T. (1971). The nature and average magnitude of the sources of transient excitation of the Schumann resonances. *J. Atmos. Terr. Phys.*, 33:557–566.

Kemp, D.T. (1971). The global radiolocation of large lightning discharges from single station observations of ELF disturbances in the Earth-ionosphere waveguide. *J. Atmos. Terr. Phys.*, 33:919–927.

Kitagawa, N. and Michimoto, K. (1994). Meteorological and electrical aspects of winter thunderclouds. *J. Geophys. Res.*, 99:10713–10722.

Lyons, W. A., Nelson, T. E., Williams, E. R., Cummer, S. A., and Stanley, M. A. (2003). Characteristics of sprite-producing positive cloud-to-ground lightning during the 19 July 2000 STEPS mesoscale convective systems. *Mon. Wea. Rev.*, 131:2417–2427.

Lyons, Walter A. (1996). Sprite observations above the U.S. High Plains in relation to their parent thunderstorm systems. *J. Geophys. Res.*, 101(D23):29641–29652.

MacGorman, D. R. and Rust, W. D. (1998). *The Electrical Nature of Storms*. Oxford Univ. Press, New York.

Madden, T. R. and Thompson, W. (1965). Low frequency electromagnetic oscillations of the earth-ionosphere cavity. *Rev. Geophys.*, 3:211–254.

Michimoto, K. (1993). A study of radar echos and their relation to lightning discharges of thunderclouds in the Hokuriku district. *J. Meteor. Soc. Japan*, 71:195–204.

Mohr, K. I. and Zipser, E. J. (1996). Mesoscale convective systems defined by the 85 GHz ice scattering signature: Size and intensity comparison over tropical oceans and continents. *Mon. Wea. Rev.*, 24:2417–2437.

Mushtak, V. C. and Williams, E. R. (2002). ELF propagation parameters for uniform models of the Earth-ionosphere waveguide. *J. Atmos. Sol.-Terr. Phys.*, 64:1989–2001.

Nickolaenko, A. P. and Hayakawa, M. (2002). *Resonances in the Earth-Ionosphere Cavity*. Kluwer Acad., Norwell, Mass.

Nickolaenko, A. P., Hayakawa, M., and Hobara, Y. (1996). Temporal variations of the global lightning activity deduced from the Schumann resonance data. *J. Atmos. Terr. Phys.*, 58:1699–1709.

Nickolaenko, A. P. and Kudintseva, I. G. (1994). A modified technique to locate the sources of ELF transients. *J. Atmos. Terr. Phys.*, 56:1493–1498.

Pasko, V. P., Inan, U. S., Taranenko, Y. N., and Bell, T. F. (1995). Heating, ionization and upward discharges in the mesosphere due to intense quasi-electrostatic thundercloud fields. *Geophys. Res. Lett.*, 22:365–368.

Rakov, V. A. and Uman, M. A. (2003). *Lightning: Physics and Effects*. Cambridge University Press.

Rust, W. D., MacGorman, D. R., and Arnold, R. T. (1981). Positive cloud-to-ground lightning flashes in severe storms. *Geophys. Res. Lett.*, 8:791–794.

Saito, M., Ishii, M., Hojo, J., Sugita, A., Idogawa, T., and Kotani, K. (2003). Development of lightning discharge observed by VHF radiation. In *Joint Technical Meeting on Electrical Discharges, Switching and High Voltage*, Okinawa. IEE Japan, HV-03-90. (in Japanese).

Sentman, D. D. (1995). Schumann resonances. In Volland, H., editor, *Handbook of Atmospheric Electrodynamics*, volume 1, pages 267–298. CRC Press Inc.

Sentman, D. D. (1996). Schumann resonance spectra in a two-scale-height Earth-ionosphere cavity. *J. Geophys. Res.*, 101:9479–9488.

Sentman, D. D. and Wescott, E. M. (1995). Red sprites and blue jets: Thunderstorm-excited optical emissions in the stratosphere, mesosphere, and ionosphere. *Phys. Plasmas*, 2:2514–2522.

Takeuti, T. and Nakano, M. (1983). Study on winter lightning activity in Hokuriku. *Tenki*, 30:13–18. (in Japanese).

Toracinta, E. R. and Zipser, E. J. (2001). Ice-scattering mesoscale convective systems in the global tropics. *J. Appl. Meteor.*, 40:983–1002.

Valdivia, J. A., Milikh, G., and Papadopoulos, K. (1997). Red sprites: Lightning as a fractal antenna. *Geophys. Res. Lett.*, 24:3169–3172.

Wait, J. R. (1996). *Electromagnetic Waves in Stratified Media.* IEEE Press, Piscataway, N.J.

Williams, E. R. (1998). The positive charge reservoir for sprite-producing lightning. *J. Atmos. Sol.-Terr. Phys.*, 60:689–692.

Williams, E. R. (2001). Sprites, elves and glow discharge tubes. *Phys. Today*, 54(11):41–47.

Williams, E. R. and Sátori, G. (2004). Lightning, thermodynamic and hydrological comparison of two tropical continental chimneys. *J. Atmos. Sol.-Terr. Phys.*, 66:1213–1231.

Wilson, C. T. R. (1925). The electric field of a thundercloud and some of its effects. *Proc. Roy. Soc. Lond.*, 37(32D).

Zipser, E. J. (1982). Use of a conceptual model of the life-cycle of mesoscale convective systems to improve very-short-range forecasts. In Browning, K., editor, *Nowcasting*, pages 191–204. Academic Press, London and New York.

CALIBRATED RADIANCE MEASUREMENTS WITH AN AIR-FILLED GLOW DISCHARGE TUBE: APPLICATION TO SPRITES IN THE MESOSPHERE

Earle Williams [1], M. Valente [2], E. Gerken [3] and R. Golka [4]

[1] *Massachusetts Institute of Technology, Cambridge, MA 02139, USA.*

[2] *East Coast Induction, Inc., Brockton, MA.*

[3] *SRI International, Menlo Park, CA.*

[4] *Golka Associates, Brockton, MA.*

Abstract Quantitative observations of a large air-filled glow discharge tube are used to shed light on the behavior of sprites in the mesosphere. The optical spectrum in the (red) positive column is shown to resemble the spectrum of the sprite body. Experimental estimates of the electrical conductivity of the positive column are reasonably consistent with theoretical estimates for the conductivity of sprites. The relationship between absolute radiance and current density in the tube are used to interpret the current flow in sprites, under the working assumption that laboratory "DC" excitation is appropriate for time-dependent sprite excitation. In general, the current inferred for sprites is modest in comparison to the current of the parent lightning flash.

11.1 Introduction

The light resulting from the flow of current through partially evacuated tubes is known as 'glow discharge'. The study of glow-type discharges in gases at sub-atmospheric pressure began in earnest with the development of vacuum pumps, in the 19th century (Harvey, 1957). The field of spectroscopy also developed rapidly in this period. A century later, it is now recognized that sprites in the mesosphere are electrodeless glow discharges at a grander scale (Dowden et al., 1996; Williams, 2001), and that much of what we now know about sprite spectroscopy (Mende et al., 1995; Hampton et al., 1996) was documented in air-filled laboratory discharge tubes nearly a century before sprites gained major recognition (Pearse and Gaydon, 1963). This study is concerned with a return to glow discharges at the laboratory scale to gain further quantitative insight about the relationship between sprite radiance and the current density characterizing the luminous sprite.

The traditional air-filled glow discharge tube in the laboratory, under DC voltage excitation, is illustrated in Figure 1. Typical internal pressures are 0.01-1 torr. The region near the (negative) cathode is blue in color, and is often called the negative glow. Its spectrum is characterized by so-called 'nitrogen first negative' emission (the first in a sequence of spectroscopic signatures observed in the negative end of the glow discharge tube), now linked with the ionized species N_2^+ (Pearse and Gaydon, 1963). The lower tendrils of sprites have been shown to emit this light e.g. (Heavner, 2000). Returning to the discharge tube, the red light is the dominant feature in the remainder of the tube. This emission fills the so-called positive column, extending from the Faraday 'dark space' to the (positive) anode. The name given this emission on the basis of spectroscopic analysis of light from the discharge tube is 'nitrogen first positive' (Pearse and Gaydon, 1963). It is now known that such emission results from the collisions of free electrons in the tube plasma with neutral nitrogen molecules. This emission was identified as the dominant spectral feature in red sprite light by Mende et al. (1995) and by Hampton et al. (1996). The current density in the laboratory tube and in sprites is the key physical parameter in the present study. Table 1 summarizes the current density observed in several earlier investigations. In comparisons with values explored in the present study, the current densities are quite large – of the order of a few amperes per square meter with DC excitation. Still larger current densities (10^3-$10^4 A/m^2$) have been documented in transient capacitive discharge experiments (Naudé, 1932; Elmore, 1973; Goto et al., 2003). In the present experiments, current densities as small as one milliampere per square meter were explored, consistent with the perceived need to replicate quantitative conditions in sprites.

11.2 Methodology

The tube used in the present experiments is substantially larger than the traditional laboratory tube, typically some centimeters in diameter and tens of centimeters in length. This Plexiglas tube (Figure 2) is 6 ft (1.8 m) long and 2 ft (0.61 m) in diameter, with a wall thickness of $^3/_4$" (1.9 cm). The purpose of the larger tube was twofold: (1) to minimize the effects of the tube wall on the plasma it contains, and (2) to minimize the effects of the two electrodes, which are sources for contaminants in the tube. Neither of these laboratory features is present in sprites. The tube itself is mounted on a hydraulic lifter (also shown in Figure 2) to enable easy access to one end of the tube for electrode replacement and for cleaning the inside walls of the tube. Three different power supplies have been used to explore a wide range of current and voltage in the tube. The initial experiments were carried out with a supply capable of

Table 1. Summary of earlier studies on glow discharge in air.

Investigator	Current Density J [A/m^2]
DC Current	
Herz (1895)	2 – 15
Wilson (1902)	0.3 – 3
Townsend (1915)	3
Günter-Schulze (1924)	4.2
Günter-Schulze (1927)	35 – 70
Found and Langmuir 1931	1.5
Capacitive Discharge	
Naudé (1932)	$1 - 2 \cdot 10^4$
Elmore (1974)	$7 \cdot 10^4$
Goto et al. (2003)	$8 \cdot 10^4$
This study	0.004 to 40

∼100 A but were limited in peak voltage to ∼1 kV. A second DC supply with higher voltage capability but lower peak current provided access to the regime most closely duplicating sprite conditions. DC current and voltage were measured with commercial digital meters. The choice of operating pressure for the tube is constrained by the observations of the altitudes where sprites are most prevalent. Figure 3 shows the pressure in the atmosphere as a function of altitude. The relationship is quasi- exponential. The chemical composition of the atmosphere does not change appreciably over this range of altitude. Ground-based observations of sprites indicate initiation near 75 km (Stanley, 2000) and subsequent upward development to the nighttime conductivity ledge near 85-90 km, and downward development to 40-50 km, and occasionally to 35 km. For convenience with our vacuum system, we have typically operated in a pressure range of 40-70 millitorr (0.5-0.9 Pa), corresponding to an altitude range of 65-70 km, also shown in Figure 3. Three vacuum pumps are used to evacuate the tube to desired operating pressure, in a time of several minutes. An adjustable needle valve is used to maintain a quasi-steady state pressure with fresh air replacing air withdrawn by the pumps. No filtering of the air (either aerosol or water vapor) is undertaken.

Reliable quantitative inter-comparisons between the red light in the tube and the red light of sprites require attention to absolute units. The device used in the laboratory for radiance measurements is the PR-650 Colorimeter, manufactured by Photo Research, Inc. This device reports absolute radiance in an

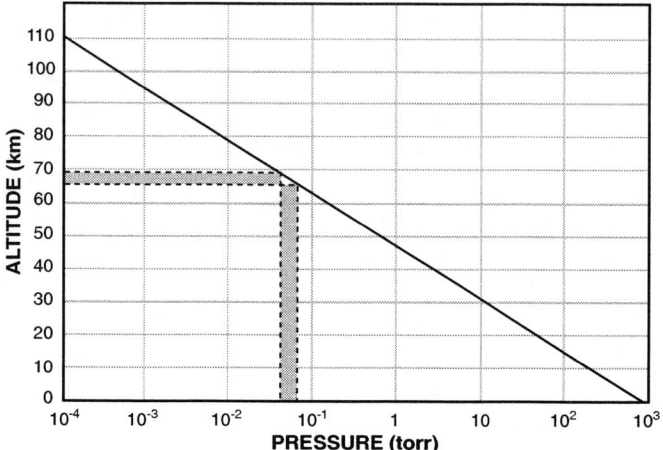

Figure 3. Atmospheric pressure versus altitude to illustrate the operating range of pressure in the laboratory tube in the context of sprites in the mesosphere.

optical band from 380 to 780 nm in units of watts/m^2/sr/nm. Following on the expertise in auroral studies (Chamberlain, 1995, pp. 704), the preferred unit of measurement for sprite radiance is the Rayleigh. The dual use of units requires the conversion from one to the other. The conversion developed here, assuming a bandwidth (BW) for N$_2$1P emission of 50 nm, and with details shown in Eq. 1, gives: radiance in Rayleighs = 2.39 10^{11} L$_\lambda$ where L$_\lambda$ is the radiance in watts/m^2/sr/nm provided by the PR-650 instrument. The Colorimeter provides spectral radiance L$_\lambda$ in absolute units (watts/m^2/sr/nm)

$$\text{Rayleighs}(\text{megaphotons/cm}^2/\sec) =$$

$$= \frac{4\pi(sr/\text{unit sphere})\ BW(nm)\ L_\lambda(watts/m^2/sr/nm)}{hc/\lambda(joule/photon)10^4 cm^2/m^2(10^6 \text{photons/megaphoton})} \qquad (1)$$

$$= (2.39 \times 10^{11})L_\lambda$$

11.3 Optical Spectrum

A demonstration that the spectral characteristics of the light of the positive column of the laboratory tube are similar to the characteristics of the light from sprites is shown in Figure 4. The lab spectrum is taken from the PR-650 Colorimeter and the sprite spectrum is taken from the field study of Hampton et al. (1996). Three broad features, all in the red region of the spectrum, are seen in both spectra – the molecular spectra of nitrogen first positive emission (N$_2$1P). In detail, the spectral shapes differ. The most pronounced difference is in the

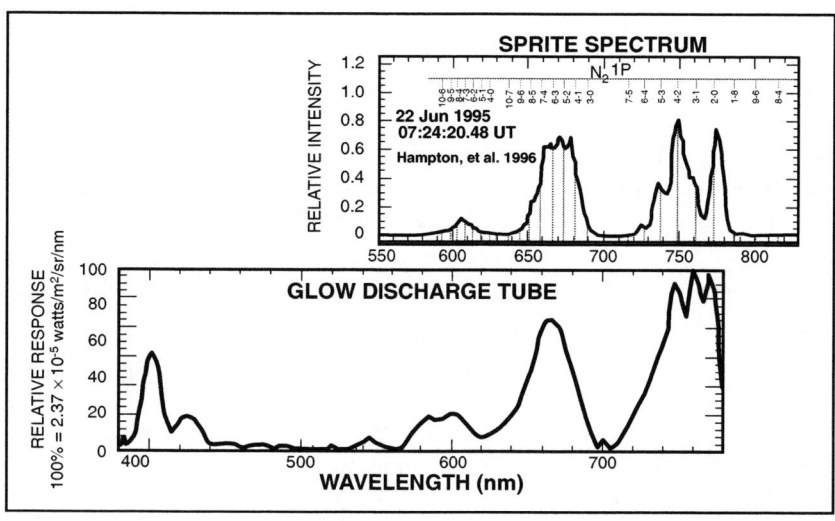

Figure 4. A comparison of optical spectra for (a) a sprite in the mesosphere (Hampton et al., 1996) and for (b) the positive column of the large glow discharge tube. See text for detailed comparisons.

vicinity of the atmospheric O_2 absorption feature centered at 762 nm (Nichols, 1993; Mende et al., 1995). Energy from this region of the $N_2$1P molecular emission is lost from the sprite spectrum due to the long path length through the atmosphere. Mende et al. (1995) also document the dark absorption feature of O_2 in their sprite spectrum. A more subtle difference in the two spectra in Figure 4 is the ~5 nm offset. The reasons for this apparent deviation are unknown, but the minimum at 766 nm in the sprite spectrum in Figure 4 suggests an upward bias. Figures 5 and 6 show examples of the illumination in the tube for conditions of relatively high pressure (1.2 torr (16 Pa)), and at a lower pressure (70 millitorr (0.92 Pa)), more in keeping with the mean altitudes (Figure 3) of sprites. The high-pressure case clearly shows both the negative cathode glow and the positive column, but the latter is somewhat constricted relative to the tube diameter. This constriction increases as the pressure increases, to the point that the discharge is "lightning-thin" (i.e., only a centimeter or so in diameter). Such a trend illustrates the difference between diffuse glow discharges at low pressure and more strongly thermalized and constricted discharges at high pressures. When the pressure is as low as 70 millitorr (Figure 6), variations in radiance along the tube known as striations are often present in discharge tubes (Donahue and Dieke, 1951), but only in certain circumstances are they visible to the eye. This photograph represents a snapshot of the tube, and so shows the striations quite clearly. They are generally not visible in neon sign tubes (of the kind used for advertising), until one examines such tubes at close range. In this mode, the dynamic nature of the positive column becomes apparent. Integration of tube light over periods of tens of seconds was undertaken to obtain radiance measurements with the PR-650 Colorimeter. The role of striations in sprites is presently unknown, though Sentman et al. (1995, 1997) and Gerken and Inan (2002) have noted the presence of 'dark bands' in sprites.

11.4 Radiance Response to Power and Current Density

Figure 7 shows the measured radiance versus input power supplied to the tube. The input power is simply the product of the measured voltage drop along the tube and the steady current flowing in the tube. Since the total voltage drop is measured, these power computations must be viewed as upper bounds for the power dissipation in the positive column, the main region of interest. Since the volume of the tube is approximately 0.5 m^3, the abscissa values can be transformed to units of mean power density by simply doubling the numerical values. Two different power supplies were used to make these measurements – one with high current (<10 amperes) capability (the cluster of data points in the upper right), and one with low current capability. It was the realization that the measured radiances at high current density (of the order Giga Rayleighs) were much too large for sprite applicability that led to the use of lower current density. Overall, despite some scatter, the radiance is proportional to the input power, consistent with earlier results (Cobine, 1958, pp. 606) that the luminous efficiency of gas discharge tubes is a stable quantity, relatively insensitive to other variables. Figure 8 shows the measured radiance in the positive column

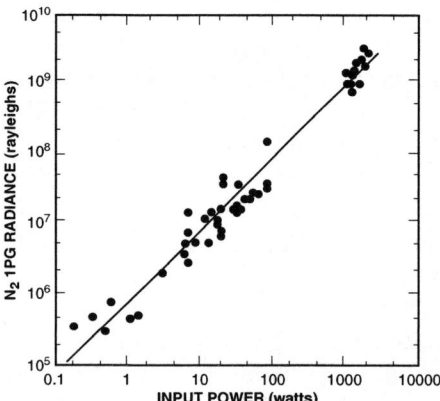

Figure 7. The measured radiance of the positive column of the large glow discharge tube versus the input power, computed as the product of the current and the voltage drop across the tube.

versus the current density flowing in the tube. Here we assumed a uniform current density equal to the measured DC current divided by the cross-sectional area of the tube. Again, a rough proportionality is observed, with somewhat greater scatter. Both current density and total power show proportionality with radiance, because the measured electric field in the tube is only weakly dependent on total current. To be more specific, measurements show that the mean electric field in the tube changes by only a factor of two or so over a two order-of-magnitude variation in current. See also Donahue and Dieke (1951).

A key reference point for sprites in Figure 8 is that a current density of 10 milliamperes per square meter corresponds with a radiance of roughly one MR, a value in the upper range for typical sprites (Gerken et al., 2000).

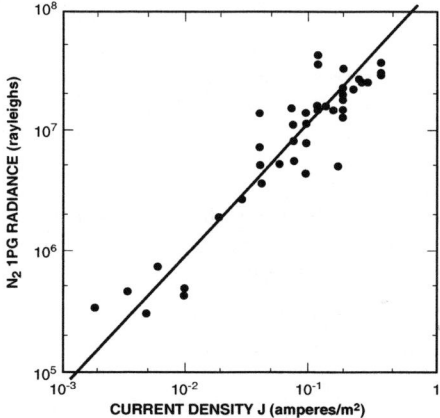

Figure 8. The measured radiance of the positive column of the large glow discharge tube versus the current density, computed as the total current divided by the cross-sectional area of the tube.

11.5 Estimates of Bulk Plasma Conductivity

Figure 9 explores the electrical conductivity of the plasma in the tube and its variation with the measured radiance. The mean conductivity σ of the plasma is computed assuming ohmic behavior ($J = \sigma E$), as the measured current density divided by the mean electric field in the tube. The latter quantity has been approximated as the measured voltage drop along the tube divided by the tube length. Given the evidence from the literature on glow discharge tubes that the voltage drop across the negative glow is greater than in the positive column, we can expect the bulk conductivity estimates to be on the low side relative to the true value in the positive column. For a radiance value in the range of sprite measurements, the inferred conductivity is of order $3 \cdot 10^{-5}$ S/m. This value is in line with theoretical predictions for sprites (Pasko et al., 1998; Liu and

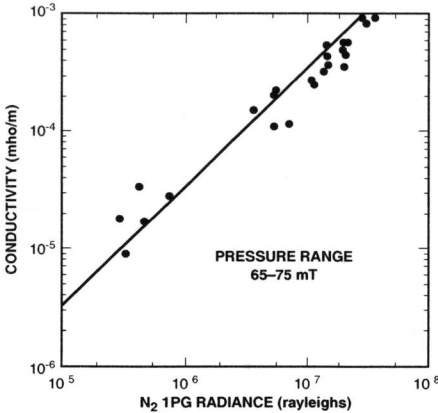

Figure 9. The inferred electrical conductivity conductivity of the plasma in the glow discharge tube versus the measured radiance of the positive column. See text for details.

Pasko, 2004; Chapter 12 by V. Pasko). It is also worth noting that this value is some three orders of magnitude greater than the ambient electrical conductivity at this simulated altitude (∼65 km), thereby providing additional evidence that sprites are ionized plasmas and represent substantial local perturbations in conductivity in the middle atmosphere.

11.6 Application to Current Flow in Sprites

As noted earlier, when the laboratory measurements are applied to sprites, it is essential that calibrated measurements of sprite radiance are used. Secondly, it is important that the sprite observations be spatially resolved. Unfortunately, neither of these conditions is fulfilled for the majority of optical observations on sprites to date, and most conventional video observations on sprites present a fuzzy picture of sprite structure. Fortunately, both these conditions are fulfilled in telescopic observations by Stanford University (Gerken et al., 2000; Gerken and Inan, 2002, 2003). Such observations expose the lightning-like filamentary structure of sprites and have been used extensively in this study.

A sprite is a decidedly unsteady phenomenon, and so questions naturally arise about the applicability of an essentially 'DC' phenomenon, the glow discharge. A more specific question is: Can we expect the relationship between current density and radiance, obtained under DC excitation, to be relevant to the transient sprite conditions? A fundamental time scale for the $N_2 1P$ emission characterizing the positive column is the relaxation time obtained from the Einstein coefficient (Rybicki and Lightman, 1979). This relaxation time is 6 microseconds (Vallance-Jones, 1974), and is the time required for the natural decay in photon emission on the cessation of the forcing. Lightning is the natural forcing agent for sprites, and is set by the charge transfer of the parent lightning. The time scale for charge transfer in the lightning return stroke is hundreds of microseconds, but for sprites, the continuing current time scale is more relevant (Huang et al., 1999), with time scales of milliseconds and longer. Sprite streamers have time scales of order 1 millisecond (V. Pasko and N. Liu, personal communication, 2004). All of the discharge time scales cited above are orders of magnitude longer than the relaxation time, lending some confidence that sprite emission in the $N_2 1P$ band is in some sense 'steady'. An important experimental test of this claim would involve the determination of the total power-radiance relationship (Figure 7) and the current density-radiance relationship (Figure 8) for capacitive discharge experiments of the kind previously cited (Naudé (1932), Elmore (1973) and most notably Goto et al. (2003)). Unfortunately, the measurements of this kind carried out to date are neither time-resolved nor calibrated, so this test is not readily achieved. An example of some column sprites captured with the Stanford telescope is shown in Figure 10. The column sprite is generally regarded as the elemental sprite form. In the right hand side of these figures we have included the scale of current density beside the calibrated color scale of radiance (given here in kR). The one-MR radiance level is at the upper limit of the scale, in red. The bright narrow columns appear to be of the order of 100 meters wide. If the filament cross-sectional area is assumed to be $10^4 m^2$ and the core radiance is 1 MR, then the inferred total current is $10^4 m^2 \times 10^{-2}$ A/m^2 = 100 amperes. Such a current is definitely small in comparison to the lightning currents observed to flow in the production of sprites (Boccippio et al., 1995; Cummer et al., 1998; Stanley et al., 2000; Füllekrug et al., 2001; Cummer and Füllekrug, 2001). The telescopic image for a larger, more spectacular sprite (Gerken et al., 2000), sometimes called "A-bomb sprite" or "angel" (see also Chapter 2 by W. A. Lyons), is shown in Figure 11. The image within the small white rectangle of the upper frame is magnified in the middle frame (Figure 10 and Figure 11), thereby exposing a myriad of filaments as substructure. Based on this magnified image it can be estimated that the total sprite cross-section contains one thousand filaments, each carrying a current of the order of 30 A, based on the observations that only a few filaments here attain the red (1 MR) level. In this situation, a total current of 30 kA might be achieved. This current is appreciable, larger than the mean peak return stroke current for ground flashes, but is still smallish in comparison with the measured peak current in the lightning that made this sprite (129 kA).

246 SPRITES, ELVES AND INTENSE LIGHTNING DISCHARGES

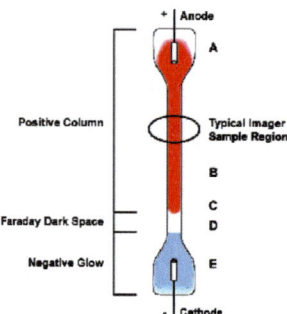

Figure 1. Illustration of DC-excited glow discharge in an air-filled tube, showing its key regions.

Figure 2. Photograph of the large discharge tube used in the present study.

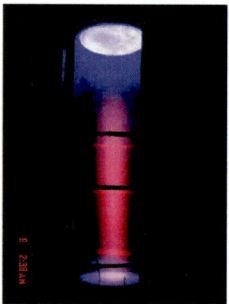

Figure 5. Photograph of the large glow discharge tube at a pressure of 1.2 Torr (160 Pa) and a current density of 6.6 A/m². At this pressure, the positive column is restricted relative to the tube diameter.

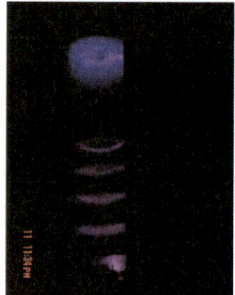

Figure 6. Snapshot photograph of the large glow discharge tube at a pressure of 70 mT (9.2 Pa) and a current density of 0.12 A/m² illustrating quasi-periodic striations, uniformly distributed across the tube at this pressure.

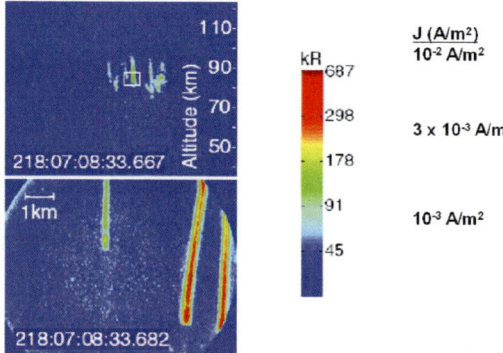

Figure 10. Telescopic imagery on column sprites (Gerken et al., 2000), showing isolated luminous channels. The scale for current density is based on results in Figure 8.

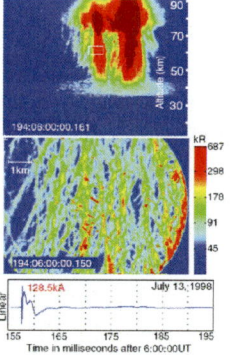

Figure 11. Telescopic imagery on an angel sprite (Gerken et al., 2000), showing detailed filamentary structure.

11.7 Conclusion

In conclusion, the quantitative measurements in the air-filled discharge tube appear to have important applicability to sprites. The measurements substantiate the idea that sprites are conductive plasma in which appreciable currents can flow. Further comparisons using spatially and temporally resolved sprite radiance measurements may help resolve controversy about where the currents are flowing in the overall lightning-sprite configuration. Fully calibrated measurements with both fine space and time resolution are best suited to this end.

Acknowledgments

The authors wish to acknowledge valuable discussions on this topic with R. Armstrong, S. Cummer, R. Dowden, M. Füllekrug, Y. Goto, M. Ishii, W. Lyons, V. Pasko, D. Suszcynsky, Y. Takahashi, D. Williams and Y. Yair. This work was bootlegged on support to MIT from the Physical Meteorology Section, U.S. National Science Foundation (Grant ATM-0003346), for ELF measurements on sprite lightning.

Bibliography

Boccippio, D. J., Williams, E. R., Heckman, S. J., Lyons, W. A., Baker, I. T., and Boldi, R. (1995). Sprites, ELF transients, and positive ground strokes. *Science*, 269:1088–1091.

Chamberlain, J. W. (1995). *Physics of the Aurora and Airglow*. Amer. Geophys. Union. Washington, D.C.

Cobine, J. D. (1958). *Gaseous Conductors—Theory and Engineering Applications*. Dover Publications.

Cummer, S. A. and Füllekrug, M. (2001). Unusually intense continuing current in lightning causes delayed mesospheric breakdown. *Geophys. Res. Lett.*, 28:495.

Cummer, S. A., Inan, U. S., Bell, T. F., and Barrington-Leigh, C. P. (1998). ELF radiation produced by electrical currents in sprites. *Geophys. Res. Lett.*, 25:1281–1284.

Donahue, T. and Dieke, G. H. (1951). Oscillatory phenomena in direct current glow discharges. *Phys. Rev.*, 81:248–261.

Dowden, R., Brundell, J., Rodger, C., Molchanov, O., Lyons, W., and Nelson, T. (1996). The structure of red sprites determined by VLF scattering. *IEEE Antennas and Propagation Magazine*, 38(3):7–15.

Elmore, W. C. (1973). Microwave experiments for an advanced laboratory. *Amer. J. Phys.*, 41:865–870.

Füllekrug, M., Moudry, D. R., Dawes, G., and Sentman, D. D. (2001). Mesospheric sprite current triangulation. *J. Geophys. Res.*, 106(17):20189.

Gerken, E. A. and Inan, U. S. (2002). A survey of streamer and diffuse glow dynamics observed in sprites using telescopic imaging. *J. Geophys. Res.*, 107(SIA):4–1.

Gerken, E. A. and Inan, U. S. (2003). Observations of decameter-scale morphologies in sprites. *J. Atmos. Sol.-Terr. Phys.*, 65:567–572.

Gerken, E. A., Inan, U. S., and Barrington-Leigh, C. P. (2000). Telescopic imaging of sprites. *Geophys. Res. Lett.*, 27:2637–2640.

Goto, Y., Sato, Y., and Ohba, Y. (2003). The optical and spectral measurements of low pressure air discharges as sprite models. In *Proc. 12th International Conference on Atmospheric Electricity*, volume 1, pages 325–328, Versailles, France.

Hampton, D. L., Heavner, M. J., Wescott, E. M., and Sentman, D. D. (1996). Optical spectral characteristics of sprites. *Geophys. Res. Lett.*, 23:89–92.

Harvey, E. N. (1957). *A History of Luminescence—From the earliest times until 1900*. The American Philosophical Society. 692 pp.

Heavner, M. J. (2000). *Optical Spectroscopic Observations of Sprites, Blue Jets and Elves: Inferred Microphysical Processes and their Macrophysical Implications*. Doctoral thesis, University of Alaska, Fairbanks, AK. 141 pp.

Huang, E., Williams, E., Boldi, R., Heckman, S., Lyons, W., Taylor, M., Nelson, T., and Wong, C. (1999). Criteria for sprites and elves based on Schumann resonance measurements. *J. Geophys. Res.*, 104:16943–16964.

Liu, N. and Pasko, V. P. (2004). Effects of photoionization on propagation and branching of positive and negative streamers in sprites. *J. Geophys. Res.*, 109(A04301):doi:10.1029/2003JA010064.

Mende, S. B., Rairden, R. L., and Swenson, G. R. (1995). Sprite spectra: N21PG band identification. *Geophys. Res. Lett.*, 22(19):2633–2636.

Naudé, S. M. (1932). Quantum analysis of the rotational structure of the first positive bands of nitrogen (N_2). *Proc. Roy. Soc. Lond.*, 136:114–144.

Pasko, V. P., Inan, U. S., Bell, T. F., and Reising, S. C. (1998). Mechanism of ELF radiation from sprites. *Geophys. Res. Lett.*, 25:3493–3496.

Pearse, R. W. B. and Gaydon, A. G. (1963). *The Identification of Molecular Spectra*. Chapman and Hall LTD.

Rybicki, G. B. and Lightman, A. P. (1979). *Radiative Processes in Astrophysics*. John Wiley and Sons.

Sentman, D. D., Wescott, E. M., Heavner, M. J., and Moudry, D. R. (1997). Horizontal banded structure in sprites. *EOS Trans. Amer. Geophys. Union*, 78(46):F71.

Sentman, D. D., Wescott, E. M., Osborne, D. L., Hampton, D. L., and Heavner, M. J. (1995). Preliminary results from the Sprites 94 campaign: 1. Red sprites. *Geophys. Res. Lett.*, 22:1205–1208.

Stanley, M. A. (2000). *Sprites and their Parent Discharges*. Doctoral thesis, New Mexico Institute of Mining and Technology, Socorro, New Mexico. 164 pp.

Stanley, M. A., Brook, M., Krehbiel, P., and Cummer, S. A. (2000). Detection of daytime sprites via a unique sprite ELF signature. *Geophys. Res. Lett.*, 27:871–874.

Vallance-Jones, A. (1974). *Aurora*. D. Reidel Publishing Co.

Williams, E. R. (2001). Sprites, elves and glow discharge tubes. *Phys. Today*, 54(11):41–47.

THEORETICAL MODELING OF SPRITES AND JETS

Victor P. Pasko
Communications and Space Sciences Laboratory, Department of Electrical Engineering, The Pennsylvania State University, University Park, PA 16802, USA.

Abstract   An overview of the recent modeling efforts directed on interpretation of observed features of transient luminous events (TLEs) termed sprites, blue jets, blue starters and gigantic jets is presented. The primary emphasis is placed on discussion of similarity properties of gas discharges and interpretation and classification of the observed features of TLEs in the context of previous experimental and theoretical studies of gas discharges of various types. Some of the currently unsolved problems in the theory of TLEs are also discussed.

12.1   Introduction

Transient luminous events (TLEs) are large scale optical events occurring at stratospheric and mesospheric/lower ionospheric altitudes, which are directly related to the electrical activity in underlying thunderstorms (e.g., Sentman et al., 1995; Neubert, 2003; Pasko, 2003b) and references cited therein. Although eyewitness reports of TLEs above thunderstorms have been recorded for more than a century, the first image of one was captured only in 1989, serendipitously during a test of a low-light television camera (Franz et al., 1990). Since then, several different types of TLEs above thunderstorms have been documented and classified. These include relatively slow-moving fountains of blue light, known as 'blue jets', which emanate from the top of thunderclouds up to an altitude of 40 km (e.g., Wescott et al., 1995, 2001; Lyons et al., 2003a); 'sprites' that develop at the base of the ionosphere and move rapidly downwards at speeds up to 10,000 km/s (e.g., Sentman et al., 1995; Lyons, 1996; Stanley et al., 1999) and 'elves', which are lightning induced flashes that can spread over 300 km laterally (e.g., Fukunishi et al., 1996a; Inan et al., 1997). Recently several observations of 'gigantic jets', which propagated upwards from thunderclouds to altitudes about 90 km, have been reported (Su et al., 2003). It appears from observations using orbiting sensors that TLEs occur over most regions of the globe (in temperate and tropical areas, over the oceans, and over the land) (Boeck et al., 1995; Yair et al., 2003, 2004; Blanc et al., 2004; Israelevich et al., 2004; Su et al., 2004; Mende et al., 2004). To date TLEs have been successfully detected from ground and airborne platforms in North America (e.g., Sentman et al., 1995), Central and South America (e.g., Heavner et al., 1995; Holzworth et al., 2003; Pinto Jr. et al., 2004), in the Caribbean region (Pasko et al., 2002a,b) in Australia (Hardman et al., 1998, 2000), over winter storms in Japan (e.g., Fukunishi et al., 1999; Hobara et al., 2001; Hayakawa et al., 2004), on the Asian continent and over the oceans around Taiwan (e.g., Su et al., 2002, 2003; Hsu et al., 2003) and in Europe (Neubert et al., 2001).

The total electrostatic energy associated with charge separation inside a thundercloud is on the order of 1-10 GJ and a substantial fraction of this energy is released in one lightning discharge on time scales less than 1 sec (e.g., Raizer, 1991, p. 372), making lightning one of the most spectacular and dangerous phenomena on our planet. One of the important aspects of lightning phenomena at low altitudes is that the energy release is happening in highly localized regions of space leading to formation of spark channels with temperatures $\sim 25000°K$ and plasma with electron densities exceeding $10^{17} cm^{-3}$ (e.g., Raizer, 1991, p. 373). Due to the exponential decrease in the atmospheric neutral density as a function of altitude even a small fraction of the thundercloud electrostatic energy released at mesospheric/lower ionospheric altitudes during TLE phenomena may have a profound effect on the thermal and chemical balance of these regions; nevertheless these important aspects of TLEs are not yet well understood and quantified. Although the bright ionized channels observed in TLEs (e.g., Gerken and Inan, 2002, 2003, 2005; Pasko et al., 2002a) may indicate high localized energy deposition rates, very little is known at present about actual microphysics of these important elements of TLEs and their effects.

Early theories of TLEs have been reviewed by Rowland (1998), Sukhorukov and Stubbe (1998) and Wescott et al. (1998). The goal of this chapter is to provide a limited overview of some of the recent modeling efforts directed on interpretation of observed features of TLEs termed sprites and jets. We primarily attempt to interpret and classify the observed features in the context of previous experimental and theoretical studies of gas discharges of various types. Some of the currently unsolved problems in theory of TLEs are also discussed.

12.1.1 Phenomenology of Sprites

Sprites are large luminous discharges, which appear in the altitude range of ~ 40 to 90 km above large thunderstorms typically following intense positive cloud-to-ground lightning discharges (e.g., Sentman et al., 1995; Boccippio et al., 1995).

The remote sensing of sprite-producing lightning discharges using Extremely-Low Frequency (ELF) waves and by utilizing a known charge moment change threshold for sprite initiation (see Section 12.3.1 for a definition of the charge moment change) provides an estimate of the global occurrence rate of sprites $\sim$720 events/day on average (Sato and Fukunishi, 2003). However, it is known that sprites are not always associated with ELF transients (e.g., Price et al., 2004), and estimates by other authors indicate that 80% of ELF signatures produced by positive lightning are related to sprites (Füllekrug and Reising, 1998), but only 20% of all sprites are associated with ELF signatures (Füllekrug et al., 2001; Reising et al., 1999), which leads to an estimate of $\sim$5 sprite events/minute or 7200 events/day globally (Füllekrug and Constable, 2000).

Recent telescopic imaging of sprites at standard video rates (i.e., with $\sim$16 ms time resolution) revealed an amazing variety of generally vertical fine structure with transverse spatial scales ranging from tens to a few hundreds of meters (Gerken et al., 2000; Gerken and Inan, 2002, 2003, 2005). First high-speed (1 ms) telescopic imaging of sprites has been reported indicating that streamer-like formations in sprites rarely persist for more than 1-2 ms (Marshall and Inan, 2005). Also recently, it has been demonstrated that sprites often exhibit a sharp altitude transition between the upper diffuse and the lower highly structured regions (Stenbaek-Nielsen et al., 2000; Pasko and Stenbaek-Nielsen, 2002; Gerken and Inan, 2002, 2003, 2005). Many sprites are observed with an amorphous diffuse glow at their tops, the so-called sprite "halo" (e.g., Barrington-Leigh et al., 2001; Wescott et al., 2001; Miyasato et al., 2002, 2003; Moudry et al., 2003; Gerken and Inan, 2003; Moore et al., 2003).

In addition to well documented optical and ELF/VLF radiation associated with sprites, well distinguishable infra-sound signatures of these events have recently been reported (Liszka, 2004; Farges et al., 2005).

The appearance of the fine structure in sprites has been interpreted in terms of positive and negative streamer coronas, which are considered as scaled analogs of small scale streamers, and which exist at high atmospheric pressures at ground level (e.g., Pasko et al., 1998a; Raizer et al., 1998; Petrov and Petrova, 1999; Pasko et al., 2001; Pasko and Stenbaek-Nielsen, 2002). These aspects of sprite phenomenology will be discussed in Section 12.3 of this chapter.

12.1.2 Phenomenology of Blue Jets, Blue Starters and Gigantic Jets

Blue jets develop upwards from cloud tops to terminal altitudes of about 40 km at speeds of the order 100 km/s and are characterized by a blue conical shape (Wescott et al., 1995, 1998, 2001; Lyons et al., 2003a). Blue starters can be distinguished from blue jets by a much lower terminal altitude. They protrude upward from the cloud top (17-18 km) to a maximum altitude of 25.5 km (Wescott et al., 1996, 2001). Blue jets were originally documented during airplane based observations (Wescott et al., 1995). Ground observations of blue jets are believed to be difficult due to severe Rayleigh scattering of blue light during its transmission through the atmosphere (Wescott et al., 1998; Heavner et al., 2000, p. 74). Several ground based video recordings of blue jets, which also electrically connected a thundercloud with the lower ionosphere, have recently been reported (Pasko et al., 2002a; Su et al., 2003; Pasko, 2003b). This type of events is now termed gigantic jets (Su et al., 2003). Recent photographic (Wescott et al., 2001) and video (Pasko et al., 2002a) observations of blue jets at close range have clearly shown the small scale streamer structure of blue jets, earlier predicted in (Petrov and Petrova, 1999), and similar to that reported in sprites. The modeling interpretation of the observed features of blue jets will be presented in Section 12.4.

12.2 Classification of Breakdown Mechanisms in Air

12.2.1 Concept of Electrical Breakdown

The subject of gas discharge physics is very broad and is extensively covered in many textbooks devoted to industrial applications of plasmas and to studies of electrical phenomena associated with naturally occurring lightning (e.g., Raizer, 1991; Roth, 1995, 2001; Lieberman and Lichtenberg, 1994; van Veldhuizen, 2000; Rakov and Uman, 2003; Cooray, 2003; Babich, 2003). In the most general sense, the electrical breakdown is the process of transformation of a non-conducting material into a conductor as a result of applying a sufficiently strong field (Raizer, 1991, p. 128). The specific observed features depend on many factors, including: gas pressure (p), inter-electrode gap size (d), applied electric field (E) magnitude, polarity and time dynamics, electrode geometry, gas composition, electrode material, external circuit resistance, levels of medium pre-ionization, initial energy of seed electrons, etc. (e.g., Raizer, 1991, Chapters 7–12). The presence of electrodes (i.e., the cathode serving as a source of secondary electrons in the discharge volume due to positive ion bombardment) represents an important component of the classical Townsend/Paschen breakdown theory based on simple exponential multiplication of electrons as a function of distance from the cathode, with no significant space charge effects (e.g., Roth, 1995, p. 275). Although the presence of such electrodes, or their analogs, is not immediately obvious in the case of TLE discharges, some of the discharge events (i.e., needle-shaped filaments of ionization, called streamers, embedded in originally cold (near room temperature) air and driven by strong fields due to charge separation in their heads (Raizer, 1991, p. 334)) can effectively proceed without any significant contribution from the secondary cathode emission. The streamers also serve as precursors to a more complicated leader phenomenon, which involves significant heating and thermal ionization of the ambient gas and which represents a well known initiation mechanism of breakdown in long gaps and in lightning discharges at near ground pressure (Raizer, 1991, p. 363). The understanding of particular parameter regimes in which streamer and leader discharges can occur may therefore still be useful for interpretation of TLE events, even though the electrode dominated Townsend/Paschen theory is not directly applicable to them. In the following subsections we provide a brief overview of a classification of breakdown mechanisms in terms of pd and E/p (or E/N, where N is the ambient gas number density) values. The usefulness of these two parameters will be further discussed in Section 12.2.4 devoted to discussion of similarity properties of gas discharges.

12.2.2 Classification of Breakdown Mechanisms in Terms of pd Values

Here we assume relatively low overvoltage conditions, with overvoltage defined as $\Delta V = V - V_b$, where V_b is the Paschen theory breakdown voltage (e.g., Roth, 1995, p. 278).

At pd values such that $pd < 10^{-3}$ Torr·cm an electron crosses the gap practically without collisions, so there is no multiplication of electrons in the discharge volume (Raizer, 1991, p. 137). In this chapter we choose Torr as a unit for pressure (760 Torr = 760 mmHg = 1 atm = 101.3 kPa = 1.013 bar) following commonly accepted practice in existing literature on gas discharges (e.g., Raizer, 1991; Lieberman and Lichtenberg, 1994; Roth, 1995).

The $10^{-1} < pd < 200$ Torr·cm is the Paschen curve range in which Townsend's breakdown model of multiplication of electron avalanches via secondary cathode emission is predominant (Raizer, 1991, p. 325). This is a parameter regime under which glow discharges typically operate. The glow discharge can be generally defined as a self-sustaining discharge at a moderately large pd ($10^{-1} < pd < 200$ Torr·cm) with a cold cathode emitting electrons due to secondary emission mostly due to positive ion bombardment (Raizer, 1991, p. 167). The discharge is self-sustained in a sense that it can maintain itself in a steady state of primitive reproduction of electrons in the discharge volume, when total outflow of electrons and ions through the boundaries is fully compensated by inflow due to the emission of secondary electrons from the cathode and production of both types of species in the discharge volume due to ionization (Roth, 1995, p. 275). Typical characteristics of stable glow discharges used in practical applications include: tube radius $R \simeq 1$ cm, tube length $L \simeq 10\text{-}100$ cm, pressure $p \simeq 10^{-2}\text{-}10$ Torr, voltage $V \simeq 10^2\text{-}10^3$ V and current $I \simeq 10^{-4}$-10^{-1} A (Raizer, 1991, p. 167; Roth, 1995, p. 284). The discussion of voltage-current characteristic, distribution of glow intensity and nomenclature of various regions appearing in low pressure DC discharges can be found in numerous textbooks on the subject, some of which are cited at the beginning of the preceding Section 12.2.1. For the purposes of this chapter we only note that the cathode region layered structures observed in low pressure DC discharges generally scale with pressure p as a mean free path of electrons λ_e ($\lambda_e \sim 1/p$) (Raizer, 1991, p. 168). At p=0.1 Torr (65 km altitude) in air $\lambda_e \sim$ 0.3 cm leading to the important conclusion that the dark horizontal bands and bright beads with sizes of tens of meters, which are sometimes observed in sprite discharges (e.g., Gerken and Inan, 2002), cannot be explained by using direct analogy with stratification observed in discharge tubes at the same pressure. It is also well known that the presence of boundaries in discharge tubes, and loss (due to ambipolar diffusion) and production (i.e., due to secondary electron emission) of charged particles on these boundaries play a definitive role in maintaining the particle and energy balance in this type of discharges and in defining the discharge properties (i.e., temperature and density of electrons) (e.g., Lieberman and Lichtenberg, 1994, p. 454). In contrast to discharge tubes in which electron energy distributions are defined by the balance of the ambipolar diffusion of discharge plasma to the walls and production of charged particles in the discharge volume, the energy distributions in large volumes of TLEs are controlled by the local electric field (see discussion and related supporting references concerning the local field approximation in Liu and Pasko, 2004). Additionally, discharge tubes most commonly operate under quasi-steady conditions, which do not directly correspond to sprite discharges having a very transient nature (see Section 12.3.1). The discharge tubes therefore cannot be used as direct analogs of sprite discharges and comparison of any visually similar aspects of these two should be done carefully and with full realization of the above discussed physical differences.

In the range 200<pd<4000 Torr·cm the Townsend and streamer breakdown can be observed depending on specific conditions, i.e., electrode geometry, applied voltage, etc. (Raizer, 1991, pp. 327, 342).

In the range 4000<pd<$\sim 10^5$ Torr·cm the Townsend/Paschen theory fails, streamer breakdown dominates. In this range of pd values the breakdown of plane gaps develops much faster than is predicted by multiplication of avalanches through cathode emission in Townsend/Paschen theory. The breakdown proceeds in this case in a streamer form dominated by space charge effects and at much lower fields than predicted by the Paschen's curve. The secondary cathode emission can be ignored because there is not enough time for ions to cross the gap (Raizer, 1991, p. 326).

The streamer breakdown theory was put forward in the 1930's to explain spark discharges (Loeb and Meek, 1940). The theory was based on the concept of a streamer. Streamers are narrow filamentary plasmas, which are driven by highly nonlinear space charge waves (e.g., Raizer, 1991, p. 327). At ground level, the streamer has a radius of $10^{-1} - 10^{-2}$ cm and propagates with a velocity of $10^5 - 10^7$ m/s. The dynamics of a streamer are mostly controlled by a highly enhanced field region, known as streamer head. A large amount of net space charges exists in the streamer head, which strongly enhances the electric field in the region just ahead of the streamer, while screening the ambient field out of the streamer channel. The peak space charge field can reach a value about 4-7 times of the conventional breakdown threshold field E_k (see definition of E_k in Section 12.2.3). This large space charge field results in a very intense electron impact ionization occurring in the streamer head. This ionization rapidly raises the electron density from an ambient value to the level in the streamer channel, resulting in the extension of the streamer channel to the head region. Therefore, streamers are often referred to as space charge waves. The streamer polarity is defined by a sign of the charge in its head. The positive streamer propagates against the direction of the electron drift and requires ambient seed electrons avalanching toward the streamer head for the spatial advancement (e.g., Dhali and Williams, 1987) The negative streamer is generally able to propagate without the seed electrons since electron avalanches originating from the streamer head propagate in the same direction as the streamer (e.g., Vitello et al., 1994; Rocco et al., 2002). At low atmospheric pressures, at sprite altitudes, streamers may be initiated from single electron avalanches in regions, where the electric field exceeds the conventional breakdown threshold field E_k (see Figure 6 and related discussion in Section 12.3.4). In this case double-headed streamers are expected to form (e.g., Loeb and Meek, 1940; Kunhardt and Tzeng, 1988; Vitello et al., 1993) with the negative head propagating upward toward the ionosphere and the positive downward toward the cloud tops, assuming the positive polarity of a typical cloud-to-ground lightning discharge producing sprites (e.g., Hu et al., 2002).

For many years, the studies of streamers in air at ground pressure have been motivated by their known ability to generate chemically active species, which can be used for treatment of hazardous and toxic pollutants (e.g., Kulikovsky, 1997; van Veldhuizen, 2000), and references cited therein, and in connection with high voltage external insulation problems (e.g., Allen and Ghaffar, 1995). The electrons in the streamer head can gain sufficiently high energy to ionize, dissociate, and excite molecules, and the chemically active excited molecules, radicals and atoms can initiate multiple chemical reactions in the ambient air (Kulikovsky, 1997).

The values $pd > \sim 10^5$ Torr·cm correspond to the leader breakdown mechanism. At ground pressure p=760 Torr the leader mechanism is predominant in gaps ≥ 1 m (Raizer, 1991, p. 363).

The above classification represents a highly condensed and simplified version of that given in (Raizer, 1991), nevertheless, when considered in conjunction with similarity laws (Section 12.2.4), it can provide useful insights into expected breakdown features at low pressures in large TLE volumes ($\geq$1000s of km^3). In particular, streamers propagating in relatively cold (i.e., 300 °K) atmospheric pressure (p=760 Torr) air are known to initiate spark discharges in gaps $d \geq$5-6 cm (Raizer, 1991, p. 327). At a typical sprite altitude of 65 km, p=0.1 Torr and an effective spatial scale needed for development of streamers can be estimated as $d \geq$400 m, which is readily available in this system. In atmospheric pressure air leader mechanism works in gaps with $d \geq$1 meter. At 65 km, p=0.1 Torr, the pd=10^5 Torr·cm gives $d \simeq$10 km, so one generally would expect that spatial scales available at sprite altitudes are also sufficient for development of the leader phenomena. Additionally, recent modeling of neutral gas heating times associated with a streamer-to-leader transition processes for the altitude range corresponding to the transient luminous events indicates a substantial relative (scaled by neutral density as discussed in Section 12.2.4) acceleration of the air heating, when compared to the ground level (Pasko, 2003a). This acceleration is attributed to strong reduction in electron losses due to three-body attachment and electron-ion recombination processes in streamer channels with reduction of air pressure. However, the very transient nature of the electric field at mesospheric altitudes (see Section 12.3.1) may prevent the streamer-to-leader transition since the driving field may not persist long enough to cause significant heating of the neutral gas. From this point of view the streamers are expected to dominate at least in the upper portions of sprites. The exact altitude range in which streamer to leader transition occurs in transient luminous events is not known at present and represents one of yet unsolved problems in current TLE research (see Section 12.5.1).

In addition to the time dynamics of the electric field mentioned in the previous paragraph, the medium pre-ionization (i.e., due to the presence of the lower ionospheric boundary) is of great importance for interpretation of discharge processes in transient luminous events. In particular, following arguments presented above we expect the streamer mechanism to be applicable at $\sim$80 km altitude in sprites. However, a relatively dense lower ionospheric electron population prevents streamers from developing, and even though pd>4000 Torr·cm, the ionization proceeds at these altitudes in the form of the simplest volumetric multiplication of electrons initially provided by the lower ionospheric boundary (Pasko et al., 1997, 1998b) (this aspect will be further discussed in Section 12.3.2).

12.2.3 Classification of Breakdown Mechanisms in Terms of Applied Electric Field

The magnitude of the applied electric field has a strong effect on the type of discharge, which is observed in gaps with the same pd values. The increase in electric field leads to streamer formation even in relatively short gaps $d<$5-6 cm at ground pressure. For instance, (Raizer, 1991, p. 342), indicates that in molecular nitrogen gas at 400 Torr in the gap with size d=3 cm the transition between Townsend mechanism of electron multiplication to streamer discharge occurs at $\simeq$17% overvoltage. Therefore, it is essential to attempt a classification of different discharge mechanisms in terms of applied electric field values.

The most important reference field for gas discharges in air is the conventional breakdown threshold field E_k which is defined by the equality of the ionization and dissociative attachment coefficients (e.g., Raizer, 1991, p. 135). We note that at large values $pd \geq 1000$ Torr·cm the breakdown voltage V_b defined by the Paschen curve is almost proportional to pd (or d if p is kept constant) and $E_k \simeq V_b/d$ (Raizer, 1991, p.135; Lieberman and Lichtenberg, 1994, p. 460). The E_k field therefore can be used as an approximate reference field needed for Townsend breakdown at relatively high pd values discussed in the previous subsection. We assume $E_k \simeq 32$ kV/ which agrees with typical figures observed in centimeters-wide gaps (Raizer, 1991, p. 135), and also with the ionization and two-body attachment models used recently for streamer modeling in (Liu and Pasko, 2004).

The minimum field required for the propagation of positive streamers in air at ground pressure has been extensively documented experimentally and usually stays close to the value E_{cr}^+=4.4 kV/cm (Allen and Ghaffar, 1995), in agreement with recent results of numerical simulations of positive streamers (Babaeva and Naidis, 1997; Morrow and Lowke, 1997). The absolute value of the similar field E_{cr}^- for negative streamers is a factor of 2-3 higher (e.g., Raizer, 1991; Babaeva and Naidis, 1997, p. 361). One estimate of this field is E_{cr}^-=-12.5 kV/cm (in accordance with Figure 7 of Babaeva and Naidis, 1997). It should be emphasized that the fields E_{cr}^+ and E_{cr}^- are the minimum fields needed for the propagation of individual positive and negative streamers, but not for their initiation (e.g., Petrov and Petrova, 1999). Streamers can be launched by individual electron avalanches in large fields exceeding the conventional breakdown threshold E_k, or by initial sharp points creating localized field enhancements, which is a typical case for point-to-plane discharge geometries (e.g., Raizer et al., 1998). The possibility of simultaneous launching (in opposite directions) of positive and negative streamers from a single midgap electron avalanche is well documented experimentally (e.g., Loeb and Meek, 1940; Raizer, 1991, p. 335) and reproduced in numerical experiments (e.g.,

Vitello et al., 1993). In addition to the E_k, E_{cr}^+ and E_{cr}^- fields discussed above there are several other important reference fields, which can most conveniently be described using a so-called dynamic friction force of electrons in air

$$F = \sum_j N_j \sigma_j(\varepsilon) \delta \varepsilon_j$$

which is shown as a function of electron energy in Figure 1. Here the summation is performed over all inelastic processes characterized by energy dependent cross sections $\sigma_j(\varepsilon)$, with the corresponding energy loss per one collision $\delta \varepsilon_j$. The N_j represents a partial density (in m^{-3}) of target molecules in air corresponding to a particular collision process defined by the cross section σ_j (in plotting F in Figure 1 we assumed partial densities $0.8 N_0$ and $0.2 N_0$ for N_2 and O_2 molecules, respectively, with $N_0 = 2.68 \times 10^{25}$ m^{-3} being a reference value corresponding to ground pressure). The total of 43 different processes were taken into account in Figure 1 using cross sectional data provided by A. V. Phelps at http://jilawww.colorado.edu/www/research/colldata.html. The electron energy losses due to non-zero energy of secondary electrons emerging from ionizing collisions with N_2 and O_2 molecules are accounted for us-

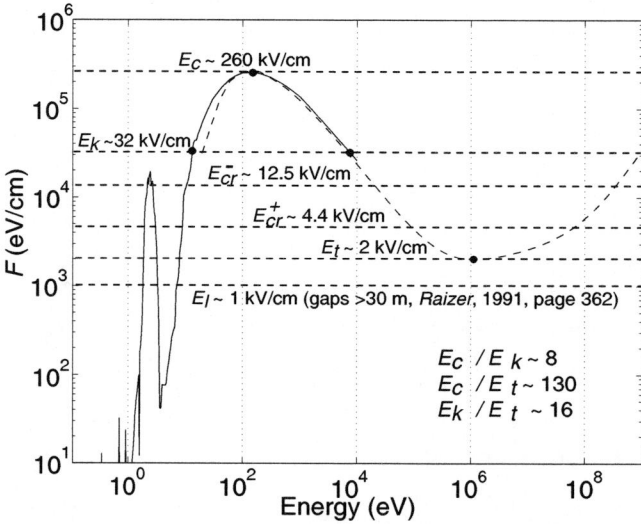

Figure 1. Dynamic friction force of electrons in air. Reference fields (shown by horizontal dashed lines) corresponding to different discharge regimes in air at ground pressure are: E_c – thermal runaway field; E_k – conventional breakdown field; E_{cr}^+ and E_{cr}^- – minimum fields required for propagation of positive and negative streamers, respectively; E_t – relativistic runaway field; and E_l – field needed for advancement of leaders in long gaps (see text for references and discussion).

ing differential ionization cross sections provided in (Opal et al., 1971). The dashed line in Figure 1 shows F values for energy range 20 eV-1 GeV taken from (ICRU, 1984), which combine radiative and collisional losses of electron energy. We note that the radiative losses exceed collisional losses at energy $\sim$100 MeV (ICRU, 1984). The friction force F shown in Figure 1 is conveniently expressed in units of eV/cm and can be directly compared to the applied electric field E in V/cm to provide intuitively simple insight into the expected dynamics of electrons at different energies. We note that F may be directly used in the equation of motion of electrons only under an approximation that electrons move along the applied electric field and scatter mostly in forward direction at small angles (a reasonable approximation above energies $\sim$50 eV). A description for electrons moving at an arbitrary angle with respect to the applied electric field can be found in (Gurevich et al., 1992; Babich, 2003, p. 65). There is a maximum in F at $\sim$150 eV, which is called the thermal runaway threshold ($E_c \simeq$260 kV/cm) and a minimum around $\sim$1 MeV, called the relativistic runaway threshold ($E_t \simeq$2 kV/cm). The maximum is created by a combined action of the ionization and excitation of different electronic states of N_2 and O_2 molecules. At higher energies >150 eV the friction force F decreases with increasing electron energy. The main reason for this is that scattering at these energies mostly governed by Coulomb's law interaction of projectile electrons with individual electrons and nuclei constituting ambient gas molecules. The corresponding scattering potential $\sim 1/r$, where r is the center-of-mass scattering radius, leads to scaling of the scattering cross section with energy $\sigma \sim 1/\varepsilon^2$ and effective scaling of $F \sim 1/\varepsilon$ (e.g., Liebermann and Lichtenberg, 1994, p. 57; Gurevich and Zybin, 2001, p. 57). At higher energies the decrease in the dynamic friction force becomes weaker due to relativistic effects; for $\varepsilon \geq 1$ MeV it reaches its minimum and then a logarithmically slow increase begins (see Figure 1 in Gurevich and Zybin, 2001). The previously discussed E_k, E_{cr}^+ and E_{cr}^- fields are also shown in Figure 1 by horizontal dashed lines. Additionally, we include a reference minimum field $E_l \simeq$1 kV/cm needed for propagation of leaders in gaps >30 m at ground pressure in air (Raizer, 1991, p. 362). It has been shown in the case of lightning discharges that the average field for stable leader propagation can be as low as 100 V/m (Gallimberti et al., 2002, and references cited therein). We note that E_k, E_{cr}^+, E_{cr}^- and E_l fields do not associate with any particular features of the F curve and are shown in Figure 1 simply for easy comparison with E_c and E_t fields.

If we apply to the weakly ionized air an electric field with magnitude above the E_c peak shown in Figure 1, electrons with any (even zero) initial energy will gain more energy from the electric field than they lose in inelastic collisions (i.e., they will become runaway electrons). Some electrons can be energized to very high energies in this way (a substantial population of electrons will still be present at low energies due to large angle scattering and low energy secondary electrons produced in ionizing collisions). In reality, however, it is not easy to produce and maintain fields above E_c, since the electron runaway is also accompanied by an avalanche multiplication of electrons and strong increase in plasma conductivity, which tends to reduce (screen out) the applied field. At lower fields, comparable to the conventional breakdown threshold field E_k, electrons starting with low energies of several eV are expected to remain at low energies and be trapped in the region of energies $\leq$20 eV, as evident from Figure 1. And finally, at very low applied fields (i.e., a fraction of E_k) electrons are expected to be trapped by the first bump in F around 1-2 eV, which is a result of strong electron energy losses in air due to excitation of vibrational degrees of freedom of nitrogen and oxygen molecules.

As evident from Figure 1, the relativistic threshold field $E_t \sim 2$ kV/cm is the minimum field needed to balance the dynamic friction force acting in air on a relativistic electron with $\sim$1 MeV energy (e.g., McCarthy and Parks, 1992; Gurevich et al., 1992; Roussel-Dupré et al., 1994; Lehtinen et al., 1999; Gurevich and Zybin, 2001). In fields above the E_t threshold the 1 MeV electrons, which are readily available in the Earth's atmosphere as cosmic ray secondaries, gain more energy from electric field than they lose in collisions with ambient neutral gas (i.e., become runaways). The possibility of acceleration of energetic cosmic ray secondary electrons in thunderclouds was first suggested by Wilson (1925), and the role of this process in charge redistribution in thunderclouds, in lightning initiation, and in production of observed X-ray fluxes in thunderstorms still remains a subject of active research (see related discussion and references in McCarthy and Parks, 1992; Marshall et al., 1995; Eack et al., 1996b,a; Moore et al., 2001; Dwyer et al., 2003, 2004a,b; Dwyer, 2003, 2004). An interesting aspect of the relativistic runaway process is that an avalanche multiplication of electrons is possible when a fraction of secondary electrons produced in ionizing collisions appear with energies high enough to become runaways themselves (Gurevich et al., 1992; Gurevich and Zybin, 2001). A substantial progress has been made in recent years in modeling of the relativistic runaway process (e.g., Symbalisty et al., 1998; Lehtinen et al., 1999; Gurevich and Zybin, 2001; Dwyer, 2003); and references cited therein. This process has also been extensively discussed in recent literature with applications to TLE phenomena (e.g., Bell et al., 1995; Roussel-Dupré and Gurevich, 1996; Taranenko and Roussel-Dupré, 1996; Lehtinen et al., 1996, 1997, 1999, 2000, 2001; Roussel-Dupré et al., 1998; Yukhimuk et al., 1998a,b, 1999; Kutsyk and Babich, 1999). The relativistic runaway air breakdown is admittedly the most viable mechanism by which terrestrial gamma ray flashes (TGFs) (Fishman et al., 1994; Smith et al., 2005) can be produced in the Earth's atmosphere (Lehtinen et al., 1999, 2001; Inan, 2005; Milikh et al., 2005) and references therein. The first one-to-one correlations of TGF events and lightning discharges have been established (Inan et al., 1996; Smith et al., 2005; Cummer et al., 2005; Inan et al., 2005). There is no strong evidence at present time of direct correlation of TGFs and sprites (e.g., Cummer et al., 2005) and the most recent modeling of energy spectrum of TGFs agrees with relativistic runaway breakdown at 16.5 km altitude (Dwyer, 2005).

We note that Figure 1 presents E_k, E_{cr}^+, E_{cr}^-, E_l, E_c and E_t fields at ground pressure. For discussion in this chapter we assume that most of these fields can be directly scaled proportionally to atmospheric pressure p (or neutral density N) to find corresponding values at TLE altitudes. This approach is generally justified by similarity laws for the electric field in gas discharges discussed in the next subsection. We note, however, that the actual scaling of E_{cr}^+ and E_{cr}^- for the altitude range of TLEs has not yet been verified experimentally (see Section 12.5.3), and the simple scaling of E_{cr}^+ and E_{cr}^- proportionally to neutral density adopted in modeling studies discussed in the subsequent sections of this chapter therefore should be considered as one of the approximations, which can be improved as more data on this subject become available. The exact details of the minimum field for leader formation (E_l) scaling with N are not known at present. The E_l field is related to Joule heating processes and may exhibit substantial deviations from the N scaling due to non-similar behavior of this process (see Section 12.2.4).

12.2.4 Similarity Relations

A problem which often arises in the design of DC glow discharge devices is the scaling in size of a glow discharge from a size which is known to work satisfactorily, to a larger or smaller size (Roth, 1995, p. 306). We have already encountered some of the similarity properties of gas discharges earlier in this chapter when we discussed pd values (i.e., discharges preserve similar properties for the same pd=const values) and the scaling of critical breakdown fields proportionally to pressure p (i.e., discharges preserve similar properties for the same E/p=const values). Here we provide a summary of useful similarity relationships for gas discharges. Note that we use neutral density N in place of p therefore assuming constant temperature of ambient neutral gas. The physical quantities with "o" subscript correspond to reference values at ground pressure.

Length (i.e., mean free path, discharge tube length or diameter, streamer radius, etc.) scales as:

$$L = L_o \frac{N_o}{N}$$

(note that this statement can also be written as NL=const, which is identical to the previously discussed pd=const).

Time (i.e., between collisions, dielectric relaxation, 2-body attachment, etc.) scales as:

$$\tau = \tau_o \frac{N_o}{N}$$

Velocity does not scale (i.e., electron or ion drift velocity, streamer velocity, etc.):

$$v = L/\tau = \text{const}$$

Temperature and energy of electrons and ions do not scale remaining the same in similar discharges (for the same reason as v=const).

Electric field (i.e., in streamer head, in streamer body, etc.) scales as:

$$E = E_o \frac{N}{N_o}$$

(note that this statement can also be written as E/N=const, which is identical to the previously discussed E/p=const).

Mobility (electrons and ions) scales as:

$$\mu = v/E = \mu_o \frac{N_o}{N}$$

Diffusion coefficient (electrons and ions) scales as:

$$D = D_o \frac{N_o}{N}$$

Plasma and charge density (i.e., electron and ion densities in the streamer body, etc.) scale as:

$$n = n_o \frac{N^2}{N_o^2}$$

(the above relationship can be easily verified using τ and μ scaling and definition of the dielectric relaxation time $\tau = \varepsilon_o/q_e n \mu$, where ε_o is the permittivity of free space and q_e is the absolute value of electron charge).

Charge (i.e., in the streamer head) scales as:

$$Q = Q_o \frac{N_o}{N}$$

Ionization and two-body attachment coefficients scale as:

$$\nu = 1/\tau = \nu_o \frac{N}{N_o}$$

Conductivity scales as:

$$\sigma = q_e n \mu = \sigma_o \frac{N}{N_o}$$

Current density scales as:

$$J = q_e n v = J_o \frac{N^2}{N_o^2}$$

Current does not scale remaining the same in similar discharges:

$$I = JL^2 = \text{const}$$

The discharge parameters do not always scale in accordance with the above described similarity laws. Examples of processes which do not obey similarity laws are discussed below.

The characteristic time of the three-body attachment of electrons in air (important at high pressures) $\tau_{a3}=1/\nu_{a3}$, where ν_{a3} is the three body attachment coefficient, scales as:

$$\tau_{a3} = \tau_{a3o} \frac{N_o^2}{N^2}$$

The joule heating characteristic time scales as

$$\tau_h = \tau_{ho} \frac{N_o^2}{N^2}$$

Electron-ion recombination is described by a quadratic term including product of electron and ion densities and therefore does not obey similarity laws (the electron-ion recombination in streamer discharges is dramatically reduced at TLE altitudes due to reduction of electron and ion densities $\sim N^2$ as discussed above).

Non-similarity of photoionization in air is introduced by quenching of N_2 excited states which give rise to UV emissions photoionizing O_2 (e.g., Liu and Pasko, 2004).

Additional discussion of similarity relationships pertinent to a particular case of streamer discharges can be found in (Pasko et al., 1998b).

12.3 Physical Mechanism and Numerical Modeling of Sprites

12.3.1 Large Scale Electrodynamics

The possibility of large-scale gas discharge events above thunderclouds, which we currently know as sprite phenomenon, was first predicted in 1925 by the Nobel Prize winner C. T. R. Wilson (Wilson, 1925). He first recognized that the relation between the thundercloud electric field which decreases with altitude r (Figure 2) as $\sim r^{-3}$ and the critical breakdown field E_k which falls more rapidly (being proportional to the exponentially decreasing atmospheric density) leads to the result that "there will be a height above which the electric

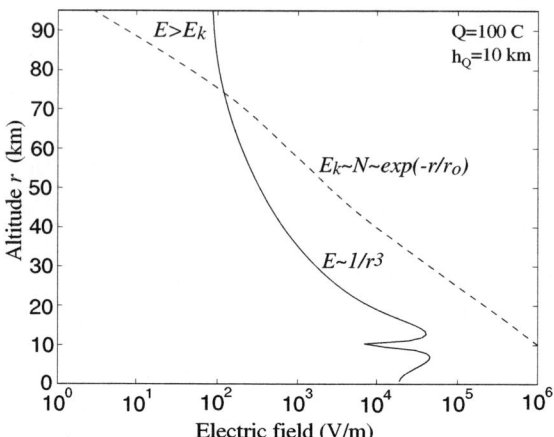

Figure 2. Physical mechanism of sprites Wilson (1925): "While the electric force due to the thundercloud falls off rapidly as r increase, the electric force required to causing sparking (which for a given composition of the air is proportional to its density) falls off still more rapidly. Thus, if the electric moment of a cloud is not too small, there will be a height above which the electric force due to the cloud exceeds the sparking limit."

force due to the cloud exceeds the sparking limit" (Wilson, 1925). It should be noted that due to the finite atmospheric conductivity above thunderclouds the dipole field configuration shown in Figure 2 is realized at mesospheric altitudes only during very transient time periods ~1-10 ms following intense lightning discharges, in part defining similarly transient nature of the observed sprite phenomenon (e.g., Pasko et al., 1997), and references cited therein.

The mechanism of the penetration of the thundercloud electric fields to the higher-altitude regions is illustrated in Figure 3. As the thundercloud charges slowly build up before a lightning discharge, high-altitude regions are shielded from the quasi-electrostatic fields of the thundercloud charges by the space charge induced in the conducting atmosphere at the lower altitudes. The appearance of this shielding charge is a consequence of the finite vertical conductivity gradient of the atmosphere above the thundercloud. When one of the thundercloud charges (e.g., the positive one as shown in Figure 3) is quickly removed by a lightning discharge, the remaining charges of opposite sign above the thundercloud produce a large quasi-electrostatic field that appears at all altitudes above the thundercloud, and endures for a time equal to approximately (see related discussion in Pasko et al., 1997) the local relaxation time ($\tau_\sigma = \varepsilon_o/\sigma$, where σ is the local conductivity and ε_o is the permittivity of free space) at each altitude. These temporarily existing electric fields lead to the heating of ambient electrons and the generation of ionization changes and optical emissions known as sprite phenomena. Figure 4 illustrates the above

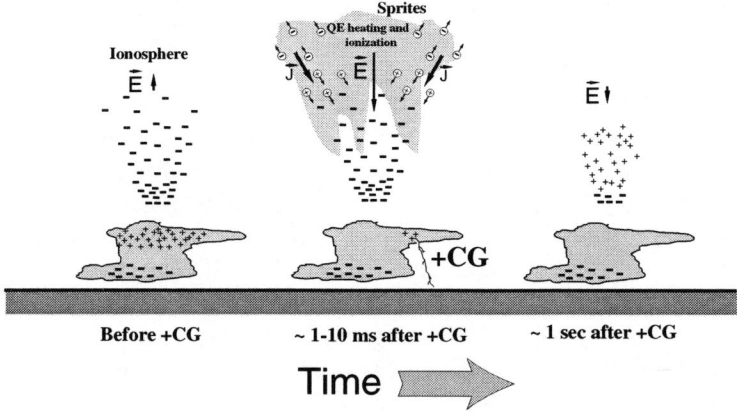

Figure 3. Illustration of the mechanism of penetration of large electric fields to mesospheric altitudes (Pasko et al., 1997). Reprinted by permission from American Geophysical Union.

discussed scenario by showing model calculations of the vertical component of the electric field at altitudes 50, 60, 70 and 80 km directly above a positive lightning discharge removing 200 C of charge from altitude 10 km in 1 ms (Pasko et al., 1997). During a very transient time period ∼1 ms, mostly defined by atmospheric conductivity profile, the electric field can reach values on the order of the critical breakdown threshold field E_k at mesospheric/lower ionospheric altitudes. The quasi-static approximation employed here is valid for relatively slow source variations with time scales >0.5 ms (Pasko et al., 1999).

It should be emphasized that the simplified schematics shown in Figure 3 is used to discuss the physical concept of penetration of large electric field transients to mesospheric altitudes and by no means reflects the complexity of charge distributions observed in thunderclouds. In cases of more realistic charge distributions in the thundercloud, which sometimes involve up to six charge layers in the vertical direction (e.g., Marshall and Rust, 1993; Shepherd et al., 1996), each of the charge centers can be viewed as generating its own polarization charge in and above the thundercloud, and the resultant configuration of the electric field and charge density can be obtained by using the principle of superposition. This consideration is helpful in visualization of the fact that the electric field appearing at mesospheric altitudes after the charge removal by cloud-to-ground lightning discharge is defined mostly by the absolute value and altitude of the removed charge and is essentially independent of the complexity of the charge configuration in the cloud. The charge removal can also be viewed as the "placement" of an identical charge of opposite sign. The initial field above the cloud is simply the free space field due to the "newly placed" charge and its image in the ground which is assumed to be perfectly conducting. The most recent observations indicate that most of the charge responsible for production of sprites can be lowered from relatively low altitudes 2-5 km, with an average height 4.1 km for one particular storm studied in (Lyons et al., 2003b), and references therein. The charge moment change Qh_Q (i.e., charge removed by lightning Q times the altitude from which it was removed h_Q) represents the key parameter which is used in current sprite literature to measure the strength of lightning in terms of sprite production potential (e.g., Cummer et al., 1998; Hu et al., 2002; Cummer, 2003). One of the major unsolved problems in current sprite research, which is also directly evident from the Figure 2 depicting the field created by a charge moment Qh_Q=1000 C·km, is the observed initiation of sprites at altitudes 70-80 km by very weak lightning discharges with charge moment changes as small as 120 C·km (e.g., Hu et al., 2002), and references cited therein. Several theories have been advanced to explain these observations, which include localized inhomogeneities created by small conducting particles of meteoric origin (e.g., Zabotin and Wright, 2001) and the formation of upwardly concave ionization regions near the lower ionospheric boundary associated with sprite halos (Barrington-Leigh et al., 2001). The problem of initiation of sprite streamers in low applied electric fields is one of unsolved problems in current TLE research (see Section 12.5.2).

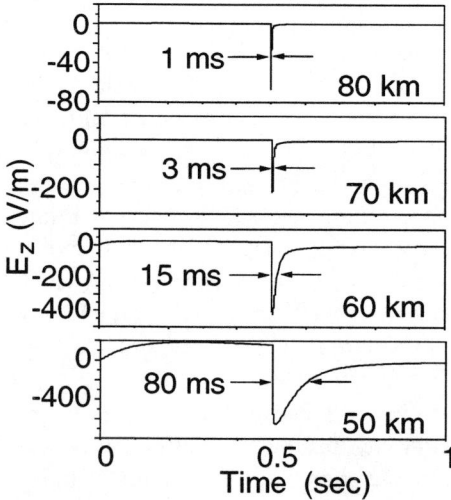

Figure 4. Time dynamics of the vertical component of the electric field at selected altitudes directly above a positive cloud to ground lightning discharge (Pasko et al., 1997). Reprinted by permission from American Geophysical Union.

12.3.2 Altitude Structuring of Optical Emissions

In spite of the apparent simplicity of the basic mechanism of penetration of large quasi-electrostatic fields to the mesospheric altitudes described in the previous section and depicted in Figure 3, the sprite morphology, and sprite altitude structure in particular, appear to be quite complex.

Pasko et al. (1998a) proposed a theory indicating that sprite structure as a function of altitude should exhibit a transition from essentially non-structured diffuse glow at altitudes $\geq$85 km to the highly structured streamer region at altitudes $\leq$75 km (Figure 5). It is proposed that the vertical structuring in sprites is created due to interplay of three physical time scales: (1) The dissociative attachment time scale τ_a (which is defined by the maximum net attachment coefficient as $1/(\nu_a-\nu_i)_{\max}$, where ν_i and ν_a are the ionization and attachment coefficients, respectively); (2) The ambient dielectric relaxation time scale $\tau_\sigma = \varepsilon_o/\sigma$; (3) The time scale for the development of an individual electron avalanche into a streamer t_s. This time is an effective time over which the electron avalanche generates a space charge field comparable in magnitude to the externally applied field (e.g., Pasko et al., 1998b). The interplay between these three parameters creates three unique altitude regions as illustrated in Figure 5: (1) The diffuse region ($\tau_\sigma < \tau_a$, $\tau_\sigma < t_s$) characterized by simple volumetric multiplication of electrons (Townsend electron multiplication mechanism); (2) The transition region ($\tau_\sigma > \tau_a$, $\tau_\sigma < \sim t_s$) characterized by strong attachment of ambient electrons before the onset of the electrical breakdown; (3) The streamer region ($\tau_\sigma > \tau_a$, $\tau_\sigma > t_s$) also characterized by the strong attachment as well as by individual electron avalanches evolving into streamers. The upper and the lower boundaries of the transition region shown in Figure 5 represent an estimate of the altitude range in which the actual transition between the diffuse and streamer regions is expected to occur. The upper boundary may shift downward under conditions of an impulsive lightning discharge which generates substantial electron density (i.e., conductivity) enhancement associated with the sprite halo at the initial stage of sprite formation (e.g., Barrington-Leigh et al., 2001). The lower boundary may shift upward due to streamers originating at lower altitudes but propagating upward toward the lower ionosphere (e.g., Stanley et al., 1999). Barrington-Leigh et al. (2001) conducted one-to-one comparison between high-speed video observations of sprites and a fully electromagnetic model of sprite driving fields and optical emissions. Sprite halos are brief descending glows with lateral extent 40-70 km, which are sometimes observed to accompany or precede more structured sprites. The analysis conducted by Barrington-Leigh et al. (2001) demonstrated a very close agreement of model optical emissions and high-speed video observations, and for the first time identified sprite halos as being produced entirely by quasi-electrostatic thundercloud fields. Sprites indeed of-

272   SPRITES, ELVES AND INTENSE LIGHTNING DISCHARGES

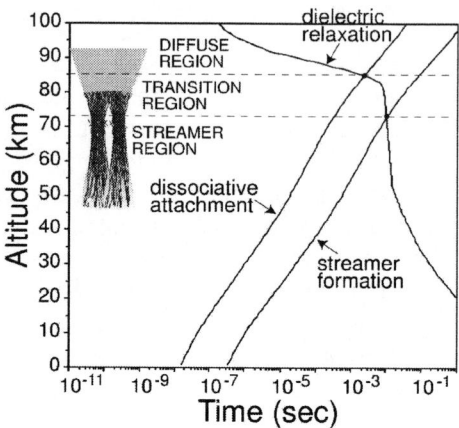

Figure 5. The altitude distribution of different time scales characterizing the vertical structuring of optical emissions in sprites (Pasko et al., 1998b). Reprinted by permission from American Geophysical Union.

ten exhibit sprite halos which appear as relatively amorphous non-structured glow at sprite tops, and which convert to highly structured regions at lower altitudes (e.g., Stanley et al., 1999; Gerken et al., 2000, and cited references therein). This vertical structure in sprites is apparent in recent high-speed video images of Stenbaek-Nielsen et al. (2000) and also was reported during telescopic observations of sprites by Gerken and Inan (2002, 2005).

12.3.3   Large Scale Fractal Models of Sprites

The large scale plasma fluid models based on continuity equations for electrons and ions coupled with either Poisson (Pasko et al., 1997) or the full set of Maxwell's equations (Pasko et al., 1998a; Veronis et al., 1999; Barrington-Leigh et al., 2001) have been successfully used to reproduce some of the observed features of sprites, including their ELF radiation (Cummer et al., 1998) and optical emissions coming from their upper diffuse ("halo") region (Barrington-Leigh et al., 2001). Such direct modeling of the lower, highly structured sprite regions poses insurmountable computational difficulties due to the dramatic differences between the scale of individual ionized channels (i.e., streamers) constituting sprites (on the order of several meters Gerken et al., 2000) and the overall spatial extent of sprites (of order 10 to 100 km). To overcome this difficulty, two-dimensional (Pasko et al., 2000) and three dimensional (Pasko et al., 2001) fractal models have been proposed, which allow the reproduction of the observed large scale volumetric shapes of sprites. One of the important questions, which can be studied with this type of models is the question about attachment of sprites to cloud tops (Pasko et al., 2001). This question is of fundamental importance for understanding a role of sprites in the global atmospheric electric circuit, since it directly relates to the possibility of establishing a highly conducting link between the Earth's surface and the lower ionosphere. The fractal type of the model simulates the propagation of branching streamers associated with sprites as a growth of fractal trees composed of a large number of line channels, and allows realistic modeling of upward, downward and quasi-horizontal propagation and branching of sprite ionization. The model is based on a phenomenological probabilistic approach, which was proposed in (Niemeyer et al., 1989) for modeling of a streamer corona and uses experimentally and theoretically documented properties of positive and negative streamers in air for a realistic determination of the propagation of multiple breakdown branches in a self-consistent electric field. The fractal model effectively follows the dynamics of highly branched streamers in large volumes of space without actually resolving the internal physics of individual streamer channels, but rather relying on demonstrated collective characteristics of streamers in air (Pasko et al., 2000). The fractal model uses E_{cr}^+=4.4 kV/cm and E_{cr}^-=12.5 kV/cm as minimum electric field magnitudes required for propagation of positive and negative streamers in air at ground pressure, respectively (see discussion in Section 12.2.3). The critical fields E_{cr}^+ and E_{cr}^- are assumed to scale with altitude proportionally to the atmospheric neutral density as shown in Figure 6.

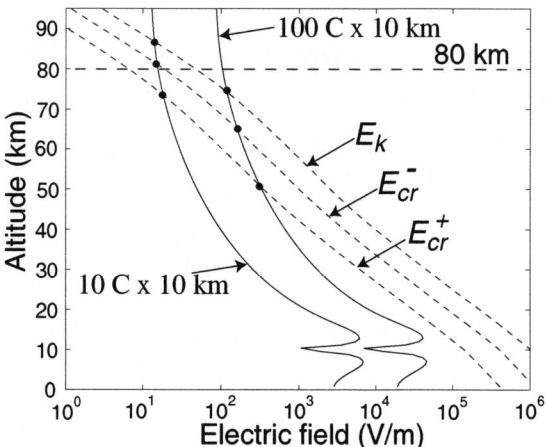

Figure 6. Altitude scans of the electrostatic field corresponding to two different charge moments. The conventional breakdown field E_k, and the minimum fields required for the propagation of positive (E_{cr}^+) and negative (E_{cr}^-) streamers are also shown for reference by the dashed lines (Pasko et al., 2000). Reprinted by permission from American Geophysical Union.

Figures 7a, 7b and 7c illustrate the development of a sprite initiated by a single electron avalanche at altitude 80 km using a two-dimensional version of the fractal model (Pasko et al., 2000). The starting field is defined by the removal of +100 C of charge by a positive cloud-to-ground (CG) lightning from 10 km altitude (Figure 6). The sprite is characterized by the upward development of negative, and downward development of positive, streamers. This behavior agrees with high speed video observations of initial sprite development (Stanley et al., 1999). We note that this type of model cannot provide information on the velocity of streamers. Figure 7b illustrates the moment of attachment of negative streamers to the lower ionosphere and Figure 7c shows the final configuration after the arresting of propagation of all streamer trees. A large "jelly fish" sprite is formed with its lower extremities reaching an altitude of ~38 km. Figure 7d illustrates the corresponding configuration of the field characterized by the expulsion of the field from the body of the sprite and the enhancement of the field on the sharp edges of streamer trees. As was already emphasized previously the field exceeding the conventional breakdown threshold E_k is needed for initiation of streamers from individual electron avalanches. The initiated streamers can propagate in fields E_{cr}^+ or E_{cr}^-, depending on the streamer polarity, which are substantially lower than E_k. We note that in Figure 7 positive streamers propagated downward well below the point at ~50 km altitude above which the initial field exceeded the E_{cr}^+ value (Figure 6). This effect is due to self-consistent focusing and enhancement of the field around lower streamer branches (Figure 7d).

12.3.4 Modeling of Small-scale Sprite Streamer Processes and Photoionization Effects

It is well established that the dynamical properties and geometry of both positive and negative streamers can be affected by the population of the ambient seed electrons, and many of the recent modeling studies have been devoted to the understanding of the role of the ambient medium pre-ionization, including effects of photoionization by UV photons originating from a region of high electric field in the streamer head, on the dynamics of negative (e.g., Babaeva and Naidis, 1997; Rocco et al., 2002) and positive (e.g., Babaeva and Naidis, 1997; Kulikovsky, 2000; Pancheshnyi and Starikovskii, 2001) streamers in different mixtures of molecular nitrogen (N_2) and oxygen (O_2) gases, and in air at ground pressure.

The importance of the photoionization effects on sprite streamers at low air pressures at high altitudes is underscored by the fact that the effective quenching altitude of the excited states $b^1\Pi_u$, $b'^1\Sigma_u^+$ and $c_4'^1\Sigma_u^+$ of N_2 that give the photoionizing radiation is about 24 km (corresponding to the air pressure $p=p_q=30$ Torr) (e.g., Zheleznyak et al., 1982). The quenching of these

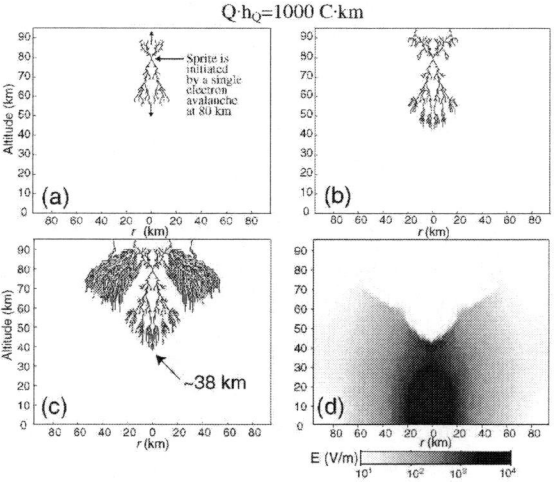

Figure 7. The dynamics of sprite development produced by 100 C of charge removed from 10 km altitude. The sequence of Figures (a)-(c) shows cross-sectional view of discharge trees; (d) A cross-sectional view of the distribution of the absolute values of the electric field corresponding to the structure shown in (c) (Pasko et al., 2000). Reprinted by permission from American Geophysical Union.

states is therefore negligible at typical sprite altitudes 40-90 km, leading to an enhancement of the electron-ion pair production ahead of the streamer tip due to the photoionization, when compared to the previous studies of streamers at ground level. This and other effects have recently been studied by Liu and Pasko (2004) using a newly developed streamer model and some principal results of these studies will be outlined below.

Figure 8 illustrates results of model calculations of electron densities corresponding to double-headed streamers developing at altitudes 0, 30 and 70 km in electric field $E_0=1.5E_k$. In accordance with the similarity laws (Section 12.2.4), the streamer time scales, the streamer spatial scales, and the streamer electron densities scale with the air density as $\sim 1/N$, $\sim 1/N$, and $\sim N^2$, respectively, and the scaled streamer characteristics remain otherwise identical for the same values of the reduced electric field E/N (Pasko et al., 1998). In order to facilitate discussion of similarity properties of streamers at different altitudes/air densities, the results presented in Figures 8b and 8c are given at the moments of time, which are obtained by scaling ($\sim 1/N$) of the ground value, 2.7 ns, specified in Figure 8a. The horizontal and vertical dimensions of the simulation boxes in Figures 8b and 8c also directly correspond to scaled ($\sim 1/N$) ground values shown in Figure 8a. The electron density scale in Figures 8b and 8c also corresponds to scaled ($\sim N^2$) values given in Figure 8a. The differences observed between model streamers at the ground and at 30 and 70 km altitudes in Figure 8 are primarily due to the reduction in photoelectron production at high atmospheric pressures through the quenching of UV emitting excited states of N_2 (Liu and Pasko, 2004). In all cases shown, the model streamers exhibit fast acceleration and expansion (Liu and Pasko, 2004). The results presented in Figure 8 correspond to a "free" (i.e., not affected by electrodes) development of double headed streamers under conditions of the same E/N reduced field values. In this context we note that discussion of similarity of streamers obtained in point-to-plane discharge geometry (e.g., Pancheshnyi et al., 2005; Briels et al., 2005) necessitates inclusion of electrode effects, and, assuming a relatively fast voltage rise time, similar streamer patterns are expected at the same E/N values (see additional discussion of the quenching effects on streamer similarity properties and branching at high pressures in Section 12.5.4) only under conditions when electrode geometry (i.e. interelectrode distance, point electrode diameter, etc.) are scaled as $\sim 1/N$. The fast expansion and acceleration are important characteristics of the considered model streamers. For instance, a positive streamer initiated in a $1.1E_k$ field at 70 km altitude would reach an effective radius of 55 m and speed of about one tenth of the speed of light (2.2×10^7 m/s) by traveling a distance of only 1 km (Liu and Pasko, 2004). Such high speeds of sprite streamers indeed have

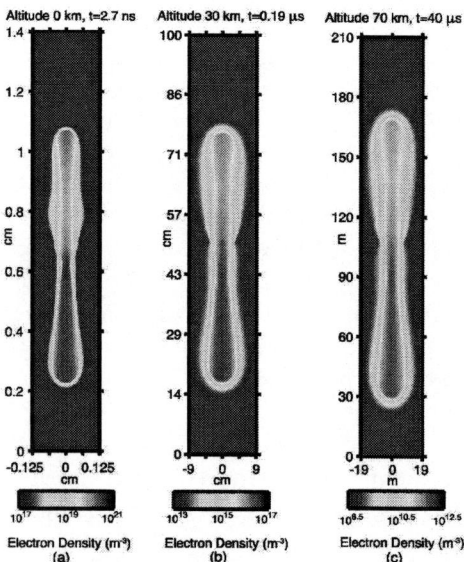

Figure 8. A cross-sectional view of the distribution of the electron number density of the model streamers at altitudes (a) 0 km, (b) 30 km, and (c) 70 km (Liu and Pasko, 2004). Reprinted by permission from American Geophysical Union.

been recently documented by high-speed video (Stanley et al., 1999; Moudry et al., 2002, 2003) and multi-channel photometric (McHarg et al., 2002) systems. The initiation of sprites at altitudes 70-75 km in a form of simultaneous upward and downward propagating streamers is also well documented (Stanley et al., 1999; Stenbaek-Nielsen et al., 2000; Moudry et al., 2002, 2003; McHarg et al., 2002). It is clear, however, that the effective streamer diameters observed by an imager zooming on sprite structures at different altitudes would inevitably depend on the geometry of the mesospheric electric fields and the history of the sprite development (i.e., the altitude of the initiation point(s)). Gerken et al. (2000) and Gerken and Inan (2002, 2003) have recently employed a novel telescoping imager to measure effective streamer diameters at different altitudes in sprites. The measured diameters are 60-145 m ($\pm$12 m), 150 m ($\pm$13 m), 19 m ($\pm$13 m), for altitude ranges 60-64 km ($\pm$4.5 km), 76-80 km ($\pm$5 km), 81-85 km ($\pm$6 km), respectively. Although the 60-145 m ($\pm$12 m) is more than one order of magnitude greater than the scaled initial diameters of streamers shown in Figure 8c (at 60 km, $2r_s \simeq 4$ m), given realistic charge moments available for the sprite initiation (Hu et al., 2002), it is likely that streamers appearing at these low altitudes were initiated at much higher altitudes and propagated long distances experiencing substantial expansion. All observed diameters by Gerken et al. (2000) and Gerken and Inan (2002, 2003) can therefore be realistically accounted for by the modeling studies presented in (Liu and Pasko, 2004).

12.3.5 Optical Emissions Associated with Sprite Streamers

Figures 9 and 10 illustrate distributions of the electric field magnitude and intensities of optical emissions (in Rayleighs) corresponding to the model streamers at 30 and 70 km altitudes, respectively, at the same instants of time as specified in Figure 8. The strong blue emissions (associated with 1st negative N_2^+ ($1NN_2^+$) and 2nd positive N_2 ($2PN_2$) band systems, Figures 10c and 10d) originating primarily in the streamer heads are expected to be produced during the early time of sprite development, as the sprite develops over its altitude extent on a time scale short with respect to the total sprite emission time. This agrees well with recent narrow-band photometric and blue-light video observations of sprites (Armstrong et al., 1998, 2000; Suszcynsky et al., 1998; Morrill et al., 2002) indicating short duration ($\sim$ms) bursts of blue optical emissions appearing at the initial stage of sprite formation. The time averaged optical emissions are expected to be dominated by red emissions associated with the first positive band system of N_2 ($1PN_2$, Figure 10b), which has the lowest energy excitation threshold ($\sim$7.35 eV) and can effectively be produced by relatively low electric fields in the streamer channels, in agreement with sprite

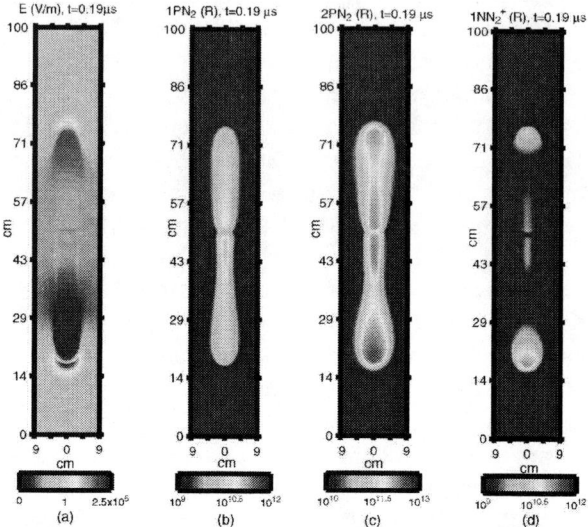

Figure 9. The magnitude of the electric field (a) and the intensity of optical emissions (in Rayleighs) in selected band systems (b)- associated with the model streamers at altitude 30 km (Liu and Pasko, 2004). Reprinted by permission from American Geophysical Union.

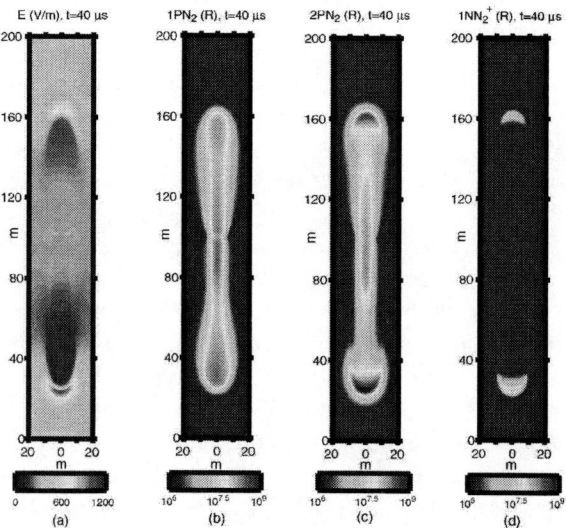

Figure 10. The same as Figure 9 only for model streamers at altitude 70 km (Liu and Pasko, 2004). Reprinted by permission from American Geophysical Union.

observations (Mende et al., 1995; Hampton et al., 1996; Morrill et al., 1998, 2002; Takahashi et al., 2000; Bucsela et al., 2003). We also note that the suppression of 1PN$_2$ emissions due to the strong quenching of the $B^3\Pi_g$ state at altitudes below 50 km (e.g., Vallance-Jones, 1974, p. 119) (note the intensity scale difference between Figures 9b and 9c) is the primary factor which is responsible for making the blue color a dominant color of streamer coronas at lower extremities of sprites (e.g., Sentman et al., 1995) and in blue jet type phenomena observed near thundercloud tops (e.g., Wescott et al., 1995). The red (1PN$_2$) emissions are not completely quenched at altitudes <50 km and have been detected in red-filtered images of sprites (Armstrong et al., 1998). A detailed calculation of the streamer color requires knowledge of the spectral range of the color TV system, the specifics of the observational geometry, allowing to account for the effects of the atmospheric transmission, and such factors as the transmission through an aircraft window (e.g., Wescott et al., 1998; Morrill et al., 1998). The interested readers can find more discussion on related topics in Section 12.4.2 of (Pasko and George, 2002) and references cited therein. The related aspects will also be discussed with relation to blue jets in Section 12.4.4.

Recently, the ISUAL instrument on FORMOSAT-2 (former ROCSAT-2) satellite has successfully observed far-UV emissions from sprites due to N$_2$ Lyman-Birge-Hopfield (LBH) band system (Mende et al., 2004; Frey et al., 2004). Modeling results on LBH emissions from sprite streamers at 70 km altitude indicate that the LBH emissions are stronger by up to a factor of 10 than those from the first negative band system of N$_2^+$ (Liu and Pasko, 2005).

12.4 Physical Mechanism and Numerical Modeling of Blue Jets, Blue Starters and Gigantic Jets

Petrov and Petrova (1999) were first to propose that blue jets correspond qualitatively to the development of the streamer zone of a positive leader and therefore should be filled with a branching structure of streamer channels. These predictions appear to be in remarkable agreement with recent experimental discoveries indicating the streamer structure of blue jets (Wescott et al., 2001; Pasko et al., 2002a). Following the suggestion of Petrov and Petrova (1999), we outline below a possible scenario of events leading to the upward launch of blue jets and blue starters, which occupy large volumes of the atmosphere above thunderstorms measured of thousands of cubic kilometers; many orders of magnitude greater than volumes typically associated with the conventional lightning processes at lower altitudes.

12.4.1 Blue Jets as Streamer Coronas

It is now well-established that electric fields measured from balloons at different altitudes in thunderstorms very rarely exceed 0.5-1 kV/cm (e.g., Winn et al., 1974; Marshall et al., 1996, 2001), and references therein. Results of one study specifically devoted to the investigation of electric field magnitudes and lightning initiation conditions in thunderstorms indicate that in most observed cases the thundercloud electric field as a function of altitude is bounded by the relativistic runaway threshold field E_t, which has a value close to 2 kV/cm at ground level (see Figure 1) and is reduced with altitude proportionally to the atmospheric neutral density (e.g., Marshall et al., 1995). The E_t field has been discussed earlier in Section 12.2.3 and is referred to as the breakeven field in some publications (e.g., Marshall et al., 1995). For the purposes of our discussion in this section we use E_t only as a reference upper bound on fields which are typically observed inside thunderclouds, making no direct association of the relativistic runaway phenomena with blue jets and blue starters. The E_t field as a function of altitude is illustrated in Figure 11. We note that the threshold field E_t appears to be very close to the documented minimum fields ($E_l \sim 1$ kV/cm, see Figure 1) required for propagation of positive and negative leaders in long gaps with sizes exceeding several tens of meters at ground pressure (Raizer, 1991, p. 362). The leader process is also a well-documented means by which conventional lightning develops in thunderstorms (Uman, 2001, p. 82). We note that the electric fields in thunderstorms

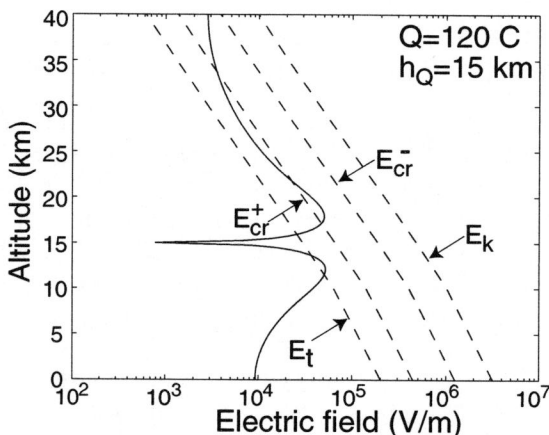

Figure 11. Altitude scan of the electrostatic field produced by a 120 C thundercloud charge placed in the center of the simulation box at altitude 15 km (solid line). Dashed lines show the characteristic fields E_t, E_{cr}^+, E_{cr}^-, and E_k (Pasko and George, 2002). Reprinted by permission from American Geophysical Union.

can occasionally exceed the E_t threshold, and in those rarely observed cases lightning usually followed, immediately destroying the electric field meters at the place where the electric field went substantially above the E_t value (Marshall et al., 1995). In two out of three such cases reported in (Marshall et al., 1995), the E_t threshold was exceeded at altitudes close to 10 km, and in both cases the electric field was upward directed, so that positive streamers would have propagated upward (Marshall et al., 1995). The maximum field enhancement observed by (Marshall et al., 1995) before the instrument was struck and destroyed by lightning is $1.6E_t$. The maximum electric field of 1.86 kV/cm at 5.77 km altitude (i.e., $\sim 1.8E_t$) in close proximity to a lightning initiation location has recently been reported in (Marshall et al., 2005). Another report of electric field observations substantially greater than E_t is that by (Winn et al., 1974), whose authors observed fields on the order of $4E_t$. A summary of other observations of maximum electric field magnitudes measured in thunderclouds is given in (Rakov and Uman, 2003, p. 83). The fact that fields greater than E_t are observed only rarely in balloon sounding data does not necessarily mean that they are uncommon in thunderclouds, if one considers a reasonable argument that regions exhibiting these fields may be localized and also that their persistence in time may be limited by the fast development of a lightning discharge which would try to reduce them. (Marshall et al., 1995) point out that the balloon soundings give the electric field only at the balloon location and that electric fields substantially larger than E_t might be present elsewhere in the cloud.

It is assumed that as soon as the electric field inside the thundercloud approaches the E_t threshold the leader process is developed. The normal role of the leader process is to initiate a discharge of the system, leading to a reduction of charge accumulation in the thundercloud responsible for the field enhancement. The leader process itself is known to be quite complex, and its initiation mechanism and internal physics are not yet fully understood (e.g., Uman, 2001, p. 79; Raizer, 1991, p. 370, Bazelyan and Raizer, 1998, p. 203, 253). For the purposes of discussion here, we do not consider specifics of the leader initiation and postulate presence of this process in high field ($\sim E_t$) regions of the thundercloud.

The head of the highly ionized and conducting leader channel is normally preceded by a streamer zone looking as a diverging column of diffuse glow and filled with highly branched streamer coronas (e.g., Bazelyan and Raizer, 1998, p. 203, 253). Due to its high conductivity, the leader channel can be considered as equipotential and therefore plays the primary role in focusing/enhancement of the electric field in the streamer zone, where the relatively weakly conducting streamer coronas propagate (e.g., Raizer, 1991, p. 364). Leaders of positive polarity attract electron avalanches, while in those of negative polarity the avalanching electrons move in the same direction as the leader head. In large experimental gaps (>100 m) and in thunderclouds, the electric fields required for propagation of leaders of the positive and negative polarity are known to be nearly identical, but the internal structure of their streamer zones, which is closely associated with the direction of electron avalanches, is very different (Raizer, 1991, p. 375; Bazelyan and Raizer, 1998, p. 253). We note that the experimentally documented electric fields E_{cr}^+ and E_{cr}^- required for propagation of streamer coronas, which constitute essential components of the leader streamer zone, are substantially higher than the ambient E_t (Figure 1), and as a result the leader streamer zone is normally confined to a limited region of space around the leader head. A remarkable feature of the streamer corona is that in spite of its internal structural complexity, involving multiple highly branched streamer channels, its macroscopic characteristics remain relatively stable under a variety of external conditions and the field measurements inside the streamer zone of positive (Petrov et al., 1994) and negative (Petrov and Petrova, 1993) leaders indicate that the minimum fields required for propagation of positive E_{cr}^+ and negative E_{cr}^- streamers discussed in Section 12.2.3 and shown in Figure 1 are also close to the integral fields established by positive and negative coronas, respectively, in regions of space through which they propagate.

Figure 11 shows an altitude scan of the electric field created by a static charge of 120 C, having a Gaussian spatial distribution with spatial scale 3 km placed at 15 km altitude between two perfectly conducting planes positioned at the ground (0 km) and at 40 km altitude, as well as the critical fields E_t, E_{cr}^+, E_{cr}^-, and E_k, which assumed to scale with altitude proportionally to the neutral atmospheric density (see additional discussion on scaling in Section 12.5.3 and additional references for physical factors defining the position of the upper boundary at the end of Section 12.4.2).

We note that the fields (E_{cr}^+, E_{cr}^-, E_k) can be comfortably exceeded at high altitudes (>70 km) following intense positive cloud-to-ground lightning discharges leading to the sprite phenomenon (see Figure 6). However, these fields (especially E_k) are much greater than the large scale fields typically observed inside thunderclouds, as discussed above and illustrated by the E_t altitude distribution shown in Figure 11. The large-scale electric field enhancement inside thunderclouds above even E_{cr}^+ (which is closest to the E_t) should be considered as an unusual and rare circumstance.

In view of the above discussion it is clear that if, due to the fast growth of the thundercloud charge, the large-scale electric field does exceed the E_{cr}^+ threshold, then positive streamer coronas, which are normally confined close to the leader head, can quickly (with propagation speeds $>10^5$ m/s, substantially exceeding typical leader speeds $\sim 2 \times 10^4$ m/s (e.g., Bazelyan and Raizer, 1998, p. 227) fill a large volume of space in the vicinity of a thundercloud. Such field distribution, which exceeds E_{cr}^+ in a relatively narrow (± 2 km) region around 20 km altitude, is illustrated in Figure 11. Although the initial volume of space occupied by streamer coronas is defined by the geometry of thundercloud charges (the volume of space in which electric fields exceed the E_{cr}^+ threshold), the streamer coronas themselves self-consistently modify the electric field distribution. Results of three-dimensional modeling of streamer coronas under these circumstances (Pasko and George, 2002) clearly demonstrate that under a variety of initial conditions the streamer coronas form upward propagating conical shapes closely resembling the experimentally observed geometry of blue jets. An example of these calculations will be shown in Section 12.4.3 below.

12.4.2 Thundercloud Charge and Current Systems Supporting Blue Jets, Blue Starters and Gigantic Jets

The formulation of large-scale charge and current systems in thunderclouds, which support upward propagation of blue jets and blue starters and which we discuss here, closely follows that given in (Pasko and George, 2002), and is based on a fast-growing positive charge at the thundercloud top. It is assumed that this charge is a primary source of the electric field which drives blue jets and starters, with no association with lightning activity. The positive (top) and the negative (bottom) thundercloud charges (Figure 12) accumulate due to the current $\vec{J}_s'$ associated with the separation of charges inside the cloud and directed opposite to the resulting electric field. We assume that the charge accumulation time scale can in some cases be very fast (fraction of a

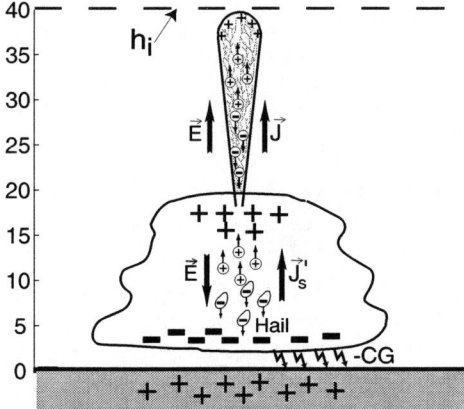

Figure 12. Currents, charges and electric fields associated with blue jets (Pasko and George, 2002). Reprinted by permission from American Geophysical Union.

second). This time scale, in combination with the middle atmospheric conductivity profile, plays a primary role in defining the upper termination altitude of blue jets (Pasko and George, 2002, Section 2.6). The current $\vec{J}'_s$ may be related to the small, light and positively charged ice splinters driven by updrafts and heavy negatively charged hail particles driven downward by gravity (e.g., Uman, 2001, p. 65). Unusually intense precipitation of large hail was indeed observed in association with blue jets and blue starters (Wescott et al., 1995, 1996) and due to negative charge associated with hail particles discussed above is a strong indication of intense electrical activity inside the cloud. The recently observed blue jet event in Puerto Rico (Pasko et al., 2002a) was produced during a fast growth stage of thunderstorm development. The electromagnetic data which was available for both the Wescott et al. (1995, 1996) and the Pasko et al. (2002a) observations indicate no direct triggering of blue jets and blue starters by a lightning event. It is assumed that a charge of $\sim$110-150 C with Gaussian spatial distribution of scale $\sim$3 km can accumulate at altitude $\sim$15 km, creating electric field magnitudes capable of crossing the E_{cr}^+ threshold as depicted in Figure 11 and discussed in the previous section. As a general note, we emphasize that although our model results presented in the following section depend on the spatial scale of the charge, the charge value and the charge altitude, and these parameters are expected to vary for conditions existing in real thunderclouds, the results appear to be very robust in terms of production of upward conical shapes of blue jets as long as the large-scale fields exceed the E_{cr}^+ threshold at the thundercloud top. In part this has to do with the general geometry of electric field lines created by a localized charge placed between two conducting plane boundaries and the exponential reduction of E_{cr}^+ as a function of altitude. Following (Pasko and George, 2002), we neglect the lower (negative) thundercloud charge due to its proximity to the ground (Figure 12). This charge also may be removed by a series of negative lightning discharges during several seconds before the appearance of jets (Wescott et al., 1995, 1996). The electric field is non-zero inside the positive streamer coronas constituting blue jets and blue starters, and an integral upward directed current always flows in the body of the jet (Figure 12). The blue jet propagates when this current is supported by a source in or near the cloud. Otherwise, the negative charge flowing towards the positive thundercloud charge would reduce the source charge and the electric field above the cloud and would eventually suppress the propagation. Thus, the blue jet or blue starter can propagate as long as $\vec{J}'_s$ can deliver sufficient positive charge to the thundercloud top. An equal amount of negative charge is accumulated at the thundercloud base so that the overall charge in the cloud-jet/starter system is conserved. For the fractal model calculations which we show in the next section, we simply assume that the source thundercloud charge remains unchanged during the development of the phenomena, which physically corresponds to a situation when thundercloud current $\vec{J}'_s$ compensates any reduction in the source charge due to the jet current. We note that in the fractal model of Pasko and George (2002) the upper terminal altitude of blue jets is defined by the location of the upper boundary of the simulation box, as discussed in the next section (i.e., 40 km, in Figure 11). The discussion of physical factors, which determine terminal altitudes of blue jets, blue starters and gigantic jets, providing effective classification of different phenomena as a function of their vertical extent, can be found in (Pasko and George, 2002, Sections 2.6 and 4.1).

12.4.3 Numerical Simulation of Blue Jets and Blue Starters

The large scale volumetric shapes of blue jets, blue starters and gigantic jets can be successfully modeled with two and three dimensional fractal models discussed in Section 12.3.3, and related results are documented in (Pasko and George, 2002). Figure 13a shows an example of the model calculation corresponding to the same charge configuration as in Figure 11, but with the upper simulation box boundary set to 70 km altitude, corresponding to the terminal altitude of the blue jet event reported in (Pasko et al., 2002a). Figure 13b shows one of the images of the blue jet phenomena, taken from the video sequence reported in (Pasko et al., 2002a) and corresponding to the moment of the attachment of the blue jet to the lower ionospheric boundary. This stage of the blue jet development is similar to the "final jump stage" of the leader process observed in laboratory experiments, when the streamer zone makes contact with the opposite electrode (Bazelyan and Raizer, 1998, p. 212). The range of observed speeds during the final jump, 5×10^4 m/s to 10^6 m/s (Bazelyan and

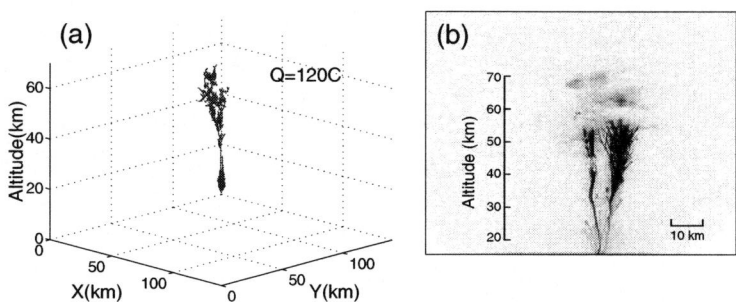

Figure 13. (a) Fractal model results for thundercloud charge Q=120 C at altitude h_Q=15 km and the upper simulation box boundary at 70 km (Pasko and George, 2002). Reprinted by permission from American Geophysical Union. (b) Image of a blue jet at the moment of attachment to the lower ionospheric boundary (Pasko et al., 2002a). Reprinted by permission from *Nature*.

Raizer, 1998, p. 212), is similar to the range of speeds, from 5×10^4 m/s to more than 2×10^6 m/s, reported in (Pasko et al., 2002a). Although this type of model cannot provide information on the velocity of streamer coronas, Figure 13 demonstrates a good agreement between model results and observations in terms of the general volumetric shape of the blue jet.

12.4.4 Modeling of Optical Emissions from Blue Jets and Blue Starters

Blue jets and blue starters have been captured by black and white and color video cameras, allowing for some important suggestions concerning optical bands responsible for the observed blue color (Wescott et al., 1995). Evidence from color TV suggesting that the blue light must have an ionized 1st negative N_2^+ component has been presented in (Wescott et al., 1998). The first conclusive evidence of 427.8 nm (1st negative N_2^+) emission in blue starters has been recently reported in (Wescott et al., 2001). Wescott et al. (2001) also analyzed color TV frames associated with blue starters and concluded that the combined red and green channel intensity constituted 7% of the total blue channel intensity.

The fractal model allows accurate determination of the macroscopic electric fields in regions of space occupied by streamers. The results reported in (Pasko and George, 2002) indicate that for a variety of input parameters these fields are very close (within several %) to the minimum electric field required for propagation of positive streamers in air, E_{cr}^+. This behavior is consistent with earlier findings (Niemeyer et al., 1989; Pasko et al., 2000, 2001) and experimental measurements of Petrov et al. (1994). We note, however, that the low fields on the order of E_{cr}^+ are generally not sufficient to excite any observable optical emissions (Pasko and George, 2002).

The fractal model does not allow resolution of microscopic properties of individual streamer channels constituting streamer coronas and therefore does not allow resolution of the regions of space around streamer tips. It is known that under a variety of conditions the electric field enhancements around streamer tips reach values $\sim 5E_k$ (e.g., Dhali and Williams, 1987; Vitello et al., 1994; Babaeva and Naidis, 1997; Kulikovsky, 1997; Pasko et al., 1998b), where E_k is the conventional breakdown threshold field discussed in Section 12.2.3 and shown in Figure 1. This property of streamers is also valid for positive streamers propagating in low ambient electric fields comparable to E_{cr}^+ (e.g., Grange et al., 1995; Morrow and Lowke, 1997), similar to the ambient conditions for propagation of streamer coronas considered in (Pasko and George, 2002) and illustrated in the previous section. Our conclusion, therefore, is that the observed optical luminosity in blue jets and starters arises from large electric fields existing in narrow regions of space around tips of small-scale corona streamers constituting them.

Figure 14 presents comparison of recent spectral observations reported by Wescott et al. (2001) and the calculated ratio of the combined red and green emissions to the total blue emission assuming the driving field to be $5E_k$, using optical model formulation documented in (Pasko and George, 2002; Liu and Pasko, 2004), and also accounting for the atmospheric transmission and aircraft window corrections pertinent to experimental conditions of (Wescott et al., 2001). The resultant ratio appears to be in good agreement with the recent analysis of color TV frames associated with blue starters reported in (Wescott et al., 2001), who concluded that the combined red and green channel intensity constituted 7% of the total blue channel intensity.

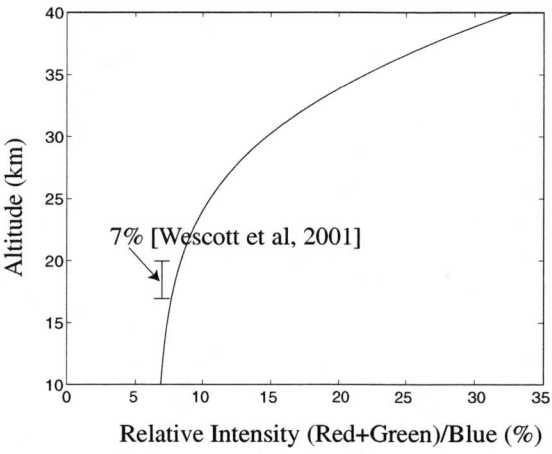

Figure 14. Ratio of the combined red and green emissions to the total blue emission as a function of altitude (Pasko and George, 2002). Reprinted by permission from American Geophysical Union.

12.5 Unsolved Problems

12.5.1 Relationship of Sprites and Jets to High Air Pressure Leader Processes

Although the streamer-to-leader transition (e.g., Raizer, 1991, p. 363); (Bazelyan and Raizer, 1998, p. 238), may be a part of the blue jet/starter and even sprite phenomena, their branched appearance is suggestive that their initial formation is due mostly to the streamer coronas expanding from the leader streamer zone at lower altitudes in case of blue jets, and from single electron avalanches, or some other, yet unknown, initiation agents in case of sprites.

The available imaging data reported recently in Wescott et al. (2001); Pasko et al. (2002a); Su et al. (2003), and some earlier observations of lightning-like phenomena at thundercloud tops reviewed recently in (Lyons et al., 2003a) indicate that the streamer-to-leader transition may be occurring in the lower parts of blue jets and at the later stages of their development (see also related discussion in Pasko and George, 2002). The transition involves a collective action of streamers leading to an increase in ambient gas temperature to several 1000s of °K sufficient for development of the thermal ionization (Raizer, 1991, p. 365). In this respect we note that the understanding of ambient gas heating processes initiated by streamers, embedded in originally cold (near room temperature) air represent a long standing problem, which is of interest for studies of long laboratory sparks and natural lightning discharges (e.g., Gallimberti et al., 2002). As we have already discussed in Section 12.3.4, many of the small-scale features observed in sprites at higher altitudes (e.g., Gerken and Inan, 2003) can be interpreted in terms of corona streamers, which, after appropriate scaling with air density, are fully analogous to those, which initiate spark discharges in relatively short (several cm) gaps at near ground pressure (Liu and Pasko, 2004, and references therein), and which constitute building blocks of streamer zones of conventional lightning leaders in long gaps (Gallimberti et al., 2002). The recent reports of infrasound bursts originating from 60-80 km altitudes in sprites, with durations consistent with the optical widths of the sprites (Farges et al., 2005), and recent modeling results indicating the relative acceleration of the air heating with reduction of pressure discussed in Section 12.2.2, provide additional motivation for studies of the heating of the ambient air and associated chemical effects caused by streamers in transient luminous events. The streamer-to-leader transition at low air pressures represents one of unsolved problems in current TLE research.

12.5.2 Initiation of Sprite Streamers in Low Applied Fields

As it has already been briefly mentioned in Section 12.3.1, the initiation mechanisms of sprites produced by lightning discharges associated with charge moment changes as small as 120 C·km (Hu et al., 2002) are not understood at present, and it is likely that details of the related physics are more complicated than simple scaling of the breakdown and thundercloud fields shown in Figure 6. In particular, Adachi et al. (2004) found that the number of sprite columns in each sprite event was proportional to the peak current intensity of positive cloud-to-ground lightning discharges (CGs) while the average vertical length of columns was proportional to the charge moment of the causative positive CGs. Recent studies of Gerken and Inan (2004) indicate that relatively faint and diffuse sprites confined to small altitude range are associated with very high peak currents and short time duration of causative lightning discharges. Ohkubo et al. (2005) reported association of sprites with clusters of VLF radio atmospherics (sferics), similar to those observed previously in association with subionospheric signal perturbations referred to as "early/fast" VLF events (Johnson and Inan, 2000). These studies indicate a possible and not yet fully understood role of the time characteristics of the lightning currents in the initiation of sprites.

High speed imaging of sprites has captured their temporal development, indicating that most sprites are initiated with a bright column or streamers and followed by the upward and downward branching (Stanley et al., 1999; Stenbaek-Nielsen et al., 2000; Gerken and Inan, 2003; Moudry et al., 2003). On the other hand, sprite halos are observed typically preceding the occurrences of sprites (Barrington-Leigh et al., 2001; Wescott et al., 2001; Miyasato et al., 2002, 2003; Moudry et al., 2003; Gerken and Inan, 2003). It has recently been speculated that an enhancement of the electric field at the bottom of the sprite halo may create favorable conditions for subsequent streamer breakdown (Barrington-Leigh et al., 2001). The recent observations of early/fast VLF events by Moore et al. (2003) provide additional evidence for the electron density disturbances associated with sprite halo events.

Other proposed approaches to initiation of sprite streamers in very weak electric fields include localized inhomogeneities in the mesospheric medium (Moudry et al., 2003), electromagnetic pulses from horizontal lightning discharges forming localized peaks of electron density in the altitude range 75-85 km (Cho and Rycroft, 2001), and external triggers such as meteors or micrometeors (Suszcynsky et al., 1999; Wescott et al., 2001; Zabotin and Wright, 2001). We note that initiation of sprite streamers in low fields ($E<E_k$) requires strong conductivity enhancements comparable to conductivity of streamers themselves. At a typical sprite initiation altitude of 70 km the required electron density perturbations should be on the order of $10^6 cm^{-3}$ (see Figure 8c). Cosmic rays contribute to the background ionization at these altitudes, however, the related electron densities are expected to be low even under conditions of relativistic runaway phenomena (i.e., $E_k>E>E_t$, Section 12.2.3) due to prohibitively long avalanche distances l_a of relativistic runaway electrons at low pressures (i.e., at 70 km l_a=50 m$\times(E_t/E)(N_o/N)\simeq$370 km assuming $E/E_t\simeq$2 (Gurevich and Zybin, 2001)).

12.5.3 Propagation of Sprite Streamers

The overall volumetric effects of sprites on the upper atmosphere can be evaluated using two and three dimensional phenomenological fractal models of streamer coronas (Pasko et al., 2000, 2001), and references therein (see Section 12.3.3). In this type of models the minimum fields required for positive and negative streamer advancement at various pressures are specified externally as input parameters. It should be emphasized that these fields have not yet been determined at air pressures corresponding to sprite altitudes and the existing models employ a highly simplified assumption by using the ground level values scaled linearly with pressure (e.g., Pasko et al., 2000) (this assumption, contradicts to a limited amount existing experimental data on the subject, as will be discussed below). The determination of the minimum fields needed for propagation of streamers at low pressures represents one of the important tasks of the current sprite research. The information about these fields is needed for the correct evaluation of the total volume of atmosphere affected by sprite phenomenon as well as for evaluation of the possible role of sprites in establishing a direct path of electrical contact between the tropospheric and mesospheric/lower ionospheric regions (see Pasko et al., 2001, for references to relevant experimental data).

As it has already been discussed in Section 12.2.3 the minimum field required for the propagation of positive streamers in air at ground pressure (E_{cr}^+) has been extensively documented experimentally and usually stays close to the value 4.4 kV/cm (Allen and Ghaffar, 1995) which agrees well with recent results of numerical simulations of positive streamers (Babaeva and Naidis, 1997; Morrow and Lowke, 1997). The information about the absolute value of the similar field (E_{cr}^-) for the negative streamers at present is very limited. The existing sources indicate that this field is a factor of 2-3 times higher than the corresponding field for the positive streamers (Raizer, 1991, p. 361; Babaeva and Naidis, 1997).

The data on minimum fields required for propagation of streamers at pressures lower than atmospheric are very limited in currently available literature (e.g., Griffiths and Phelps, 1976; Aleksandrov and Bazelyan, 1996; Bazelyan and Raizer, 1998, p. 216). Although from the streamer similarity laws one generally would expect the minimum fields required for streamer propagation to scale with altitude proportionally to the air neutral density N, the actual scaling for the altitude range of sprites has not yet been verified experimentally. In experiments reported in Griffiths and Phelps (1976) and Bazelyan and Raizer (1998), p. 216, the measurements were performed for a set of relatively high pressures corresponding to altitudes <12 km, and the E_{cr}^+ was found to drop with altitude faster than N. These deviations may be a manifestation of the effects, which are able to destroy the streamer similarity at high pressures (i.e., possibly related to the neutral gas heating and the three-body electron attachment) and which are not yet well understood at this time. On the other hand, the data presented by different authors remain controversial, and some studies of streamers indicate nearly linear scaling of E_{cr}^+ with pressure (Aleksandrov and Bazelyan, 1996, and references cited therein).

12.5.4 Branching of Sprite Streamers

The complex morphology of sprites is well documented in existing literature (e.g., Stenbaek-Nielsen et al., 2000; Gerken and Inan, 2003; Moudry et al., 2003). Early video observations have revealed upward and downward branching structures in sprites (Sentman et al., 1996; Taylor and Clark, 1996; Stanley et al., 1996; Fukunishi et al., 1996a). Complex time dynamics with upward and downward propagating tree-like structures has also been reported during more recent high-speed video observations of sprites (Stanley et al., 1999; Stenbaek-Nielsen et al., 2000). The observed tree-like structures are likely to be related to the branching property of individual streamer channels. The upward (negative streamers) and downward (positive streamers) branching has been clearly seen in some of the sprites (Moudry et al., 2003; Gerken and Inan, 2003).

The branching phenomenon has received increasing attention in recent experimental and modeling studies of streamers at ground pressure (Popov, 2002; Arrayas et al., 2002; Rocco et al., 2002; Yi and Williams, 2002; van Veldhuizen and Rutgers, 2002; Ono and Oda, 2003; Akyuz et al., 2003; Hallac et al., 2003; Arrayas and Ebert, 2004; Pancheshnyi et al., 2005; Briels et al., 2005). Interest in these studies is largely motivated by numerous chemical applications of streamers, including treatment of large gas volumes for the purposes of pollution control and ozone production (e.g., van Veldhuizen, 2000, and references therein). The branching features of sprites (Gerken and Inan, 2003) and ground (van Veldhuizen and Rutgers, 2002) streamers are very similar, although their spatial scales are different by several orders of magnitude (i.e., decameter streamers in comparison with a fraction of millimeter ground streamers, as discussed in Section 12.3.4 and illustrated in Figure 8).

Finding the exact physical factors, which define the transverse spatial scale of a streamer, is a difficult task since simplified streamer models do not usually provide a characteristic spatial scale for the streamer radius (e.g., Raizer and Simakov, 1998, Bazelyan and Raizer, 1998, p. 277; Kulikovsky, 2000, and references cited therein). It has recently been demonstrated that negative streamers developing in high ambient fields, when no is pre-ionization available ahead of the streamer, are reaching an unstable "ideal conductivity" state with approximately equipotential and weakly curved head (Arrayas et al., 2002; Rocco et al., 2002). It was proposed that this new state exhibits a Laplacian instability, like that in viscous fingering, which leads to branching of the streamer (Arrayas et al., 2002; Rocco et al., 2002). However, the authors also pointed out that the branching of streamers does depend on the numerical discretization to certain degree. There still exists disagreement on whether the branching of model streamers is a pseudo-branching and is a consequence of a numerical instability (Pancheshnyi and Starikovskii, 2001, 2003), or correctly reflects the real splitting physics of streamers.

A limited set of modeling studies conducted by Liu and Pasko (2004) indicate that the maximum radius of the expanding streamers is predominantly controlled by the combination of the absorption cross section $\chi_{min}=3.5\times10^{-2}$ cm^{-1}Torr^{-1} of the molecular oxygen (O$_2$) at 1025 Å and the partial pressure of O$_2$ in air, p_{O_2}, and streamers exhibit branching when their radius becomes greater than $1/\chi_{min}p_{O_2}$. Model results indicate a lower branching threshold radius for positive streamers in comparison with negative streamers, under otherwise identical ambient conditions. These results are in good agreement with recent results of high-speed photography of laboratory streamers in near-atmospheric pressure N$_2$/O$_2$ mixtures (Yi and Williams, 2002; van Veldhuizen and Rutgers, 2002; Ono and Oda, 2003) and similar morphology documented during recent telescopic and high-speed video observations of sprites (Moudry et al., 2003; Gerken and Inan, 2003). The importance of photoionization effects for the branching of streamers has also been noticed in the recent numerical study of ground streamers by Hallac et al. (2003). Following arguments presented in (Liu and Pasko, 2004) we note that non-similar reduction in photoelectron production at high pressures due to quenching of UV emitting excited states of N$_2$ is expected to lead to enhanced branching activity and smaller transverse scales of streamers at high pressures. Such behavior indeed was noted in recent experiments (Pancheshnyi et al., 2005; Briels et al., 2005).

The investigation of streamer splitting characteristics is still in the very preliminary stage. Although some pre-branching features, such as extremely high peak field and electron density in the head with a weak curvature, have been commonly observed in recent studies by several authors (Arrayas et al., 2002; Rocco et al., 2002; Liu and Pasko, 2004), the results on branching morphology reported by different research groups remain highly controversial (Kulikovsky, 2000; Pancheshnyi and Starikovskii, 2001; Kulikovsky, 2001, 2002; Ebert and Hundsdorfer, 2002; Hallac et al., 2003; Liu and Pasko, 2004). The documentation of exact branching conditions of positive and negative streamers in sprites constitutes one of the important tasks of the current TLE research.

12.5.5 Thermal Runaway Electrons in Streamer Tips in Sprites

As it has been already discussed in Section 12.3.4 the observed filamentary structures in sprites (Gerken and Inan, 2003) have recently been interpreted in terms of thin channels of ionization called streamers, which exhibit acceleration, expansion and branching (Liu and Pasko, 2004). The expanding streamers can reach a branching state for a wide range of applied electric fields and an extremely high peak field can be generated in the streamer tip immediately preceding the branching (Liu and Pasko, 2004) (see also discussion on streamer branching in the previous section). The acceleration of electrons in tips of highly overvolted streamers (Babich, 1982) has been proposed for interpretation of X-ray radiation observed in experiments reported by Tarasova et al. (1974). It has also been proposed that the high electric fields in streamer tips can accelerate electrons to energies of several keV, initiating electron runaway in relatively low ambient electric fields $E{\sim}E_k$ (Pasko et al., 1998b), and references cited therein, where E_k is the conventional breakdown threshold field defined in Section 12.2.3. The peak charge moment changes of 3070 C·km (Hu et al., 2002) and 5950 C·km (Cummer and Füllekrug, 2001) corresponding to $E{\simeq}2.3E_k$ and $E{\simeq}4.4E_k$, respectively, at 70 km altitude, may create favorable conditions for the streamer tip runaway phenomena. It is known that X-ray emissions produced by corona discharges in high voltage equipment can be a serious personnel hazard (Roth, 1995, p. 256; Roth, 2001, pp. 48-49). The recently reported X-ray emissions observed in association with natural lightning stepped leaders and triggered lightning dart leaders (Moore et al., 2001; Dwyer et al., 2003, 2004b, 2005) may be related to the enhancement of electric field in the leader streamer zone (Dwyer et al., 2004a) leading to the generation of runaway electrons in streamer tips. The quantitative calculation of fluxes of thermal runaway electrons generated in streamer tips and evaluation of their overall importance in conventional lightning as well as in sprite phenomenon represents an example of another outstanding problem in sprite research. Since streamers can represent a robust source of runaway electrons, the related studies may also be relevant to gamma ray flashes of terrestrial origin (e.g., Smith et al., 2005), and references therein.

Acknowledgments

This research was supported by NSF ATM-0134838 grant to Penn State University.

Bibliography

Adachi, T., Fukunishi, H., Takahashi, Y., and Sato, M. (2004). Roles of the EMP and QE field in the generation of columniform sprites. *Geophys. Res. Lett.*, 31(4):doi:10.1029/2003GL019081.

Akyuz, M., Larsson, A., Cooray, V., and Strandberg, G. (2003). 3D simulations of streamer branching in air. *J. Electrostatics*, 59(115–141):doi: 10.1016/S0304–3886(03)00066–4.

Aleksandrov, N. L. and Bazelyan, E. M. (1996). Temperature and density effects on the properties of a long positive streamer in air. *J. Phys. D: Appl. Phys.*, 29:2873–2880.

Allen, N. L. and Ghaffar, A. (1995). The conditions required for the propagation of a cathode-directed positive streamer in air. *J. Phys. D: Appl. Phys.*, 28:331–337.

Armstrong, R. A., Shorter, J. A., Taylor, M. J., Suszcynsky, D. M., Lyons, W. A., and Jeong, L. S. (1998). Photometric measurements in the SPRITES' 95 and 96 campaigns of nitrogen second positive (399.8 nm) and first negative (427.8 nm) emission. *J. Atmos. Sol.-Terr. Phys.*, 60:787–799.

Armstrong, R. A., Suszcynsky, D. M., Lyons, W. A., and Nelson, T. E. (2000). Multi-color photometric measurements of ionization and energies in sprites. *Geophys. Res. Lett.*, 27:653–657.

Arrayas, M. and Ebert, U. (2004). Stability of negative ionization fronts: Regularization by electric screening? *Phys. Rev. E*, 69:doi:10.1103/PhysRevE.69.036214.

Arrayas, M., Ebert, U., and Hundsdorfer, W. (2002). Spontaneous branching of anode-directed streamers between planar electrodes. *Phys. Rev. Lett.*, 88:doi:10.1103/PhysRevLett.88.174502.

Babaeva, N. Y. and Naidis, G. V. (1997). Dynamics of positive and negative streamers in air in weak uniform electric fields. *IEEE Trans. Plasma Sci.*, 25:375–379.

Babich, L. P. (1982). A new type of ionization wave and the mechanism of polarization self-acceleration of electrons in gas discharge at high overvoltages. *Sov. Phys. Dokl.*, 27:215.

Babich, L. P. (2003). *High-energy phenomena in electric discharges in dense gases: Theory, experiment and natural phenomena*, volume 2 of *ISTC Science and Technology Series*. Futurepast, Arlington, Virginia.

Barrington-Leigh, C. P., Inan, U. S., and Stanley, M. (2001). Identification of sprites and elves with intensified video and broadband array photometry. *J. Geophys. Res.*, 106(A2):doi: 10.1029/ 2000JA000073.

Bazelyan, E. M. and Raizer, Y. P. (1998). *Spark discharge*. CRC Press, Boca Raton.

Bell, T. F., Pasko, V. P., and Inan, U. S. (1995). Runaway electrons as a source of red sprites in the mesosphere. *Geophys. Res. Lett.*, 22:2127–2131.

Blanc, E., Farges, T., Roche, R., Brebion, D., Hua, T., Labarthe, A., and Melnikov, V. (2004). Nadir observations of sprites from the International Space Station. *J. Geophys. Res.*, 109(A2):doi:10.1029/2003JA009972.

Boccippio, D. J., Williams, E. R., Heckman, S. J., Lyons, W. A., Baker, I. T., and Boldi, R. (1995). Sprites, ELF transients, and positive ground strokes. *Science*, 269:1088–1091.

Boeck, W. L., O. H. Vaughan, Jr. and R. J. Blakeslee, Vonnegut, B., Brook, M., and McKune, J. (1995). Observations of lightning in the stratosphere. *J. Geophys. Res.*, 100:1465–1475.

Briels, T. M. P., van Veldhuizen, E. M., and Ebert, U. (2005). Branching of positive discharge streamers in air at varying pressures. *IEEE Trans. Plasma Sci.*, 33(2):264–265.

Bucsela, E., Morrill, J., Heavner, M., Siefring, C., Berg, S., Hampton, D., Moudry, D., Wescott, E., and Sentman, D. (2003). $N_2(B^3\Pi_g)$ and $N_2^+(A^2\Pi_u)$ vibrational distributions observed in sprites. *J. Atmos. Sol.-Terr. Phys.*, 65:583–590.

Cho, M. and Rycroft, M. J. (2001). Non-uniform ionisation of the upper atmosphere due to the electromagnetic pulse from a horizontal lightning discharge. *J. Atmos. Sol.-Terr. Phys.*, 63:559–580.

Cooray, G. V. (2003). *The lightning flash*. IEE, London, UK.

Cummer, S. A. (2003). Current moment in sprite-producing lightning. *J. Atmos. Sol.-Terr. Phys.*, 65(499–508):doi:10.1016/S1364–6826(02)00318–8.

Cummer, S. A. and Füllekrug, M. (2001). Unusually intense continuing current in lightning produces delayed mesospheric breakdown. *Geophys. Res. Lett.*, 28:495–499.

Cummer, S. A., Inan, U. S., Bell, T. F., and Barrington-Leigh, C. P. (1998). ELF radiation produced by electrical currents in sprites. *Geophys. Res. Lett.*, 25:1281–1285.

Cummer, S. A., Zhai, Y., Hu, W., Smith, D. M., Lopez, L. I., and Stanley, M. A. (2005). Measurements and implications of the relationship between lightning and terrestrial gamma-ray flashes. *Geophys. Res. Lett.*, 32(L08811):doi:10.1029/2005GL022778.

Dhali, S. K. and Williams, P. F. (1987). Two-dimensional studies of streamers in gases. *J. Appl. Phys.*, 62:4696–4707.

Dwyer, J. R. (2003). A fundamental limit on electric fields in air. *Geophys. Res. Lett.*, 30(20):doi:10.1029/2003GL017781.

Dwyer, J. R. (2004). Implications of X-ray emission from lightning. *Geophys. Res. Lett.*, 31(L12102):doi: 10.1029/2004GL019795.

Dwyer, J. R. (2005). Gamma-ray events from thunderclouds. In *2005 Seminar Series on Terrestrial Gamma Ray Flashes and Lightning Associated Phenomena*, Space Science Laboratory, University of California, Berkeley.

Dwyer, J. R., Rassoul, H. K., Al-Dayeh, M., Caraway, L., Chrest, A., Wright, B., Kozak, E., Jerauld, J., Uman, M. A., Rakov, V. A., Jordan, D. M., and Rambo, K. J. (2005). X-ray bursts associated with leader steps in cloud-to-ground lightning. *Geophys. Res. Lett.*, 32:doi:10.1029/2004GL021782.

Dwyer, J. R., Rassoul, H. K., Al-Dayeh, M., Caraway, L., Wright, B., Chrest, A., Uman, M. A., Rakov, V. A., Rambo, K. J., Jordan, D. M., Jerauld, J., and Smyth, C. (2004a). A ground level gamma-ray burst observed in association with rocket-triggered lightning. *Geophys. Res. Lett.*, 31(L05119):doi:10.1029/2003GL018771.

Dwyer, J. R., Rassoul, H. K., Al-Dayeh, M., Caraway, L., Wright, B., Chrest, A., Uman, M. A., Rakov, V. A., Rambo, K. J., Jordan, D. M., Jerauld, J., and Smyth, C. (2004b). Measurements of X-ray emission from rocket-triggered lightning. *Geophys. Res. Lett.*, 31(L05118):doi: 10.1029/2003GL018770.

Dwyer, J. R., Uman, M. A., Rassoul, H. K., Al-Dayeh, M., Caraway, L., Jerauld, J., Rakov, V. A., Jordan, D. M., Rambo, K. J., Corbin, V., and Wright, B. (2003). Energetic radiation produced during rocket-triggered lightning. *Science*, 299:694–697.

Eack, K. B., Beasley, W. H., Rust, W. D., Marshall, T. C., and Stolzenburg, M. (1996a). Initial results from simultaneous observation of X rays and electric fields in a thunderstorm. *J. Geophys. Res.*, 101:29637–29640.

Eack, K. B. and W. H. Beasley, Rust, W. D., Marshall, T. C., and Stolzenburg, M. (1996b). X-ray pulses observed above a mesoscale convective system. *Geophys. Res. Lett.*, 23:2915–2918.

Ebert, U. and Hundsdorfer, W. (2002). Comment on "Spontaneous branching of anode-directed streamers between planar electrodes" - Reply. *Phys. Rev. Lett.*, 89(22):doi: 10.1103/ PhysRevLett. 89.229402.

Farges, T., Blanc, E., Pichon, A. Le, Neubert., T., and Allin, T. H. (2005). Identification of infrasound produced by sprites during the Sprite2003 campaign. *Geophys. Res. Lett.*, 32(1):doi:10.1029/2004GL021212.

Fishman, G. J., Bhat, P. N., Mallozzi, R., Horack, J. M., Koshut, T., Kouveliotou, C., Pendleton, G. N., Meegan, C. A., Wilson, R. B., Paciesas, W. S., Goodman, S. J., and Christian, H. J. (1994). Discovery of intense gamma-ray flashes of atmospheric origin. *Science*, 264:1313–1316.

Franz, R. C., Nemzek, R. J., and Winckler, J. R. (1990). Television image of a large upward electrical discharge above a thunderstorm system. *Science*, 249:48–51.

Frey, H., Mende, S., Hsu, R. R., Su, H. T., Chen, A., Lee, L. C., Fukunishi, H., and Takahashi, Y. (2004). The spectral signature of transient luminous events (TLE, sprite, elve, halo) as observed by ISUAL. *Eos Trans. AGU - Fall Meet. Suppl.*, 85(47):AE51A–05. Abstract.

Fukunishi, H., Takahashi, Y., Fujito, M., Wanatabe, Y., and Sakanoi, S. (1996a). Fast imaging of elves and sprites using a framing/streak camera and a multi-anode array photometer. *Eos Trans. AGU - Fall Meet. Suppl.*, 77(46):F60.

Fukunishi, H., Takahashi, Y., Kubota, M., Sakanoi, K., Inan, U. S., and Lyons, W. A. (1996b). Elves: Lightning-induced transient luminous events in the lower ionosphere. *Geophys. Res. Lett.*, 23(16):2157–2160.

Fukunishi, H., Takahashi, Y., Uchida, A., Sera, M., Adachi, K., and Miyasato, R. (1999). Occurrences of sprites and elves above the Sea of Japan near Hokuriku in winter. *Eos Trans. AGU - Fall Meet. Suppl*, 80:F217.

Füllekrug, M. and Constable, S. (2000). Global triangulation of intense lightning discharges. *Geophys. Res. Lett.*, 27:333–336.

Füllekrug, M., Moudry, D. R., Dawes, G., and Sentman, D. D. (2001). Mesospheric sprite current triangulation. *J. Geophys. Res.*, 106:20189–20194.

Füllekrug, M. and Reising, S. C. (1998). Excitation of Earth-ionosphere cavity resonances by sprite-associated lightning flashes. *Geophys. Res. Lett.*, 25:4145–4148.

Gallimberti, I., Bacchiega, G., Bondiou-Clergerie, A., and Lalande, P. (2002). Fundamental processes in long air gap discharges. *Compt. Rend. Phys.*, 3:1335–1359.

Gerken, E. A. and Inan, U. S. (2002). A survey of streamer and diffuse glow dynamics observed in sprites using telescopic imagery. *J. Geophys. Res.*, 107(A11):doi:10.1029/2002JA009248.

Gerken, E. A. and Inan, U. S. (2003). Observations of decameter-scale morphologies in sprites. *J. Atmos. Sol.-Terr. Phys*, 65(567–572):doi:10.1016/S1364–6826(02)00333–4.

Gerken, E. A. and Inan, U. S. (2004). Comparison of photometric measurements and charge moment estimations in two sprite-producing storms. *Geophys. Res. Lett.*, 31(L03107):doi: 10.1029/ 2003GL018751.

Gerken, E. A. and Inan, U. S. (2005). Streamers and diffuse glow observed in upper atmospheric electrical discharges. *IEEE Trans. Plsama Sci.*, 33(2):282–283.

Gerken, E. A., Inan, U. S., and Barrington-Leigh, C. P. (2000). Telescopic imaging of sprites. *Geophys. Res. Lett.*, 27:2637–2640.

Grange, F., Soulem, N., Loiseau, J. F., and Spyrou, N. (1995). Numerical and experimental-determination of ionizing front velocity in a DC point-to-plane corona discharge. *J. Phys. D: Appl. Phys.*, 28:1619–1629.

Griffiths, R. F. and Phelps, C. T. (1976). The effects of air pressure and water vapour content on the propagation of positive corona streamers, and their implications to lightning initiation. *Quart. J. Roy. Met. Soc.*, 102:419–426.

Gurevich, A. V., Milikh, G. M., and Roussel-Dupré, R. (1992). Runaway electron mechanism of air breakdown and preconditioning during a thunderstorm. *Phys. Lett. A*, 165:463–468.

Gurevich, A. V. and Zybin, K. P. (2001). Runaway breakdown and electric discharges in thunderstorms. *Phys.-Uspekhi*, 44:1119–1140.

Hallac, A., Georghiou, G. E., and Metaxas, A. C. (2003). Secondary emission effects on streamer branching in transient non-uniform short-gap discharges. *J. Phys. D: Appl. Phys.*, 36:doi:10.1088/0022–3727/36/20/011.

Hampton, D. L., Heavner, M. J., Wescott, E. M., and Sentman, D. D. (1996). Optical spectral characteristics of sprites. *Geophys. Res. Lett.*, 23:89–93.

Hardman, S. F., Dowden, R. L., Brundell, J. B., Bahr, L., Kawasaki, Z., and Rodger, C. J. (1998). Sprites in Australia's Northern Territory. *Eos Trans. AGU - Fall Meet. Suppl.*, 79:F135.

Hardman, S. F., Dowden, R. L., Brundell, J. B., Bahr, L., Kawasaki, Z., and Rodger, C. J. (2000). Sprite observations in the Northern Territory of Australia. *J. Geophys. Res.*, 105(D4):4689–4698.

Hayakawa, M., Nakamura, T., Hobara, Y., and Williams, E. (2004). Observations of sprites over the Sea of Japan and conditions for lightning-induced sprites in winter. *J. Geophys. Res.*, 109:doi:10.1029/2003JA009905.

Heavner, M. J., Hampton, D., Sentman, D., and Wescott, E. (1995). Sprites over Central and South America - Fall Meet. Suppl. *Eos Trans. AGU*, 76(46):F115.

Heavner, M. J., Sentman, D. D., Moudry, D. R., and Wescott, E. M. (2000). Sprites, blue jets, and elves: Optical evidence of energy transport across the stratopause. In Siskind, D.E., Eckermann, S.D., and Summers, M.E., editors, *Atmospheric Science Across the Stratopause*, volume 123 of *Geophysical Monograph Series*, pages 69–82. AGU.

Hobara, Y., Iwasaki, N., Hayashida, T., Hayakawa, M., Ohta, K., and Fukunishi, H. (2001). Interrelation between ELF transients and ionospheric disturbances in association with sprites and elves. *Geophys. Res. Lett.*, 28(5):doi:10.1029/2000GL003795.

Holzworth, R. H., McCarthy, M. P., Thomas, J. N., Chinowsky, T. M., Taylor, M. J., and Pinto, O. (2003). Strong electric fields from positive lightning strokes in the stratosphere: Implications for sprites - Fall Meet. Suppl., Abstract. *Eos Trans. AGU*, 84(46):AE51A–01.

Hsu, R. R., Su, H. T., Chen, A. B., Lee, L. C., Asfur, M., Price, C., and Yair, Y. (2003). Transient luminous events in the vicinity of Taiwan. *J. Atmos. Sol.-Terr. Phys.*, 65(5):561–566.

Hu, W. Y., Cummer, S. A., and Lyons, W. A. (2002). Lightning charge moment changes for the initiation of sprites. *Geophys. Res. Lett.*, 29(8):doi:10.1029/2001GL014593.

ICRU (1984). International commission on radiation units and measurements, stopping powers for electrons and positrons. ICRU Rep. 37. Tables 8.1 and 12.4, Bethesda, Md.

Inan, U. S. (2005). Gamma rays made on Earth. *Science*, 307:1054–1055.

Inan, U. S., Barrington-Leigh, C., Hansen, S., Glukhov, V. S., Bell, T. F., and Rairden, R. (1997). Rapid lateral expansion of optical luminosity in lightning-induced ionospheric flashes referred to as 'elves'. *Geophys. Res. Lett.*, 24(5):583–586.

Inan, U. S., Cohen, M., and Said, R. (2005). Terrestrial gamma ray flashes and VLF radio atmospherics. In *2005 Seminar Series on Terrestrial Gamma Ray Flashes and Lightning Associated Phenomena*, Space Science Laboratory, University of California, Berkeley.

Inan, U. S., Reising, S. C., Fishman, G. J., and Horack, J. M. (1996). On the association of terrestrial gamma-ray bursts with lightning and implications for sprites. *Geophys. Res. Lett.*, 23:1017–1020.

Israelevich, P. L., Yair, Y., Devir, A. D., Joseph, J. H., Levin, Z., Mayo, I., Moalem, M., Price, C., Ziv, B., and Sternlieb, A. (2004). Transient airglow enhancements observed from the Space Shuttle Columbia during the MEIDEX sprite campaign. *Geophys. Res. Lett.*, 31(L06124):doi: 10.1029/2003GL019110.

Johnson, M. P. and Inan, U. S. (2000). Sferic clusters associated with Early/Fast VLF events. *Geophys. Res. Lett.*, 27:1391–1394.

Kulikovsky, A. A. (1997). Production of chemically active species in the air by a single positive streamer in a nonuniform field. *IEEE Trans. Plasma Sci.*, 25:439–446.

Kulikovsky, A. A. (2000). The role of photoionization in positive streamer dynamics. *J. Phys. D: Appl. Phys.*, 33:1514–1524.

Kulikovsky, A. A. (2001). Reply to comment on 'The role of photoionization in positive streamer dynamics'. *J. Phys. D: Appl. Phys.*, 34:251–252.

Kulikovsky, A. A. (2002). Comment on "Spontaneous branching of anode-directed streamers between planar electrodes". *Phys. Rev. Lett.*, 89(22):doi: 10.1103/ PhysRevLett.89.229401.

Kunhardt, E. E. and Tzeng, Y. (1988). Development of an electron avalanche and its transition into streamers. *Phys. Rev. A.*, 38:1410–1421.

Kutsyk, I. M. and Babich, L. P. (1999). Spatial structure of optical emissions in the model of gigantic upward atmospheric discharges with participation of runaway electrons. *Phys. Lett. A*, 253:75–82.

Lehtinen, N. G., Bell, T. F., Pasko, V. P., and Inan, U. S. (1997). A two-dimensional model of runaway electron beams driven by quasi-electrostatic thundercloud fields. *Geophys. Res. Lett.*, 24:2639–2642.

Lehtinen, N. G., Inan, U. S., and Bell, T. F. (1999). Monte Carlo simulation of runaway MeV electron breakdown with application to red sprites and terrestrial gamma ray flashes. *J. Geophys. Res.*, 104:24699–24712.

Lehtinen, N. G., Inan, U. S., and Bell, T. F. (2000). Trapped energetic electron curtains produced by thunderstorm driven relativistic runaway electrons. *Geophys. Res. Lett.*, 27:1095–1098.

Lehtinen, N. G., Inan, U. S., and Bell, T. F. (2001). Effects of thunderstorm-driven runaway electrons in the conjugate hemisphere: Purple sprites, ionization enhancements, and gamma rays. *J. Geophys. Res.*, 106:28841–28856.

Lehtinen, N. G., Walt, M., Inan, U. S., Bell, T. F., and Pasko, V. P. (1996). Gamma-ray emission produced by a relativistic beam of runaway electrons accelerated by quasi-electrostatic thundercloud fields. *Geophys. Res. Lett.*, 23:2645–2648.

Lieberman, M. A. and Lichtenberg, A. J. (1994). *Principles of plasma discharges and materials processing*. John Wiley & Sons Inc.

Liszka, L. (2004). On the possible infrasound generation by sprites. *J. Low Freq. Noise, Vibration and Active Cont.*, 23(2):85–93.

Liu, N. and Pasko, V. P. (2004). Effects of photoionization on propagation and branching of positive and negative streamers in sprites. *J. Geophys. Res.*, 109:A04301, doi:10.1029/2003JA010064. (see also Liu, N., and V. P. Pasko, Correction to "Effects of photoionization on propagation and branching of positive and negative streamers in sprites," *J. Geophys. Res., 109*, A09306, doi:10.1029/2004JA010692, 2004).

Liu, N. and Pasko, V. P. (2005). Molecular nitrogen LBH band system far-UV emissions of sprite streamers. *Geophys. Res. Lett.*, 32:doi:10.1029/2004GL022001.

Loeb, L. B. and Meek, J. M. (1940). The mechanism of spark discharge in air at atmospheric pressure. *J. Appl. Phys.*, 11:438–447.

Lyons, W. A. (1996). Sprite observations above the U.S. high plains in relation to their parent thunderstorm systems. *J. Geophys. Res.*, 101(29):641.

Lyons, W. A., Nelson, T. E., Armstrong, R. A., Pasko, V. P., and Stanley, M. A. (2003a). Upward electrical discharges from thunderstorm tops. *Bull. Am. Met. Soc.*, 84(4):445–454.

Lyons, W. A., Nelson, T. E., Williams, E. R., Cummer, S. A., and Stanley, M. A. (2003b). Characteristics of sprite-producing positive cloud-to-ground lightning during the 19 July 2000 STEPS mesoscale convective systems. *Mon. Wea. Rev.*, 131:2417–2427.

Marshall, R. A. and Inan, U. S. (2005). High-speed telescopic imaging of sprites. *Geophys. Res. Lett.*, 32(L05804):doi:10.1029/2004GL021988.

Marshall, T. C., McCarthy, M. P., and Rust, W. D. (1995). Electric-field magnitudes and lightning initiation in thunderstorms. *J. Geophys. Res.*, 100:7097–7103.

Marshall, T. C. and Rust, W. D. (1993). Two types of vertical electrical structures in stratiform precipitation regions of mesoscale convective systems. *Bull. Am. Met. Soc.*, 74:2159.

Marshall, T. C., Stolzenburg, M., Maggio, C. R., Coleman, L. M., Krehbiel, P. R., Hamlin, T., Thomas, R. J., and Rison, W. (2005). Observed electric fields associated with lightning initiation. *Geophys. Res. Lett.*, 32(L03813):doi:10.1029/2004GL021802.

Marshall, T. C., Stolzenburg, M., and Rust, W. D. (1996). Electric field measurements above mesoscale convective systems. *J. Geophys. Res.*, 101:6979–6996.

Marshall, T. C., Stolzenburg, M., Rust, W. D., Williams, E. R., and Boldi, R. (2001). Positive charge in the stratiform cloud of a mesoscale convective system. *J. Geophys. Res.*, 106:1157–1163.

McCarthy, M. P. and Parks, G. K. (1992). On the modulation of X-ray fluxes in thunderstorms. *J. Geophys. Res.*, 97:5857–5864.

McHarg, M. G., Haaland, R. K., Moudry, D. R., and Stenbaek-Nielsen, H. C. (2002). Altitude-time development of sprites. *J. Geophys. Res.*, 107(A11):doi:10.1029/2001JA000283.

Mende, S., Frey, H., R. R. Hsu, H. T. Su, Chen, A., Lee, L. C., Fukunishi, H., and Takahashi, Y. (2004). Sprite imaging results from the ROCSAT-2 ISUAL instrument. *Eos. Trans. AGU, Fall Meet. Suppl., Abstract AE51A-02*, 85(47).

Mende, S. B., Rairden, R. L., Swenson, G. R., and Lyons, W. A. (1995). Sprite spectra: $N_2$1PG band identification. *Geophys. Res. Lett.*, 22:2633–2637.

Milikh, G. M., Guzdar, P. N., and Sharma, A. S. (2005). Gamma ray flashes due to plasma processes in the atmosphere: Role of whistler waves. *J. Geophys. Res.*, 110(A02308):doi: 10.1029/ 2004JA010681.

Miyasato, R., Fukunishi, H., Takahashi, Y., and Taylor, M. J. (2003). Energy estimation of electrons producing sprite halos using array photometer data. *J. Atmos. Sol.-Terr. Phys.*, 65:573–581.

Miyasato, R., Taylor, M. J., Fukunishi, H., and Stenbaek-Nielsen, H. C. (2002). Statistical characteristics of sprite halo events using coincident photometric and imaging data. *Geophys. Res. Lett.*, 29(21):doi:10.1029/2001GL014480.

Moore, C. B., Eack, K. B., Aulich, G. D., and Rison, W. (2001). Energetic radiation associated with lightning stepped-leaders. *Geophys. Res. Lett.*, 28:2141–2144.

Moore, R. C., Barrington-Leigh, C. P., Inan, U. S., and Bell, T. F. (2003). Early/fast VLF events produced by electron density changes associated with sprite halos. *J. Geophys. Res.*, 108(A10):doi:10.1029/2002JA009816.

Morrill, J., Bucsela, E., Siefring, C., Heavner, M., Berg, S., Moudry, D., Slinker, S., Fernsler, R., Wescott, E., Sentman, D., and Osborne, D. (2002). Electron energy and electric field estimates in sprites derived from ionized and neutral N_2 emissions. *Geophys. Res. Lett.*, 29(10):doi: 10.1029/ 2001GL014018.

Morrill, J. S., Bucsela, E. J., Pasko, V. P., Berg, S. L., Benesch, W. M., Wescott, E. M., and Heavner, M. J. (1998). Time resolved N_2 triplet state vibrational populations and emissions associated with red sprites. *J. Atmos. Sol.-Terr. Phys.*, 60:811–829.

Morrow, R. and Lowke, J. J. (1997). Streamer propagation in air. *J. Phys. D: Appl. Phys.*, 30:614–627.

Moudry, D. R., Stenbaek-Nielsen, H. C., Sentman, D. D., and Wescott, E. M. (2002). Velocities of sprite tendrils. *Geophys. Res. Lett.*, 29(20):1992doi:10.1029/2002GL015682.

Moudry, D. R., Stenbaek-Nielsen, H. C., Sentman, D. D., and Wescott, E. M. (2003). Imaging of elves, halos and sprite initiation at 1 ms time resolution. *J. Atmos. Sol.-Terr. Phys.*, 65(509–518):doi:10.1016/S1364-6826(02)00323-1.

Neubert, T. (2003). On sprites and their exotic kin. *Science*, 300:747–749.

Neubert, T., Allin, T. H., Stenbaek-Nielsen, H., and Blanc, E. (2001). Sprites over Europe. *Geophys. Res. Lett.*, 28(18):doi:10.1029/2001GL013427.

Niemeyer, L., Ullrich, L., and Wiegart, N. (1989). The mechanism of leader breakdown in electronegative gases. *IEEE Trans. Electr. Insul.*, 24:309–324.

Ohkubo, A., Fukunishi, H., Takahashi, Y., and Adachi, T. (2005). VLF/ELF sferic evidence for in-cloud discharge activity producing sprites. *Geophys. Res. Lett.*, 32(L04812):doi: 10.1029/ 2004GL021943.

Ono, R. and Oda, T. (2003). Formation and structure of primary and secondary streamers in positive pulsed corona discharge effect of oxygen concentration and applied voltage. *J. Phys. D: Appl. Phys.*, 36(1952–1958):doi:10.1088/0022–3727/36/16/306.

Opal, C. B., Peterson, W. K., and Beaty, E. C. (1971). Measurements of secondary-electron spectra produced by electron impact ionization of a number of simple gases. *J. Chem. Phys.*, 55:4100–4106.

Pancheshnyi, S. V., Nudnova, M., and Starikovskii, A. Y. (2005). Development of a cathode-directed streamer discharge in air at different pressures: Experiment and comparison with direct numerical simulation. *Phys. Rev. E*, 71(016407):doi:10.1103/PhysRevE.71.016407.

Pancheshnyi, S. V. and Starikovskii, A. Y. (2001). Comments on 'The role of photoionization in positive streamer dynamics'. *J. Phys. D: Appl. Phys.*, 34:248–250.

Pancheshnyi, S. V. and Starikovskii, A. Y. (2003). Two-dimensional numerical modelling of the cathode-directed streamer development in a long gap at high voltage. *J. Phys. D: Appl. Phys.*, 36(2683–2691):doi:10.1088/0022–3727/36/21/014.

Pasko, V. P. (2003a). Dynamics of streamer-to-leader transition in transient luminous events between thunderstorm tops and the lower ionosphere. *Eos Trans. AGU - Fall Meet. Suppl.*, 84(46):AE41B–05. Abstract.

Pasko, V. P. (2003b). Electric jets. *Nature*, 423:927–929.

Pasko, V. P. and George, J. J. (2002). Three-dimensional modeling of blue jets and blue starters. *J. Geophys. Res.*, 107(A12):doi:10.1029/2002JA009473.

Pasko, V. P., Inan, U. S., and Bell, T. F. (1998a). Mechanism of ELF radiation from sprites. *Geophys. Res. Lett.*, 25:3493–3496.

Pasko, V. P., Inan, U. S., and Bell, T. F. (1998b). Spatial structure of sprites. *Geophys. Res. Lett.*, 25:2123–2126.

Pasko, V. P., Inan, U. S., and Bell, T. F. (1999). Mesospheric electric field transients due to tropospheric lightning discharges. *Geophys. Res. Lett.*, 26:1247–1250.

Pasko, V. P., Inan, U. S., and Bell, T. F. (2000). Fractal structure of sprites. *Geophys. Res. Lett.*, 27:497–500.

Pasko, V. P., Inan, U. S., and Bell, T. F. (2001). Mesosphere-troposphere coupling due to sprites. *Geophys. Res. Lett.*, 28:3821–3824.

Pasko, V. P., Inan, U. S., Bell, T. F., and Taranenko, Y. N. (1997). Sprites produced by quasi-electrostatic heating and ionization in the lower ionosphere. *J. Geophys. Res.*, 102:4529–4561.

Pasko, V. P., Stanley, M. A., Mathews, J. D., Inan, U. S., and Wood, T. G. (2002a). Electrical discharge from a thundercloud top to the lower ionosphere. *Nature*, 416:152–154. http://pasko.ee.psu.edu/Nature/.

Pasko, V. P., Stanley, M. A., Mathews, J. D., Inan, U. S., Wood, T. G., Cummer, S. A., Williams, E. R., and Heavner, M. J. (2002b). Observations of sprites above Haiti/Dominican Republic thunderstorms from Arecibo Observatory, Puerto Rico. *Eos Trans. AGU - Fall Meet. Suppl.*, 83(47):A62D–01. Abstract.

Pasko, V. P. and Stenbaek-Nielsen, H. C. (2002). Diffuse and streamer regions of sprites. *Geophys. Res. Lett.*, 29(A10):doi:10.1029/2001GL014241.

Petrov, N. I., Avanskii, V. R., and Bombenkova, N. V. (1994). Measurement of the electric field in the streamer zone and in the sheath of the channel of a leader discharge. *Tech. Phys.*, 39:546–551.

Petrov, N. I. and Petrova, G. N. (1993). Physical mechanisms for intracloud lightning discharges. *Tech. Phys.*, 38:287.

Petrov, N. I. and Petrova, G. N. (1999). Physical mechanisms for the development of lightning discharges between a thundercloud and the ionosphere. *Tech. Phys.*, 44:472–475.

Pinto Jr., O., Saba, M. M. F., Pinto, I. R. C. A., Tavares, F. S. S., Naccarato, K. P., Solorzano, N. N., Taylor, M. J., Pautet, P. D., and Holzworth, R. H. (2004). Thunderstorm and lightning characteristics associated with sprites in Brazil. *Geophys. Res. Lett.*, 31(L13103):doi:10.1029/2004GL020264.

Popov, N. A. (2002). Spatial structure of the branching streamer channels in a corona discharge. *Plasma. Phys. Rep.*, 28(7):doi:10.1134/1.1494061.

Price, C., Greenberg, E., Yair, Y., Satori, G., Bor, J., Fukunishi, H., Sato, M., Israelevich, P., Moalem, M., Devir, A., Levin, Z., Joseph, J. H., Mayo, I., Ziv, B., and Sternlieb, A. (2004). Ground-based detection of TLE-producing intense lightning during the MEIDEX mission on board the Space Shuttle Columbia,. *Geophys. Res. Lett.*, 31(L20107):doi:10.1029/2004GL020711.

Raizer, Y. P. (1991). *Gas discharge physics*. Springer-Verlag, Berlin Heidelberg.

Raizer, Y. P., Milikh, G. M., Shneider, M. N., and Novakovski, S. V. (1998). Long streamers in the upper atmosphere above thundercloud. *J. Phys. D: Appl. Phys.*, 31:3255–3264.

Raizer, Y. P. and Simakov, A. N. (1998). Main factors determining the radius of the head of a long streamer and the maximum electric field near the head. *Plasma. Phys. Rep.*, 24:700–706.

Rakov, V. A. and Uman, M. A. (2003). *Lightning physics and effects*. Cambridge University Press.

Reising, S. C., Inan, U. S., and Bell, T. F. (1999). ELF sferic energy as a proxy indicator for sprite occurrence. *Geophys. Res. Lett.*, 26:987–990.

Rocco, A., Ebert, U., and Hundsdorfer, W. (2002). Branching of negative streamers in free flight. *Phys. Rev. E*, 66:doi:10.1103/PhysRevE.66.035102.

Roth, R. J. (1995). *Industrial plasma engineering*, volume 1: Principles. IOP Publishing Ltd.

Roth, R. J. (2001). *Industrial plasma engineering*, volume 2: Applications to nonthermal plasma processing. IOP Publishing Ltd.

Roussel-Dupré, R. and Gurevich, A. V. (1996). On runaway breakdown and upward propagating discharges. *J. Geophys. Res.*, 101:2297–2311.

Roussel-Dupré, R., Gurevich, A. V., Tunnell, T., and Milikh, G. M. (1994). Kinetic-theory of runaway air breakdown. *Phys. Rev. E*, 49:2257–2271.

Roussel-Dupré, R., Symbalisty, E., Taranenko, Y., and Yukhimuk, V. (1998). Simulations of high-altitude discharges initiated by runaway breakdown. *J. Atmos. Sol.-Terr. Phys.*, 60:917–940.

Rowland, H. L. (1998). Theories and simulations of elves, sprites and blue jets. *J. Atmos. Sol.-Terr. Phys.*, 60:831–844.

Sato, M. and Fukunishi, H. (2003). Global sprite occurrence locations and rates derived from triangulation of transient Schumann resonance events. *Geophys. Res. Lett.*, 30(16):doi: 10.1029/ 2003GL017291.

Sentman, D. D., Wescott, E. M., Heavner, M. J., and Moudry, D. R. (1996). Observations of sprite beads and balls. *Eos Trans. AGU - Fall Meet. Suppl*, 77.

Sentman, D. D., Wescott, E. M., Osborne, D. L., Hampton, D. L., and Heavner, M. J. (1995). Preliminary results from the Sprites94 campaign: Red sprites. *Geophys. Res. Lett.*, 22:1205–1208.

Shepherd, T. R., Rust, W. D., and Marshall, T. C. (1996). Electric fields and charges near 0°C in stratiform clouds. *Mon. Wea. Rev.*, 124:919–938.

Smith, D. M., Lopez, L. I., Lin, R. P., and Barrington-Leigh, C. P. (2005). Terrestrial gamma-ray flashes observed up to 20 MeV. *Science*, 307:1085–1088.

Stanley, M., Krehbiel, P., Brook, M., Moore, C., Rison, W., and Abrahams, B. (1999). High speed video of initial sprite development. *Geophys. Res. Lett.*, 26:3201–3204.

Stanley, M., Krehbiel, P., Brook, M., Rison, W., Moore, C., and Vaughan, O. H. (1996). Observations of sprites and jets from Langmuir Laboratory, New Mexico. *Eos Trans. AGU - Fall Meet. Suppl*, 77(46):F69.

Stenbaek-Nielsen, H. C., Moudry, D. R., Wescott, E. M., Sentman, D. D., and Sabbas, F. T. Sao (2000). Sprites and possible mesospheric effects. *Geophys. Res. Lett.*, 27:3827–3832.

Su, H., Huang, T., Kuo, C., Chen, A. C., Hsu, R., Mende, S. B., Frey, H. U., Fukunishi, H., Takahashi, Y., and Lee, L. (2004). Global distribution of TLEs based on the preliminary ISUAL data. *Eos. Trans. AGU - Fall Meet. Suppl.*, 85(47):AE51A–03.

Su, H. T., Hsu, R. R., Chen, A. B., Lee, Y. J., and Lee, L. C. (2002). Observation of sprites over the Asian continent and over oceans around Taiwan. *Geophys. Res. Lett.*, 29(4):doi:10.1029/2001GL013737.

Su, H. T., Hsu, R. R., Chen, A. B., Wang, Y. C., Hsiao, W. S., Lai, W. C., Lee, L. C., Sato, M., and Fukunishi, H. (2003). Gigantic jets between a thundercloud and the ionosphere. *Nature*, 423:974–976.

Sukhorukov, A. I. and Stubbe, P. (1998). Problems of blue jet theories. *J. Atmos. Sol.-Terr. Phys.*, 60:725–732.

Suszcynsky, D. M., Roussel-Dupré, R., Lyons, W. A., and Armstrong, R. A. (1998). Blue-light imagery and photometry of sprites. *J. Atmos. Sol.-Terr. Phys.*, 60:801–809.

Suszcynsky, D. M., Strabley, R., Roussel-Dupré, R., Symbalisty, E. M. D., Armstrong, R. A., Lyons, W. A., and Taylor, M. (1999). Video and photometric observations of a sprite in coincidence with a meteor-triggered jet event. *J. Geophys. Res.*, 104(D24):doi: 10.1029/ 1999JD900962.

Symbalisty, E. M. D., Roussel-Dupré, R. A., and Yukhimuk, V. A. (1998). Finite volume solution of the relativistic Boltzmann equation for electron avalanche studies. *IEEE Trans. Plasma Sci.*, 26:1575–1582.

Takahashi, Y., Fujito, M., Watanabe, Y., Fukunishi, H., and Lyons, W A. (2000). Temporal and spatial variations in the intensity ratio of N-2 1st and 2nd positive bands in SPRITES. *Advances in Space Research*, 26(8):1205–1208.

Taranenko, Y. N. and Roussel-Dupré, R. (1996). High altitude discharges and gamma ray flashes: A manifestation of runaway air breakdown. *Geophys. Res. Lett.*, 23:571–574.

Tarasova, L. V., Khudyakova, L. N., Loiko, T. V., and Tsukerman, V. A. (1974). Fast electrons and X rays from nanosecond gas discharges at 0.1-760 Torr. *Sov. Phys. Tech. Phys.*, 19:351.

Taylor, M. J. and Clark, S. (1996). High resolution CCD and video imaging of sprites and elves in the N_2 first positive band emission. *Eos Trans. AGU - Fall Meet. Suppl.*, 77(46):F60.

Uman, M. A. (2001). *The Lightning Discharge*. Dover Publications, New York.

Vallance-Jones, A. V. (1974). *Aurora*. D. Reidel Publishing Co.

van Veldhuizen, E. M., editor (2000). *Electrical discharges for environmental purposes: Fundamentals and applications*. Nova Science, New York.

van Veldhuizen, E. M. and Rutgers, W. R. (2002). Pulsed positive corona streamer propagation and branching. *J. Phys. D: Appl. Phys.*, 35:2169–2179.

Veronis, G., Pasko, V. P., and Inan, U. S. (1999). Characteristics of mesospheric optical emissions produced by lightning discharges. *J. Geophys. Res.*, 104:12,645–12,656.

Vitello, P. A., Penetrante, B. M., and Bardsley, J. N. (1993). Multidimensional modeling of the dynamic morphology of streamer coronas. In *Non-Thermal Plasma Techniques for Pollution Control*, volume G34, part A of *NATO ASI Ser.*, pages 249–271. Springer-Verlag, New York.

Vitello, P. A., Penetrante, B. M., and Bardsley, J. N. (1994). Simulation of negative-streamer dynamics in nitrogen. *Phys. Rev. E*, 49:5574–5598.

Wescott, E. M., Sentman, D., Osborne, D., Hampton, D., and Heavner, M. (1995). Preliminary results from the Sprites94 aircraft campaign: 2. Blue jets. *Geophys. Res. Lett.*, 22:1209–1212.

Wescott, E. M., Sentman, D. D., Heavner, M. J., Hampton, D. L., Osborne, D. L., and Vaughan Jr., O. H. (1996). Blue starters: Brief upward discharges from an intense Arkansas thunderstorm. *Geophys. Res. Lett.*, 23:2153–2156.

Wescott, E. M., Sentman, D. D., Heavner, M. J., Hampton, D. L., and Vaughan Jr., O. H. (1998). Blue jets: their relationship to lightning and very large hailfall, and their physical mechanisms for their production. *J. Atmos. Sol.-Terr. Phys.*, 60:713–724.

Wescott, E. M., Sentman, D. D., Stenbaek-Nielsen, H. C., Huet, P., Heavner, M. J., and Moudry, D. R. (2001). New evidence for the brightness and ionization of blue starters and blue jets. *J. Geophys. Res.*, 106:21549–21554.

Wilson, C. T. R. (1925). The electric field of a thundercloud and some of its effects. *Proc. Phys. Soc. London*, 37(32D–37D).

Winn, W. P., Schwede, G. W., and Moore, C. B. (1974). Measurements of electric fields in thunderclouds. *J. Geophys. Res.*, 79:1761–1767.

Yair, Y., Israelevich, P., Devir, A. D., Moalem, M., Price, C., Joseph, J. H., Levin, Z., Ziv, B., Sternlieb, A., and Teller, A. (2004). New observations of sprites from the space shuttle. *J. Geophys. Res.*, 109:doi:10.1029/2003JD004497.

Yair, Y., Price, C., Levin, Z., Joseph, J., Israelevitch, P., Devir, A., Moalem, M., Ziv, B., and Asfur, M. (2003). Sprite observations from the space shuttle during the Mediterranean Israeli dust experiment (MEIDEX). *J. Atmos. Sol.-Terr. Phys.*, 65(5):635–642,doi:10.1016/S1364–6826(02)00332–2.

Yi, W. J. and Williams, P. F. (2002). Experimental study of streamer in pure N_2 and N_2/O_2 mixtures and a ≈ 13 cm gap. *J. Phys. D: Appl. Phys.*, 35:205–218.

Yukhimuk, V., Roussel-Dupré, R. A., and Symbalisty, E. M. D. (1999). On the temporal evolution of red sprites: Runaway theory versus data. *Geophys. Res. Lett.*, 26:679–682.

Yukhimuk, V., Roussel-Dupré, R. A., Symbalisty, E. M. D., and Taranenko, Y. (1998a). Optical characteristics of blue jets produced by runaway air breakdown, simulation results. *Geophys. Res. Lett.*, 25:3289–3292.

Yukhimuk, V., Roussel-Dupré, R. A., Symbalisty, E. M. D., and Taranenko, Y. (1998b). Optical characteristics of red sprites produced by runaway air breakdown. *J. Geophys. Res.*, 103:11473–11482.

Zabotin, N. A. and Wright, J. W. (2001). Role of meteoric dust in sprite formation. *Geophys. Res. Lett.*, 28(13):doi:10.1029/2000GL012699.

Zheleznyak, M. B., Mnatsakanyan, A. Kh., and Sizykh, S. V. (1982). Photoionization of nitrogen and oxygen mixtures by radiation from a gas discharge. *High Temp.*, 20:357–362.

ON THE MODELING OF SPRITES AND SPRITE-PRODUCING CLOUDS IN THE GLOBAL ELECTRIC CIRCUIT

Eugene A. Mareev, A. A. Evtushenko and S. A. Yashunin
Institute of Applied Physics, Russian Academy of Science, Nizhny Novgorod, Russian Federation.

Abstract   Quasi-stationary and fast transient processes connected with powerful lightning discharges and large-scale thunderstorm systems are analyzed. The main physical ideas serving as the foundation for sprite and sprite-producing cloud modeling are discussed with simple examples. Special attention is paid to the adequate description of the field sources and appropriate set of boundary conditions. The features of electron heating and ionization in the middle atmosphere are briefly discussed. The importance of the global circuit for the modeling of sprites and sprite-producing clouds is recognized. Along with the setting of boundary conditions, its role is connected with the importance of large thunderstorm complexes, including mesoscale convective systems (MCSs). It was shown recently that MCS stratiform regions make an especially large current contribution to the global circuit, serving either as an effective generator or as a discharger of the circuit depending on the polarity, magnitude and thickness of the MCS layers. On the other hand, stratiform regions of MCS are characterized by an enhanced rate of positive flashes, which are known to correlate with sprites. In the case of MCS the big narrow layers, generated near the $0°$ C isotherm serve as the source of electric charges for positive CG flashes. We suggest a model of a positive charge layer near the $0°$ C isotherm, based on the hypothesis that the melting-charging mechanism plays a principal role in the formation of the layer. We illustrate how microphysical considerations result in electric currents for the use in modeling of the global circuit and discharge processes. Further we address two aspects of the global electric circuit conception, particularly important from the viewpoint of sprites and sprite-producing cloud research. First is a classical aspect of the global circuit as the quasi-stationary current contour supported by the operation of thunderstorm generators over the globe. Another aspect is connected to the energy deposition and dissipation into the circuit, treated as an open dissipative system. Simple energetic estimates of sprite occurrence are presented. Nonlinear aspects of modeling throughout the chapter are emphasized.

13.1 Introduction

Recent observations of transient luminous events in the atmosphere encouraged new efforts in the modeling of intensive lightning discharges, their parent thunderstorm systems and the global atmospheric electric circuit. It becomes now obvious that these fields of research are closely related to each other. Indeed, the understanding of the physical nature of sprites, jets and elves is impossible without detail studies of thunderstorm and CG/IC discharge processes. The modern progress in the global electric circuit research depends on the study of quasi-stationary and fast transient processes connected with powerful lightning discharges and large-scale thunderstorm systems. The correct modeling of the processes described above requires a set of proper boundary and initial conditions determined, in turn, by the state of the global electric circuit. As a result, success in the modeling depends to a high degree on taking into account the spatial and temporal structure of different subsystems forming the global electric circuit. Another important factor of model development is the use of modern (nonlinear) methods of analytical and numerical consideration.

In the present chapter we show with simple examples the main physical ideas being the foundation for sprite and sprite-producing cloud modeling. General approaches for the modeling of quasi-stationary and transient electric fields in the conducting atmosphere are explained. Special attention is paid to the adequate description of field sources and an appropriate set of boundary conditions. The features of electron heating and ionization are discussed. An importance of global circuit conception in terms of the modeling of sprites and sprite-producing clouds is recognized. One of the reasons is the role of mesoscale convective systems (MCS). It was shown recently that MCS stratiform regions make an especially large current contribution to the global circuit (Davydenko et al., 2004). On the other hand, just stratiform regions of MCS are characterized by an enhanced rate of positive flashes, which are known to correlate with sprites (Lyons, 1996). In the case of a MCS, the source of positive electric charges for positive CG flashes is situated in large horizontal narrow layers, generated near the 0° C isotherm (Marshall and Rust, 1993; Marshall et al., 2001). We suggest a model of a positive charge layer near the 0° C isotherm, serving as a source for positive flashes associated with sprites. Our model is based on the hypothesis that the melting-charging mechanism plays a principal role in the formation of the layer. By this mechanism previously uncharged precipitation particles acquire negative charge as they melt by shedding smaller, cloud-size particles. We show how microphysical consideration provides electric currents for further use in the modeling of the global circuit and discharge processes.

13.2 Time-Dependent Electric Field in the Conducting Atmosphere

When modeling electrodynamic processes in the atmosphere, the choice of the equations depends on a particular on the problem considered. From the viewpoint of Maxwell's equations a classification of different problems of electric field dynamics in the atmosphere was presented by Bostrom and Fahleson (1977). It is based on the comparison of parameters formed by the frequency of the process ω and the conductivity σ. If the wavelength $\lambda = 2\pi/\omega\sqrt{\varepsilon_0\mu_0}$ and the skin-depth $2/\sqrt{\omega\mu_0\sigma}$ are large enough compared to the characteristic scale of the problem, the electric field can be treated as a potential one (or quasi-stationary field): $rot\mathbf{E} = 0$, and therefore, $\mathbf{E} = -\nabla\varphi$. Assuming the spatial scale of the analyzed process to be less than 100 km, and conductivity less than 10^{-8} S/m, which is valid below 70 km, one finds the condition: $\omega << 2\cdot 10^4$ s^{-1}. In the framework of a quasi-stationary approach one can either neglect the displacement current for extremely slow processes (like quasi-stationary currents, charging the global circuit), or take it into account depending on the relation of characteristic frequency ω and the inverse relaxation time σ/ε_0. In most considerations we should keep the displacement current when modeling transient electric fields in the atmosphere. There is a class of problems, however, where both time derivatives (of electric and magnetic field) in the Maxwell equations must be kept. A typical example is the electromagnetic radiation of lightning, which leads in particular to the elve generation.

To solve Maxwell's equations, one has to calculate electric charges and currents in the medium of interest and connect them to motion of the medium. In a rigorous formulation, the description of the conducting fluid implies solution of kinetic equations for the distribution functions of all species (neutral, positive and negative) over velocities (e.g., Klimontovich, 1982; Holzworth, 1995). Also, the distribution over the electric charges in precipitation and cloud particles should be taken into account. Due to an extremely wide variety of spatial and temporal scales of processes involved (from microphysical to global-scale and many other complications (inhomogeneous profile of the conductivity, non-stationary character of processes, nonlinear dynamics of systems, such as the lightning channel, thunderstorm cloud, global electric circuit etc.) different simplified approaches are used instead of the full and rigorous analysis. Air-dynamic equations are solved as a rule instead of kinetic equations. If multi-component medium (like a cloud) is analyzed, it can lead to many equations for different species (e.g., Schuur and Rutledge, 2000). Even in most simplified models of thundercloud electrification some equations for heavy and light hydrometeors and for light ions are required. On the other hand, for the analysis of the global circuit (including their sources and sinks) the balance of electric currents is the most important point. Therefore, many problems of this type are simplified using a two-scale approximation, taking into account the presence of at least two very different characteristic scales – the microphysical scale, determined by charge separation processes, and the large scale, i.e. the size of electric field which is generated. Using this terminology, it is easy to give a general formulation of the electric dynamo problem, which is in fact the most fundamental problem of atmospheric electricity and unifies the problems of fair weather, cloud electrification, and global electric circuit operation. *Electric dynamo* by definition is the generation of large-scale quasi-stationary electric field and space charge due to the motion (laminar or turbulent) of a weakly ionized medium (Mareev and Trakhtengerts, 1996; Mareev, 1999). The latter is typically considered as a multi-component system, providing both separation and dissipation currents. Several examples will be analyzed below.

But in a very simplified approach the electric dynamo may be considered even for a one-component medium. In this case the set of equations includes the equation for space charge density ρ, and hydrodynamic (aerodynamic) equations of a conducting medium:

$$\frac{\partial \rho}{\partial t} + \frac{\sigma}{\varepsilon_0}\rho = -div\rho \mathbf{v} + D\Delta\rho - \sigma\mathbf{B}rot\mathbf{v} \qquad (13.1)$$

$$\frac{\partial \rho_a}{\partial t} + div\rho_a \mathbf{v} = 0; \quad P = P(\rho_a, T) \qquad (13.2)$$

$$\rho_a \frac{\partial \mathbf{v}}{\partial t} + \rho_a(\mathbf{v}\nabla)\mathbf{v} = -\nabla P + \rho E - \frac{1}{2}\nabla E^2 + \varepsilon(\mathbf{E}\nabla)\mathbf{E} \qquad (13.3)$$

The medium is characterized: by the density ρ_a, the velocity $\mathbf{v}$, the pressure P, the conductivity σ, the temperature T, the charge diffusion coefficient D and permittivity ε, which is unity in many atmospheric applications. Space charge density and electric field are connected by the Poisson equation: $\varepsilon_0 div\mathbf{E} = \rho$. In Equation (13.1) the latter term is caused by the account of the correction $[\mathbf{vB}]$ in Ohm's law expression for the conductivity current in the moving medium: $\mathbf{j} = \sigma(\mathbf{E} + [\mathbf{vB}])$. This term leads to the space charge density growth for the vortex motion of a weakly conducting medium in a magnetic field. If electric currents are not too big, the magnetic field induction $\mathbf{B}$ in the lower atmosphere may be considered as a given function of time and coordinates. Otherwise it is found from the equation:

$$\frac{\partial \mathbf{B}}{\partial t} + \frac{\varepsilon_0}{\sigma}\frac{\partial^2 \mathbf{B}}{\partial t^2} = \frac{1}{\mu_0 \sigma}\Delta\mathbf{B} + rot[\mathbf{vB}] + \frac{1}{\sigma}rot\rho\mathbf{v} \qquad (13.4)$$

Here μ_0 is the magnetic permeability in vacuum. Equations (13.1)-(13.4) together with the Poisson equation form a full set of electro-magneto-hydrodynamic equations of a conducting medium. In most practical applications Equation (13.4) is not required (Mareev, 1999), but many complications are concerned with the inhomogeneity, turbulent motion and multi-component composition of the medium. Neglecting the magnetic field, charge diffusion and convective current, one can write Equation (13.1) for quasi-stationary time-dependent electric fields in the form:

$$\text{div}\left(\frac{\partial \mathbf{E}}{\partial t} + \frac{\sigma}{\varepsilon_0}\mathbf{E}\right) = 0 \qquad (13.5)$$

If the conductivity depends on the coordinate z only, the 1D equation for the electric field is easily integrated:

$$\frac{\partial E_z}{\partial t} + \frac{1}{\varepsilon_0}j_z = \frac{\partial E_z}{\partial t} + \frac{\sigma}{\varepsilon_0}E_z = f(t), \qquad (13.6)$$

where the function $f(t)$ does not depend on the spatial coordinates and is determined by the external sources (boundary conditions). The general solution is obvious for arbitrary functions $\sigma(z)$:

$$E = e^{-\sigma t/\varepsilon_0}\left(C + \int f(t)e^{\sigma t/\varepsilon_0}dt\right). \qquad (13.7)$$

Here the constant C is chosen according to the initial condition. Choosing for the illustration the exponential profile of the conductivity $\sigma = \sigma_0 e^{z/H}$ and switching the arbitrary field E_0 at the boundary $z = 0$ instantaneously (it corresponds to the instantaneous deposition of the charge), one gets the expression presented in (Pasko et al., 1997):

$$E = E_0\left(1 - e^{-z/H}\right)e^{-\sigma t/\varepsilon_0} + E_0 e^{-z/H}. \qquad (13.8)$$

Figure 1 shows the height profiles of the electric field (normalized by E_0 and presented on a logarithmic scale) at different instants of time. The latter is normalized by the relaxation time at $z = 0$: $\tau_{r0} = \varepsilon_0/\sigma_0$. The solid line starting from the point $E = 10$, corresponds to the breakdown field, depending on the altitude exponentially as well. It is assumed for simplicity, that the characteristic scales of neutral gas density decrease (determining the breakdown field) and conductivity growth are equal. It is seen from the figure that the region occupied by the electric field and exceeding the critical discharge level, extends from the definite level H_c equal about 20 km in this particular case, but its upper boundary moves down with the speed $V{\sim}H/t$, inversely proportional to the time until it reaches the height H_c.

Our goal however is to model the transient field corresponding to the sprite generation, which appears to be associated with the strokes followed by the stage of continuing current, lasting usually from several ms to hundreds of ms. This current provides monotonically increasing space charge, distributed

SPRITES, ELVES AND INTENSE LIGHTNING DISCHARGES

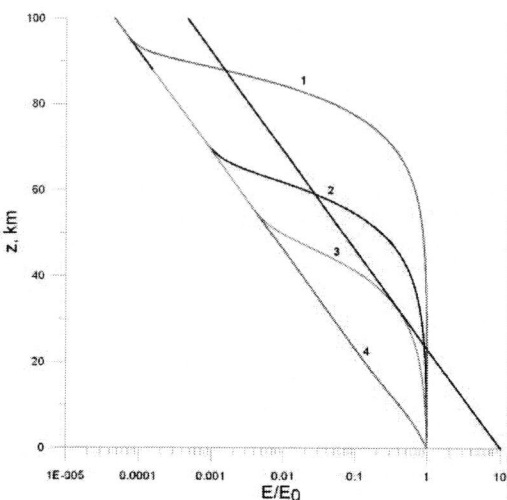

Figure 1. Transition process after instant switching of the electric field at the boundary, 1: $t/t_r = 0.001$, 2: $t/t_r = 0.01$, 3: $t/t_r = 0.04$, 4: $t/t_r = 1$.

over the cloud region, connected to the Earth by the stroke. For illustration one can consider the charge linearly growing in time $Q = It$ and distributed over the layer with the horizontal scale big enough so that the problem can be treated as one-dimensional again. It should be noted that the electric field can be correctly determined even in the vicinity of this layer if taking into account respective boundary conditions, i.e. in the 1D problem the proper voltage supported by the global electric circuit (e.g., Smirnova et al., 2000). It will give the dimensionless factor a of order of unity in the formula for the electric field strength above the layer: $E_0 = aIt/\varepsilon_0 S$, where S is the surface square of the layer, so that Q/S is the surface space charge density. The solution of Equation (13.6) for the electric field is then written in the form:

$$E = \frac{Ia}{S\sigma_0} e^{-z/H} \left(1 - e^{-z/H}\right) \left(1 - e^{-\sigma t/\varepsilon_0}\right) + \frac{Ita}{S\varepsilon_0} e^{-z/H} \quad (13.9)$$

Figure 2 shows the height profiles of the electric field (normalized by $E_a = Ia/\sigma_0 S$ and presented in logarithmic scale) at different instants of time. The latter is normalized by the relaxation time at $z = 0$: $\tau_{r0} = \varepsilon_0/\sigma_0$. The field at the lower boundary E_0 is growing linearly. The solid line starting from the point $E_c = 0.4$, corresponds to the breakdown field (chosen to be equal $0.4E_a$ at the lower boundary for illustration), depending on the altitude exponentially.

One essential feature of the field behavior, different from the field of instantaneously switching charge on the plate, is that at the given point z of the order of z_c and bigger, where the point z_c is defined from the equation

$\sigma(z_c)t = \varepsilon_0$, the field $E \approx E_a$ (which is proportional to the current and does not depend directly from the total charge transferred by the current) is constant in time during the flash continuing current stage. If this field exceeds the critical breakdown value E_c, the region where the discharge is available, extends from high enough levels (infinity in a simplest model) to the bottom boundary moving down with the speed $V \sim H/t$, inversely proportional to the time, during the continuing current stage. We refer in this connection to an interesting paper by Barrington-Leigh et al. (2002). The authors speculate on the reasons for exponential relaxation of optical emissions of sprites, which appeared in experiments. They deal with the "moving capacitor" model to show that the electric field can be constant in sprites. As we see from the consideration above this conclusion stems directly from the simplest model of the continuing current stage (while the "moving capacitor" could also be introduced in this consideration).

We turn now to the calculation of the electric field profile in a more realistic conducting atmosphere, taking as an example an experimentally measured profile of the conductivity during the night conditions (Holzworth et al., 1985). As a source of the field, also a more typical (dipole) charge distribution will be examined:

$$\frac{\partial E_z}{\partial t} + \frac{\sigma(E_z)}{\varepsilon_0} E_z = \frac{hI}{\pi \varepsilon_0 z^3}. \tag{13.10}$$

Here h is the mean height of the region, where the electric charge, transferred by the stroke current, is distributed. Equation (13.10) is convenient because

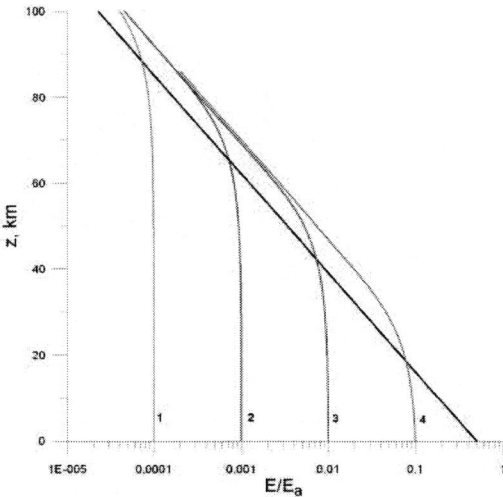

Figure 2. Transition process during the continuous current stage, 1: $t/t_r = 0.0001$, 2: 1: $t/t_r = 0.001$, 3: $t/t_r = 0.01$, 4: $t/t_r = 0.1$.

at the time small enough it gives the dipole field in the vicinity of the source. It should be noted however that it is an approximate equation, not allowing for the horizontal components of the electric field. Focusing on the search for breakdown conditions, we will not take into account in the following calculations the change of conductivity due to electron density perturbation, which is a slower process compared to electron heating in the field. Therefore, in Equation (13.10) we imply the dependence $\sigma(E_z)$ caused only by the collision frequency change due to electron heating.

To estimate the heating of electrons, one can use the equation for the electron temperature T_e written in the form (Gurevich, 1978):

$$\frac{dT_e}{dt} = \frac{2}{3}\frac{e^2}{mk}\frac{E^2}{\omega_H^2 + \nu_e^2}\left[\nu_e + \frac{\omega_H^2}{\nu_e}\cos^2\beta\right] - \delta(T_e)\nu_e(T_e)(T - T_e) \quad (13.11)$$

Here β is the angle between the magnetic field and vertical direction, ν_e is the frequency of electron collisions, δ is the fraction of electron energy lost per collision, ω_H is the electron gyro-frequency, m is the mass of the electron, k is Bolzmann's constant. The magnetic field may be neglected at heights below 75 km for any β, and for higher altitudes for sufficiently small angles or for sufficiently strong electric fields. The latter circumstance is caused by the increase of the collision frequency of electrons: $\nu_e \propto T_e^{5/6}$ in the strong electric field.

If the characteristic time of the electric field change is large compared to the time of electron heating $\tau \approx (\delta\nu_e)^{-1}$, Equation refmareev-eq-11 reduces to the corresponding quasi-stationary equation:

$$\left(\frac{E}{E_p}\right)^2 \frac{\nu_e^2(T)\delta(T)}{\nu_e^2(T_e)\delta(T_e)} = \frac{T_e}{T} - 1. \quad (13.12)$$

Here $E_p = \sqrt{3\delta(T)\nu_e^2(T)mkT/2e^2}$. It is seen from Equation (13.12) that electrons are weakly heated if $E \ll E_p$, but in the opposite case $E \gg E_p$ the electron temperature grows substantially.

Taking the expression $\nu_e = 1.84 \cdot 10^9 \cdot (N_m/10^{17}cm^{-3}) \cdot (T_e/10^3 K)^{5/6} s^{-1}$ valid for electron temperature of order of 1 eV (Borisov et al., 1986), and dependence $\delta(T_e)$ at the temperatures exceeding 1 eV, one can find

$$(E/E_p)^{\frac{3}{4}} = T/T_e, \quad T_e < 10850 \text{ K};$$
$$(E/E_p)^{\frac{3}{4}} = (T_e/T)\exp[(T_e - 10850\,\text{K})/6500\,\text{K}]\,, T_e > 10850 \text{ K}$$
$$(13.13)$$

Here the characteristic temperature values in the exponent follow from the exponential approximation of known dependence $\delta(T)$ at high temperatures (for more details see Yashunin, 2004). Respective dependence of electron temperature on the electric field strength is presented in Figure 3a. To estimate the breakdown threshould, we shall use the rate equation for electrons, which in a simplest case takes the form:

$$\frac{\partial N_e}{\partial t} = \nu_{ion}^{eff} N_e, \quad (13.14)$$

where $\nu_{ion}^{eff} = \nu_{ion} - \nu_a$ is the effective ionization frequency, determined by the difference of the actual ionization frequency ν_{ion} and the electron attachment coefficient ν_a. To allow for the dependence $\nu_{ion}^{eff}(E)$, different approximation formulas may be used (Pasko et al., 1997; Fernsler and Rowland, 1996).

We will use the expressions presented in (Kossy et al., 1994); they are obtained from kinetic solutions for $N_2{:}O_2$ mixtures. For electron energy $\geq 1 eV$, dissociative attachment of electrons to the oxygen molecules dominates: $O_2 + e \Rightarrow O^- + O$, with the attachment frequency $\nu_a = N_{O_2} k$, where the reaction coefficient k is expressed in the common way through the relation of the electric field and air density: $\lg k = -9.3 - 12.3/\theta$ at $\theta < 8$ and $\lg k = -10.2 - 5.7/\theta$ at $\theta > 8$; $\theta = 10^{16}(E/N)$, E is measured in V/cm, the air density N is measured in cm^{-3} (the generally accepted units in plasma-chemical studies).

Ionization rates are determined by expressions (Kossy et al., 1994):
$\lg k(N_2) = -8.3 - 36.5/\theta$ for the reaction $N_2 + e \to N_2^+ + e + e$, and
$\lg k(O_2) = -8.8 - 28.1/\theta$ for the reaction $O_2 + e \to O_2^+ + e + e$.

Using the expressions given above, for the nitrogen-oxygen mixture ($[N_2] \div [O_2] = 4 \div 1$) we have the effective ionization rate curve presented in Figure 3b. In this figure the critical field, measured in kV/cm, is equal $E_k = 32\left(N/2.7 \cdot 10^{19}\right)$, the air density N is measured in cm^{-3}. For estimates the following expression is convenient: $E_k/E_p = 238(\delta T)^{-1/2}(T/10^3 \text{K})^{-5/6}$, where the temperature is measured in K. For instance, for T=190 K we find

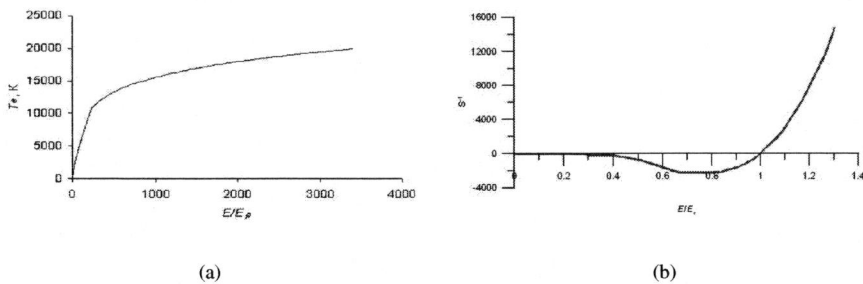

(a)   (b)

Figure 3. Electron temperature (a) and effective ionization rate (b) depending on the electric field strength.

$E_k/E_p \approx 1540$. Note that decrease of ν_{ion}^{eff} in Figure 3b is associated with the dependence of attachment on the field, while we did not take into account the effect of depletion of electron density due to this dependence (it was investigated by Pasko et al., 1997) considering here the problem of discharge initiation.

Having now the necessary estimates for electron heating and ionization, we can turn to the solution of Equation (13.10). Results of its numerical analysis for different values of the electric current during the continuing current stage of the flash are presented in Figures 4-6. It is seen that the region where the electric field exceeds the critical field strength is limited to the height range around 75 km. Boundaries of this interval slightly change during the continuing current stage, but the altitudes appear where breakdown conditions are satisfied for sufficiently long time (similar to the case, considered before). This peculiarity may be important when analyzing the conditions for sprite initiation. As we

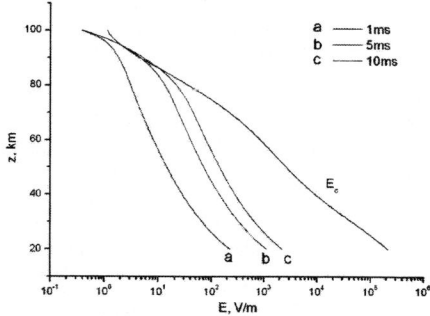

Figure 4. Electric field profiles during the continuous current stage at selected instants for the current I = 10 kA.

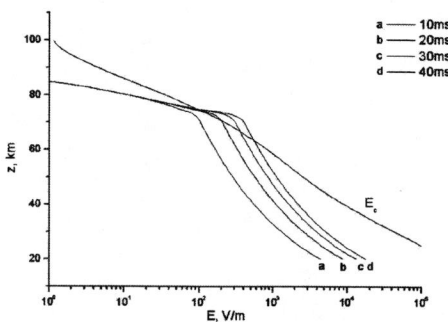

Figure 5. Electric field profiles during the continuous current stage at selected instants for the current I = 20 kA.

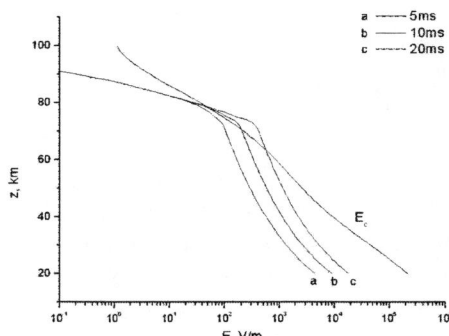

Figure 6. Electric field profiles during the continuous current stage at selected instants for the current I = 40 kA.

see from telescoping imaging of sprites, they have a pronounced fine structure (Gerken et al., 2000). Probably, it is stipulated by the streamer developing of the discharge as argued by Pasko (see Chapter 12 and References therein). But the nature of triggering mechanisms initiating the avalanches still remains questionable. The above modeling of a transient field at the stage of continuing current shows that the situation resembles a *Wilson chamber*. By this term we imply that there is supersaturation on the electric field as compared to the critical field, similar to the real Wilson chamber where supersaturation on the water vapor takes place. Also, similar to the real Wilson chamber, where condensation nuclei are needed for a fast condensation process, some centers for avalanches are needed for discharge development in the mesosphere.

There is a height interval, where discharge conditions are kept long enough, but a trigger is needed to initiate the process. An analysis of imaging photographs suggests that similar to the actual Wilson chamber, the ionized trails of high-energy particles could be triggering elements. In the problem considered taking into account runaway electron beams (e.g., Gurevich et al., 1992; Lehtinen et al., 1999) seems promising to clarify the initiation process.

The next step in our exploration of transient fields which produce high-altitude discharges is a more attentive treatment of the lightning flash which is particularly important to estimate the parameters of the electric dipole used in our previous analysis.

A key point of the model suggested in (Smirnova et al., 2000) is the assumption that a highly conducting channel arises due to a CG discharge, bringing the ground (zero) potential to some region near the cloud bottom due to the return stroke. As a result, a substantial increase in the electric field strength above the thundercloud occurs promptly. Another important point is the comparison of two types of the lightning channel geometry corresponding to different modes of leader propagation. Namely, we estimated the dipole moment of the charge distribution at the return stroke stage, using two forms of the lightning channel, corresponding to the bidirectional and simplest unipolar leader propagation models. For the bidirectional model, similar to Heckman and Williams (1989), we approximated the branching part of the channel by a horizontal stick (Figure 7). We placed the horizontal stick at the height of the potential maximum, because this geometry gives the biggest charge redistribution at the very first stages of the discharge. This geometry was found to give the biggest field transition at the return stroke stage. It should be mentioned that Coleman et al. (2003) have recently demonstrated experimentally that this choice was a good one; they showed that the horizontal lightning branches of negative polarity breakdown occur at the altitudes of potential maxima. Since the

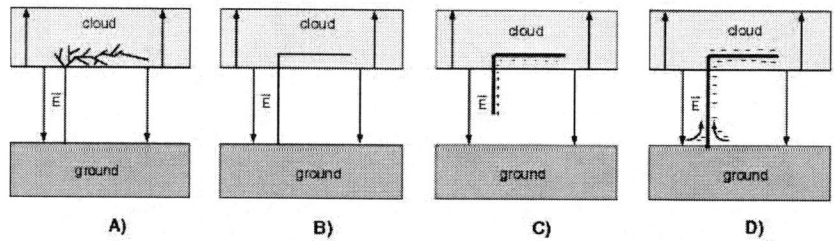

Figure 7. The actual trace of lightning (A) and the simple model we used for the calculations (B). Charge density distribution in the stepped leader channel (C). Negative charge flowing into the channel connected to the ground (D) (from Smirnova et al., 2000).

typical duration of the flash is much shorter than the time of cloud charge redistribution, we considered the channel as a thin conductor in a fixed ambient field. We calculated numerically the charge distribution in a grounded long thin conductor, consisting of two sticks located horizontally and vertically in the external field (taking into account the ground reflection). We followed the paper (Heckman and Williams, 1989) to obtain the estimates for the field on the channel surface and the channel radius.

According to the unipolar model, the leader propagates only downwards from a point located near the lower cloud boundary. For this case we considered the lightning channel as a long vertical conducting stick between the lower cloud boundary and the ground. It was shown, that the value of the transferred charge is twice as much for the bipolar model than for the unipolar one. To estimate the electric field perturbation in the middle atmosphere, we calculated the dipole moment of charges distributed along the lightning channel by the moment of the stroke completion. Respective values in dependence of the voltage between the cloud and the ground are presented in Figure 8. One can see that the bidirectional model gives values more than twice

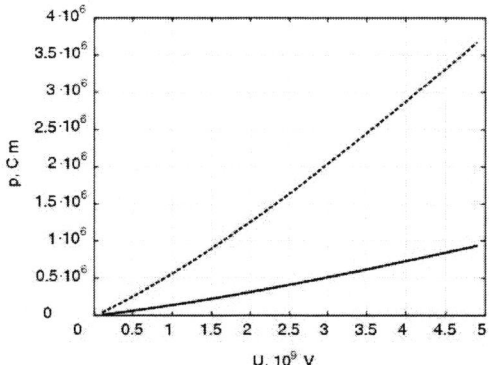

Figure 8. Dipole moment of the lightning channel as dependent on the voltage for unipolar (lower curve) and bidirectional (higher curve) models (from Smirnova et al., 2000).

as large compared to the dipole moment. Common voltages for thunderstorms are smaller than those shown in the figure, but the main conclusion is also valid. A model developed can be used for the calculation of charge moments during the flash with the following analysis of electric field profiles similar to the analysis presented above. In conclusion, it was demonstrated in this chapter by several examples of models, that transition processes in thunderclouds (including stratiform precipitation regions) can produce an extended volume of large quasi-homogeneous electric field above the thundercloud. Such a field is believed to cause the appearance of sprites and other transient luminous events in the middle atmosphere. It should be noted that only the conditions of the discharge initiation were examined. Regarding the discharge development, its scenario was numerically analyzed in some papers (in more detail, see Chapter 12 of this book and References therein), but analytical treatment, including possible quasi-stationary self-consistent solutions, is left for the future.

13.3 Modeling of the Lower Positive Charge Layer in the Stratified Region

There is evidence now that virtually all sprites are associated with positive lightning flashes, and new experimental data are available concerning the thunderstorm systems providing these flashes more regularly. Complicated multi-layer charge structures are frequently observed in stratiform precipitation regions of MCSs (Marshall and Rust, 1993; Stolzenburg et al., 1994, 1998; Shepherd et al., 1996). But the narrow layer of positive electric charge near the 0° C isotherm serving as "reservoir" for positive ground flashes appears to be a characteristic feature (Stolzenburg et al., 1994; Shepherd et al., 1996; Williams, 1998; Marshall et al., 2001). We suggest a quantitative model treating the formation of the positive charge layer near the 0° C isotherm as a result of melting-charging process. It is one of the mechanisms, which has been discussed for a long time, especially when applied to MCS stratiform regions (e.g., respective discussion by Stolzenburg et al., 1994). There is experimental evidence for its significance, while the detailed microphysics of charge separation during melting is not completely understood to date.

We assume that previously uncharged precipitation particles (large aggregates of vapor-grown crystals or smaller precipitation particles) acquire negative charge as they melt by shedding smaller, cloud-size particles (either liquid or solid). The set of equations for modelling this process is the following:

$$\frac{\partial \rho_-}{\partial t} + V_Q \cdot \frac{\partial \rho_-}{\partial z} = -\nu \cdot R(z,t) - \frac{\rho_-}{\tau_1} \qquad (13.15)$$

$$\frac{\partial \rho_+}{\partial t} + V_q \cdot \frac{\partial \rho_+}{\partial z} = \nu \cdot R(z,t) - \frac{\rho_+}{\tau_2} \qquad (13.16)$$

$$\frac{\partial E}{\partial z} = \frac{4\pi}{\varepsilon_0} \cdot (\rho_+ + \rho_-) \qquad (13.17)$$

$$V_q = V_{q0} - \alpha \cdot E \cdot \Delta q_0/\sqrt{2} \qquad (13.18)$$

$$R(z,t) = \Delta q_0 \cdot (1 - \beta \cdot \rho_+/N) \cdot (1 - \gamma E) \cdot F(z,t) \cdot N \qquad (13.19)$$

$$\tau_1(E) = \frac{\tau_{10} \cdot E}{E_C \cdot (\exp(E/E_C) - 1)}; \quad \tau_2(E) = \frac{\tau_{20} \cdot E}{E_C \cdot (\exp(E/E_C) - 1)}. \qquad (13.20)$$

Here ρ_+, N, V_Q are the charge density, the concentration and the velocity of large (precipitation) particles; ρ_-, n, V_q are the charge density, the concentration and the velocity of small (cloud) particles; ν is the frequency of small particle shedding; τ_1 and τ_2 are the lifetimes of large and small particles respectively. Finite lifetimes may be determined by the presence of light ions in the air, while other reasons for particle sinks are plausible – evaporation (for smaller particles) or, more important, removal by larger quickly falling precipitation particles (Bateman et al., 1995). The latter process can be an important factor of support for stationary layer of positive charge as argued by Bateman et al. (1995). But its detailed analysis requires some complication of the suggested model. As one example, we will examine below a model of the characteristic rate for the electric field growth and the dynamics of parameters of the system during the melting process.

$F(z,t)$ is a nondimensional function, characterizing the melting region. We considered it to have Gaussian shape with the maximum at a definite height $\langle z \rangle$. It is known from the experimental data that the hydrometeor melting occurs most effectively when the temperature is equal 1-2° C. For the linear profile of temperature in the atmosphere with a lapse rate of 20° C per 4 km, the level of +1° C, corresponding to the height $\langle z \rangle$, takes place about 200 m below the 0° C isotherm. The axis z was directed downward.

Equations 13.15 and 13.16 of this set describe the production of positive and negative particles respectively, taking into account their relative motion. Equation (13.17) is the Poisson equation. Equation (13.18) takes into account a possible dependence of the velocity of small particles upon the electric field strength by the factor α, depending on the mobility of small particles. The factor $1/\sqrt{2}$ corresponds to the effective charge to be separated during the elementary act of electrification. Equation (13.19) models the dependence of a separated charge on the external field and the charge of the melting hydrometeor with help of the parameters β and γ being determined by the microphysical processes. Equations (13.20) take into account the corona discharge around the highly charged particles and as a consequence substantially decrease the relaxation time in the cloud. Under the framework of this model we were not able to describe the dynamics of charge layers situated above the 0° C isotherm and resulting (assumingly) from other charging processes. But for the numerical calculations we added "by hand" positive and negative charge layers, commonly observed in the MCSs. This was necessary to produce a realistic set of boundary conditions, determined by the total voltage between the top and the bottom of the system, in agreement with the global circuit concept.

The coordinate z in the calculations varied from -20 km to 4 km, where the zero point corresponds to the 0° C isotherm. All the calculations have been performed with the use of the function $F(z,t)$, characterized by a width of 50 m and a maximum at $\langle z \rangle$=150 m. The process started at the time $t = 0$. Space charge densities with the Gauss distribution were added: positive one from -3 km to -12 km, and negative one from -300 m to -3 km, with the magnitude $2.4 \cdot 10^{-11}$ C/m^3 and $8 \cdot 10^{-11}$ C/m^3, respectively. The separated elementary charge was equal $\Delta q_0 = 8 \cdot 10^{-16}$ C.

The relaxation times were taken τ_1=200 s, τ_2=300 s, the frequency ν = 10 s^{-1}. The density of melting hydrometeors N=1000 m^{-3}, the particle velocities were equal $V_q = -0.5$ m/s, $V_Q = 5$ m/s. Some results of numerical analysis are presented in Figures 9-11. All the quantities are measured in SI units.

First we considered a simple model ignoring the dependence of a separated charge on the charge of a melting hydrometeor and external electric field (β=0, γ=0). At the initial moment, the generated field is already above the 0° C isotherm due to ice-ice collision charging (Figure 9a). Further, the melting charging mechanism starts to operate which leads to the formation of the field "hump" at 300 s (Figure 9b). The field magnitude grows almost linearly, reaching about 200 kV/m in 300 s (Figure 10a). At 500 s, a field magnitude as large as 300 kV/m was reached. The field profile at the last moment of the analyzed interval of time, presented on Figure 9b, allows a comparison with experimental data. The charge density profile is presented on Figure 11a. A narrow layer of the positive charge near the 0° C isotherm and more extended negative-charge layer below are clearly seen. Two additional layers above the 0° C isotherm included "artificially" into the model are seen also. It was shown that during the first 150 s the density of positively charged water drops grows linearly reaching $3 \cdot 10^6$ m^{-3} in 300 s. Note that for the bigger velocity of small particles (determined by the convective up-draft) this growth is followed by the saturation; for example when $V_q = -2$ m/s, it is saturated at the level of order of $1.5 \cdot 10^6$ m^{-3}. As to the charge of a melting particle, it is growing almost linearly up to 20 s, and it is saturated at about 70 s on the level of $-1.7 \cdot 10^{-13}$ C, which agrees well with experimental data (Bateman et al., 1995).

The electric current profiles generated by positively and negatively charged particles (polar currents) in 300 s are presented in Figure 10b. The contribution of large (negatively charged) particles obviously dominates due to their larger fall velocity. These calculations clearly illustrate how the electric currents in thunderstorm clouds may be calculated on the basis of microphysical consideration, it is necessary for further analysis of large-scale problems like the global circuit operation. We present here also for comparison the results obtained with allowance for corona discharge on charged particles. This effect leads to a growth of the conductivity growth and can result in field strengths on the order of 50 kV/m. The numerical analysis performed for $\beta=0$, $\gamma=0$ showed that during initial 50 s the field dynamics is the same as in the base model. Further the charges on large particles decrease. During 200 s the charge density of positive particles reaches $8.5 \cdot 10^{-8}$ C/m^3 and then decreases down to $8 \cdot 10^{-8}$ C/m^3.

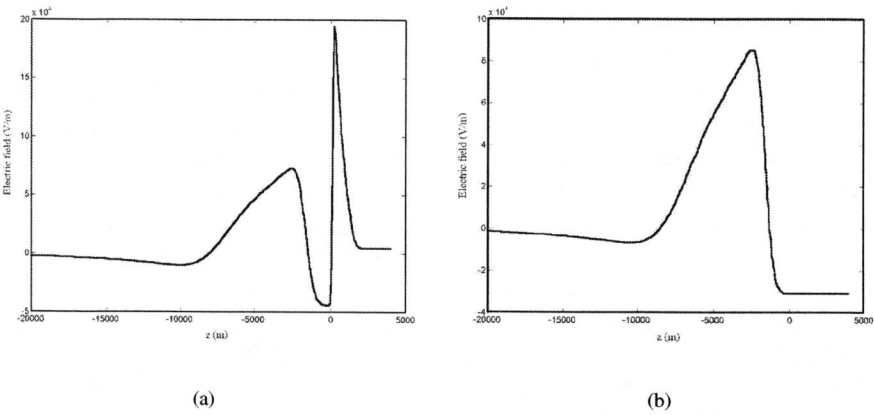

(a)   (b)

Figure 9. Electric field at the initial time (a) and at the moment 300 s (b).

SPRITES IN THE GLOBAL CIRCUIT 329

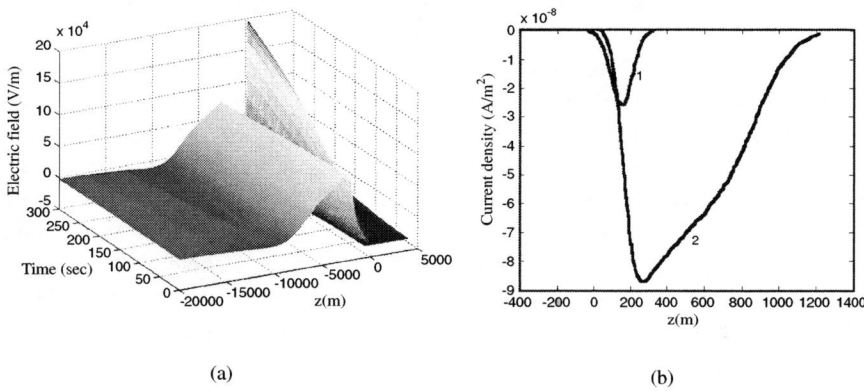

Figure 10. Dynamics of electric field distribution (a); polar currents at 300 s (b): for light particles (1), for heavy particles (2).

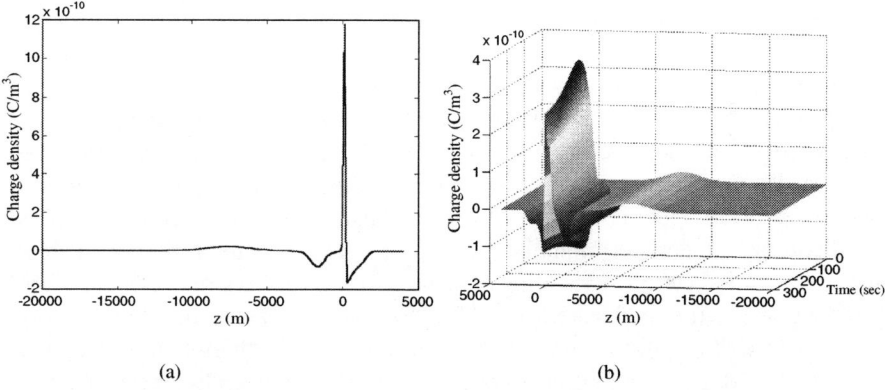

Figure 11. Charge density distribution at 300 s without (a) and with corona discharge on precipitation particles (b). Critical corona field is 50 kV/m.

The field magnitude reaches the value 120 kV/m. As an example, space charge density dynamics with account for corona discharge is presented in Figure 11b (but for V_q=-2 m/s; for small up-draft velocity the profile is rather narrow and not convenient for representation on the figure).

A detailed comparison of the developed model with experimental data is beyond the scope of this chapter. In our opinion, the main problem with the melting-charging mechanism is the absence of reliable microphysical data illustrating a possibility of accumulation of positive charges on tiny cloud particles. Existing laboratory measurements and theoretical considerations (e.g., Takahashi, 1969; Knight, 1979; Stolzenburg et al., 1994; Kochin, 1995; MacGorman and Rust, 1998; Schuur and Rutledge, 2000) are not able to explain similar processes. As to the problem with the localization of the positive charge layer above the radar bright band, it is easily explained with allowance for mesoscale updraft. Moreover, the melting-charging mechanism explains the very small vertical scale of the positive layer and its position very close to the 0°C isotherm, the point well illustrated by the respective figures from Shepherd et al. (1996), presented by Williams and Yair in the Chapter 3 of this book. This point seems to pose a problem for an explanation of the positive charge layer formation with the ice-ice collision mechanism, suggested by Williams and Yair. Nevertheless, we plan to include it in our model in the course of its generalization.

13.4   Global Electric Circuit Implications

In this chapter, we are concerned with two aspects of the global electric circuit (GEC) concept, which is particularly important for the research on sprites and sprite-producing clouds. First is a classical aspect of GEC as a quasi-stationary current contour supported by the operation of thunderstorm generators over the globe. A second aspect is related to the energy deposition and dissipation into the circuit, allowing its analysis as an open dissipative system.

The classical conception of GEC is well represented in many textbooks and reviews (e.g., Bering et al., 1998; Rycroft et al., 2000), and is not reproduced here in detail. Wilson (1920) proposed that thunderstorms are the principal generators in the global circuit. This hypothesis was subsequently supported by theoretical and experimental estimates of thunderstorm currents contributing to the global circuit. Modeling of thunderstorms as generators in the global circuit began with Holzer and Saxon (1952), who modeled the thunderstorm as positive and negative current sources of equal strength embedded in an atmosphere with an exponential conductivity profile between two perfectly conducting planes. During the next 30 years, many improvements were made in models of thunderstorms as generators in the global circuit (see Roble and Tzur, 1986, for a review), but there was still little information about the charge structure inside storms. In the last 20 years, much has been learned about the internal charge structure of storms from balloon observations of vertical profiles of electric field inside storms (e.g., Marshall and Rust, 1993; Stolzenburg et al., 1998, 2001). Also prior models of the global circuit did not include mesoscale thunderstorm systems. The charge structures of these storms called MCSs, are now fairly well known, again from balloon E measurements along vertical paths (e.g., Marshall and Rust, 1993; Stolzenburg et al., 1994, 1998; Marshall et al., 2001).

Recently a model describing the contribution of an MCS to the global electric circuit has been developed (Davydenko et al., 2004). The model explores the constant current sources embedded in the conducting atmosphere. It is a general, analytical, 2D axisymmetric model from first principles that uses recent in situ observations of the vertical profiles of E. To represent the MCS, with its very different convective and stratiform regions, two model solutions are coupled together. One solution used E data from a typical MCS convective region, and the other solution used E data from a standard Type B trailing stratiform region. The current sources in each solution were chosen so that E within the two regions matched the measured vertical profiles of E. The following equation for the electric potential with respective boundary conditions has been solved:

$$div(-\sigma \nabla \varphi + \mathbf{j}_{ex}(\mathbf{r}, z)) = 0 \quad \varphi(r \to \infty, z) \to 0; \; \varphi(r, z \to \infty) \to 0 \tag{13.21}$$

$$\sigma(r, z) = \begin{cases} \sigma_1 = \text{const}, & z \leq 0; \\ \sigma_0 \exp z/H, & z \geq 0. \end{cases} \quad \mathbf{j}_{ex}(\mathbf{r}, z) = \begin{cases} 0, & z < z_-, z > z_+; \\ \mathbf{j}_{ex}(\mathbf{r}), z_- < z < z_+. \end{cases} \tag{13.22}$$

Along with vanishing of the electric potential at large distances (Equation 13.22) the continuity of the electric potential and the vertical component of the total current density were set at the Earth's surface $z = 0$ and at the boundaries of the external current layer. The two-dimensional distribution of the electric potential φ corresponding to the given external currents (chosen with respect to the experimental profiles of the electric field inside the MCS) is presented in Figure 12. The figure plane passes through the centers of stratiform and convective regions of the MCS. One can see that the pattern of the electric potential in the vicinity of the MCS manifests a number of specific features as compared to the "ordinary" single thunderstorm. One of the most obvious peculiarities is the significant lateral scale of the potential distribution, exceeding the height of the base of the ionosphere. It should also be noticed that the potential (and, correspondingly, the vertical electric field) over the stratiform region is significantly disturbed at heights up to about 25 km (see Davydenko et al., 2004, for details). Another specific feature of the distribution shown in Figure 12 concerns the local nature of the disturbance of the electric potential associated with the currents in the convective region. Despite the fact that the magnitude of the potential inside the convective region amounts to about 210 MV and usually exceeds the potential within the stratiform precipitation region, the volume influenced by the convective region is much smaller than the volume influenced by the stratiform region. This finding is a direct result of the small lateral extent of the convective region and the small scale height of atmospheric conductivity, both of which are relatively small compared to the

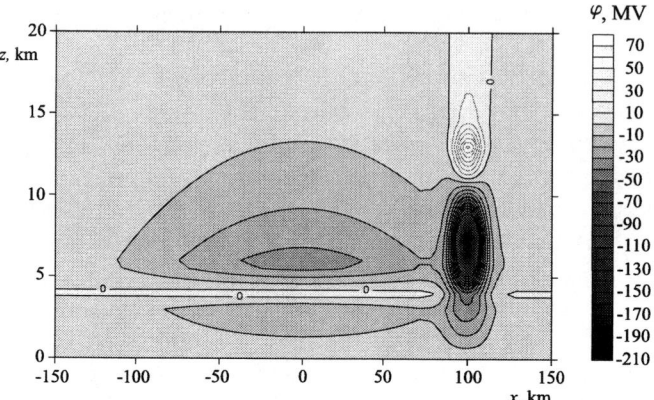

Figure 12. Two-dimensional distribution of the electric potential along a vertical plane through the centers of the stratiform and convective regions of the MCS (from Davydenko et al., 2004).

lateral extent of the stratiform region. The results show the magnitude of the current driven between the Earth and the ionosphere. Davydenko et al. (2004) found that the great lateral extent of the MCS yields a current contribution to the global circuit that is one to two orders of magnitude larger than the contribution of an ordinary storm which is commonly believed to be about 1 A (while MCSs are more rare than ordinary storms). Depending on the polarity, magnitude, and thickness of the layers of external current, an MCS can either serve as a generator or a discharger in the global circuit. The main contribution to the total upward current of the MCS is dominated by the stratiform precipitation region.

In general, to make modelling of the global circuit quasi-stationary currents, one should use either experimental data (for example, to use currents corresponding to the experimental profiles of the electric field as was demonstrated above), or take some expressions for currents derived from microphysical consideration (as was demonstrated in the previous paragraph).

Recently we have developed a simplified approach of self-consistent field-current analysis based on the solution of a nonlinear diffusion equation for electric field evolution in a cloud:

$$\frac{\partial E}{\partial t} = \varepsilon_0^{-1}(j_{sep} - j_{dis}) + D_c \frac{\partial^2 E}{\partial z^2} \quad (13.23)$$

In this equation, the charging current j_{sep} describes the field growth due to the charge separation on cloud and precipitation particles, the dissipation current j_{dis} is caused mainly by the conductivity, the charge-diffusion coefficient $D_c \cong V_c L_c$ is determined by the characteristic velocity V_c and scale

L_c of the turbulent air parcels (eddies). In our analysis (Mareev and Sorokin, 2001) we have taken into account different parameterizations of inductive and non-inductive mechanisms, leading to different expressions for the charge separation current, as well as corona discharge dissipation current and turbulent diffusion current were accounted for. At the stage of intense electrification, the charging current is dominant. However, a sharp increase in conductivity occurs in a sufficiently strong field due to the corona effect around the strongly charged particles. To allow for this effect, one can use the corresponding empirical dependence $j_{dis} = \sigma_c E_c \left[\exp\left(E/E_c\right) - 1\right]$, where σ_c is the unperturbed conductivity of the cloud medium and E_c is the critical electric field leading to the sharp growth of the corona current. Stationary states and their stability have been investigated with this model. Of particular interest is the solution in the form of a traveling front of electric field and space charge perturbation, separating the stable and unstable equilibrium states. The asymptotically stable velocity of the front is easily estimated on the basis of the classical results of Kolmogorov, Petrovsky, and Piskunov for the pseudo-wave of a spreading biological population: $V_f \approx 2\sqrt{f'(0)D_c} = 2\sqrt{\varepsilon_0^{-1} D_c \left(\gamma_s - \sigma_c\right)}$, where $f = \varepsilon_0^{-1}(j_{sep} - j_{dis})$, f' means the derivative of f over the electric field strength, γ_s is the increment of j_{sep} growth for the inductive mechanism, when $j_{sep} = \gamma_s E$. That means that the velocity depends on the diffusion coefficient. In this case, the characteristic thickness of the front is equal $L_f \approx \sqrt{D_c/f'(0)}$. Combining in the one-dimensional formulation two such waves traveling in the opposite directions, we obtain a solution in the form of a bounded region, which is occupied by the field of intensity E_1 and expands to both sides with the velocity determined by the diffusion and the intensity of charging processes. This solution can describe the growth of the thunderstorm cell at certain stages of its development. With the results of numerical calculations for actual thundercloud conditions (Mareev and Sorokin, 2001), we estimated the "front" velocity V_f of this diffusion wave. It is possible to use $f'(0) \approx 2 \cdot 10^{-2}$ s^{-1} and $D_c = V_c L_c \approx 4 \text{m}^2/\text{s}$ to obtain a velocity $V_f \approx 2$ m/s and for the front characteristic thickness $L_f = 14$ m. It is easy to estimate the growth rate of the thunderstorm-cell electrostatic energy using the above-mentioned wave velocity and the electric field in the equilibrium state: $P = \varepsilon_0 V_f S_c E_1^2$, where S_c is the area of the interaction region, i.e. thunderstorm cell. Putting $E_1 \approx 150$ kV/m and $S_c \approx 5$ km^2, we have $P \approx 2$ MW. For the electrification process with a duration of about 10 min, we obtain an accumulated electric energy (subsequently particularly released in lightning) of about 1 GJ. We can conclude that the evolution of electric field and space charge of a thunderstorm cloud can be studied in the framework of a diffusion equation for the electric field, which has auto-wave (diffusion-wave) solutions, describing the dynamics of electric charge regions separated in the cloud space.

With the use of respective statistics of thunderstorms, similar estimates seem to be very useful for studies of current and energy balance in the global electric circuit.

On the other hand, recent experimental and theoretical studies allow us to analyze the global electric circuit as a hierarchy of multi-scale dissipative systems with the atmospheric part of the global electric circuit as a thermodynamically open system driven by the external sources of energy (Mareev and Anisimov, 2003). From our point of view, this approach develops and supports substantially the classical paradigm of the global electrical circuit as the current contour formed by the bottom ionosphere (60-70 km), the terrestrial surface and the conducting layers of the atmosphere.

The energy flow from global to local scales through the atmospheric electric circuit is accompanied by the generation of multi-scale dissipative structures, including fine electrical structure of thunderstorms, clouds and aero-electric structures in the boundary layer (Anisimov et al., 2002; Anisimov and Mareev, 2003). The estimates of electrical energy stored in these structures (including highly charged hydrometeors, cloud particles, electric field fine structure) seem important for the studies of the total balance of energy in the circuit.

We will discuss briefly electric energy accumulation in thunderstorms served as generators for the global electric circuit. To date a lot of research has been performed concerning thunderstorm electrification and lightning morphology. Much less studies were devoted to the dynamics of electrical energy stored in the thunderstorm clouds and its further dissipation. Recently a one-dimensional model has been used for simple estimates of the total electrostatic energy stored in the stratiform regions of mesoscale convective systems and anvil clouds (Marshall and Stolzenburg, 2002). The calculated values of total energy ($5 \cdot 10^{11}$ J and $2 \cdot 10^{12}$ J for the MCSs and $2 \cdot 10^{11}$ J for the anvil cloud) are sufficient to support hundreds or thousands of typical lightning flashes, but only 10-100 of the energetic positive cloud-to-ground flashes. An important role of MCSs in the global electric circuit was appreciated by recent model calculations by Davydenko et al. (2004), who developed a 3-D model, which allows calculating of the electric field and current distribution in the mesoscale convective system environment according to experimental results. Estimates based on this model, give a value of order of 10^{11} J for the total energy accumulated in the system of 200 km scale. The mean dissipation rate due to ohmic losses was estimated to be $3 \cdot 10^9$ J/s.

Finally, simplest energetic estimates connected to sprite occurrence, are of interest. According to estimates by (Cummer et al., 1998; Cummer and Füllekrug, 2001), sprites can transfer the charge on the order of several C (occasionally up to tens of Coulombs), while the maximum value of the vertical electric current due to a sprite event is estimated to be 3 kA. For the voltage on the order of several MV between the cloud-top and the ionosphere it gives a characteristic value of a mean dissipation energy from units to tens (maybe occasionally as big as hundred) MJ. With allowance for recent estimations of a mean rate of sprite generation over the globe about 720 events per day (Sato and Fukunishi, 2003) the sprite contribution to the total dissipation energy turned out to be not substantial – less than 1 MW. However, using higher estimates for sprite rate occurrence based on satellite-borne observations (up to 1-10 per minute), one can get energies as large as 10 MJ for a mean sprite contribution to the total dissipation energy in the global circuit.

13.5 Conclusion and Outlook for Promising Future Work

It is obvious from the examples presented that the modeling studies of intensive lightning discharges, their parent thunderstorm systems and the global atmospheric electric circuit are closely coupled. Success in the modeling depends to a high degree on taking into account the spatial and temporal structure of different subsystems forming the global electric circuit, in particular on the study of quasi-stationary and fast transient processes connected with powerful lightning discharges and large-scale thunderstorm systems. Another important factor of model development is the use of modern (nonlinear) methods of analytical and numerical consideration. We would like to especially note as a promising direction development of fractal models of sprite and lightning discharges (e.g., Pasko et al., 2000, 2002; Iudin et al., 2003), other new methods of nonlinear modeling – auto-wave and dissipative structure generation, stochastic equation facility, etc. We are awaiting a fast development of theories describing high-energy processes (including cosmic ray and runaway electron implications) during lightning flash generation and thunderstorm evolution. In this context the concept of a Wilson chamber could be useful for the study of sprite initiation. We think that in the nearest future new theories will appear, build on the basis of concrete experiments and the actual spatio-temporal dynamics of the processes to shed more light on the complicated, but extremely interesting phenomenon of lightning discharges in the global electric circuit.

Acknowledgments

We thank S. Anisimov, S. Davydenko, V. Klimenko, T. Marshall, M. Shatalina, and M. Stolzenburg for valuable discussions. This work was partially supported by the Russian Foundation for Basic Research (grant #04-02-16634), Russian Academy of Science Program "Physics of the atmosphere: electric processes, radio physical methods", and CRDF grant # RUP1-2625-NI-04.

Bibliography

Anisimov, S. V. and Mareev, E. A. (2003). Fine structure of the global electric circuit. In *Proc. 12th Int. Conf. on Atmospheric Electricity*, pages 781–784, Versailles, France.

Anisimov, S. V., Mareev, E. A., Shikhova, N. M., and Dmitriev, E. M. (2002). Universal spectra of electric field pulsations in the atmosphere. *Geophys. Res. Lett.*, 29(24):doi:10.1029/2002GL015765.

Barrington-Leigh, C. P., Pasko, V. P., and Inan, U. S. (2002). Exponential relaxation of optical emissions in sprites. *J. Geophys. Res.*, 107(A5):1065.

Bateman, M. G., Rust, W. D., Smull, B. F., and Marshall, T. C. (1995). Precipitation charge and size measurements in the stratiform region of two mesoscale convective systems. *J. Geophys. Res.*, 100(D8):16341–16356.

Bering, E. A. III, Few, A. A., and Benbrook, J. R. (1998). The global electric circuit. *Phys. Today*, 51(10):24–30.

Borisov, N. D., Gurevich, A. V., and Milikh, G. M. (1986). *Artificial Ionization in the Atmosphere*. Academy of Science, Moscow. (in Russian).

Bostrom, R. and Fahleson, U. (1977). Vertical propagation of time-dependent electric fields in the atmosphere and ionosphere. In Dolezalek, H. and Reiter, R., editors, *Electrical Processes in Atmospheres*, pages 529–535. Dietrich Steinkopff Verlag, Darmstadt.

Coleman, L. M., Marshall, T. C., Stolzenburg, M., Hamlin, T., R.Krehbiel, P., Rison, W., and Thomas, R. J. (2003). Effects of charge and electrostatic potential on lightning propagation. *J. Geophys. Res.*, 108(D):doi:10.1029/-2002JD002718.

Cummer, S. A. and Füllekrug, M. (2001). Unusually intense continuing current in lightning produces delayed mesospheric breakdown. *Geophys. Res. Lett.*, 28(3):495–498.

Cummer, S. A., Inan, U. S., Bell, T. F., and Barrington-Leigh, C. P. (1998). ELF radiation produced by electrical currents in sprites. *Geophys. Res. Lett.*, 25(8):1281–1284.

Davydenko, S. S., Mareev, E. A., Marshall, T. C., and Stolzenburg, M. (2004). On the calculation of electric fields and currents of mesoscale convective systems. *J. Geophys. Res.*, 109:doi:10.1029/2003JD003832.

Fernsler, R. F. and Rowland, H. L. (1996). Models of lightning-produced sprites and elves. *J. Geophys. Res.*, 101(D23):29653–29662.

Gerken, E. A., Inan, U. S., and Barrington-Leigh, C. P. (2000). Telescoping imaging of sprites. *Geophys. Res. Lett.*, 27(17):2637–2640.

Gurevich, A. V. (1978). *Nonlinear phenomena in the ionosphere*. Springer-Verlag, Berlin.

Gurevich, A. V., Milikh, G. M., and Roussel-Dupré, R. (1992). Runaway electron mechanism of air breakdown. *Phys. Lett. A*, 465:463–468.

Heckman, S. J. and Williams, E. R. (1989). Corona envelopes and lightning currents. *J. Geophys. Res.*, 94:13287–13294.

Holzer, R. E. and Saxon, D. S. (1952). Distribution of electrical conduction currents in the vicinity of thunderstorms. *J. Geophys. Res.*, 57:207.

Holzworth, R. H. (1995). Quasistatic electromagnetic phenomena in the atmosphere and ionosphere. In Volland, H., editor, *Handbook on Atmospheric Electrodynamics*, volume 1, pages 65–109. CRC Press, Boca Raton, FL.

Holzworth, R. H., Kelley, M. C., Siefring, C. L., Hale, L. C., and Mitchell, J. D. (1985). Electrical measurements in the atmosphere and ionosphere over an active thunderstorm: II. Direct current electric fields and conductivity. *J. Geophys. Res.*, 90:9824.

Iudin, D. I., Trakhtengerts, V. Yu., and Hayakawa, M. (2003). Fractal dynamics of electric discharges in a thundercloud. *Phys. Rev. E*, 68:Art. no. 016601.

Klimontovich, Yu. L. (1982). *Statistical physics*. Nauka, Moscow. (in Russian).

Knight, C. A. (1979). Observations of the morphology of melting snow. *J. Atmos. Sci.*, 36(6):1123–1130.

Kochin, A. V. (1995). A mechanism of electric charge generation in nimbostratus and cumulonimbus clouds. *Meteorology and Hydrology*, N10:42–49. (in Russian).

Kossy, I. A., Kostinskiy, A. Yu., Matveev, A. A., and Silakov, V. P. (1994). Plasmo-chemical processes in nonequilibrium nitrogen-oxygen mixtures. In *Proc. IOFAN*, volume 47, pages 37–57. (in Russian).

Lehtinen, N. G., Bell, T. F., and Inan, U. S. (1999). Monte Carlo simulation of runaway MeV electron breakdown with application to red sprites and terrestrial gamma ray flashes. *J. Geophys. Res.*, 104:24699.

Lyons, W. A. (1996). Sprite observations above the U.S. High Plains in relation to their parent thunderstorm systems. *J. Geophys. Res.*, 101:29641.

MacGorman, D. R. and Rust, W. D. (1998). *The electrical nature of storms*. Oxford University Press.

Mareev, E. A. (1999). Turbulent electric dynamo in thunderstorm clouds. In *Proc. 11th Int. Conf. on Atmospheric Electricity*, pages 272–275, Guntersville, USA.

Mareev, E. A. and Anisimov, S. V. (2003). Global electric circuit as an open dissipative system. In *Proc. 12th Int. Conf. on Atmospheric Electricity*, pages 797–800, Versailles, France.

Mareev, E. A. and Sorokin, A. E. (2001). Autowave regimes of thunderstorm electrification. *Radiophys. Quant. Electr.*, 44(1-2):148–162.

Mareev, E. A. and Trakhtengerts, V. Yu. (1996). On electric dynamo problem. *Radiophys. Quant. Electr.*, 39:797–814.

Marshall, T. C. and Rust, W. D. (1993). Two types of vertical electrical structures in stratiform precipitation regions of mesoscale convective regions. *Bull. Am. Met. Soc.*, 74:2159–2170.

Marshall, T. C. and Stolzenburg, M. (2002). Electrical energy constraints on lightning. *J. Geophys. Res.*, 107(D7):doi:10.1029/2000JD000024.

Marshall, T. C., Stolzenburg, M., Rust, W. D., Williams, E. R., and Boldi, R. (2001). Positive charge in the stratiform cloud of a mesoscale convective system. *J. Geophys. Res.*, 106:1157–1164.

Pasko, V. P., Inan, U. S., and Bell, T. F. (1997). Sprites produced by quasi-electrostatic heating and ionization in the lower ionosphere. *J. Geophys. Res.*, 102(A3):4529–4561.

Pasko, V. P., Inan, U. S., and Bell, T. F. (2000). Fractal structure of sprites. *Geophys. Res. Lett.*, 27(23):497–500.

Pasko, V. P., Inan., U. S, and Bell, T. F. (2002). Diffuse and streamer regions of sprites. *Geophys. Res. Lett.*, 29(10):82–1–82–4.

Roble, R.G. and Tzur, I. (1986). The global atmospheric electrical circuit. In Krider, E.P. and Roble, R.G., editors, *The Earth's Electrical Environment*, pages 206–231. National Academy Press, Washington, D.C.

Rycroft, M., Israelsson, S., and Price, C. (2000). The global atmospheric electric circuit, solar activity and climate change. *J. Atmos. Sol.-Terr. Phys.*, 62:1563–1576.

Sato, M. and Fukunishi, H. (2003). Global sprite occurrence locations and rates derived from triangulation of transient Schumann resonance events. *Geophys. Res. Lett.*, 30(16):doi:10.1029/2003 GL017291.

Schuur, T. J. and Rutledge, S. A. (2000). Electrification of stratiform regions in mesoscale convective systems. Part II: Two-dimensional numerical model simulations of a symmetric MCS. *J. Atmos. Sci.*, 57(23):1983–2006.

Shepherd, R. T., Rust, W. D., and Marshall, T. C. (1996). Electric fields and charges near 0^0C in stratiform clouds. *Mon. Wea. Rev.*, 124.

Smirnova, E. I., Mareev, E. A., and Chugunov., Yu. V. (2000). Modeling of lightning generated electric field transitional processes. *Geophys. Res. Lett.*, 27(23):3833–3836.

Stolzenburg, M., Marshall, T., Rust, D., and Smull, B. (1994). Horizontal distribution of electrical and meteorological conditions across the stratiform region of a mesoscale convective system. *Mon. Wea. Rev.*, 1228:1777–1797.

Stolzenburg, M., Marshall, T. C., and Rust, W. D. (2001). Serial soundings of electric field through a mesoscale convective system. *J. Geophys. Res.*, 106:14079–14097.

Stolzenburg, M., Rust, W. D., Smull, B. F., and Marshall, T. C. (1998). Electrical structure in thunderstorm convective regions: 1. Mesoscale convective systems. *J. Geophys. Res.*, 103:14059–14078.

Takahashi, T. (1969). Electric potential of liquid water on an ice surface. *J. Atmos. Sci.*, 26(6):1253–1258.

Williams, E. R. (1998). The positive charge reservoir for sprite-producing lightning. *J. Atmos. Sol.-Terr. Phys.*, 60:689–692.

Wilson, C. T. R. (1920). Investigations on lightning discharges and on the electric field of thunderstorms. *Philos. Trans. Roy. Soc. Lond. - Ser. A*, 221:73–115.

Yashunin, S.A. (2004). Peculiarities of electric field generation in different atmospheric layers. Bachelor thesis. Nizhny Novgorod University.

ACTUAL PROBLEMS OF THUNDERCLOUD ELECTRODYNAMICS

Victor Y. Trakhtengerts and Dmitry I. Iudin
Institute of Applied Physcis, Russian Academy of Sciences, Nizhny Novgorod, Russian Federation.

Abstract   The electrodynamics of a thunderstorm cloud (TC) is considered with recirculation and multi-flow motion of charged cloud particles taken into account. In this consideration, the TC large-scale electric field is generated due to the charge separation during the convection of air and goes through oscillation phases during the initial and decaying stages of the TC. These oscillations explain qualitatively the observed behavior of TC electric fields. The multiflow character of convection introduces new important features of electric field generation in a TC. Multiflow convection is unstable and leads to the generation of slowly varying small-scale electrostatic fields with wavelength from $\sim 1 \div 100$ m. The amplitude of these electrostatic fields can reach the conventional breakdown value. Such instability can initiate intracloud micro-discharges at the preliminary stage of a lightning discharge and between separate lightning strokes. The development of instability scales with micro-discharges in a multiplicative way and can reach the percolation threshold for the electrical conductivity inside a TC. In this way, a drainage system for cloud space charges is formed, which initiates the leader channel of the lightning discharge. Another important process associated with the development of short-scale electric fields is the acceleration of relativistic electrons due to runaway breakdown. Unlike large-scale electric fields, slowly varying short-scale electric fields can support the acceleration of electrons in a large volume of a TC.

14.1   Introduction

The last decade was marked by fascinating discoveries in the field of atmospheric electricity, and high-altitude lightning discharges were among of these. It is clear that high-altitude electrical discharges have their roots in tropospheric thundercloud (TC) systems, which sometimes produce strong transient electric fields in the middle atmosphere. To answer how this occurs it is necessary to have a satisfactory model of electric field generation and lightning initiation in a TC.

Essential progress was achieved in experimental investigations of TC electricity on the basis of complex ground-based and balloon (*in situ*) measurements. The excellent books by MacGorman and Rust (1998) and Rakov and Uman (2003) summarise the modern state of TC electricity and lightning.

At the same time many important problems of TC electricity remain to be solved. The first concerns some dynamic features of the large-scale electric field during all stages of the evolution of a TC. These features include electric field oscillations with a temporal scale $T_{osc} \sim$ 5- 15 min, which are seen in the initial and dying stages of a TC (MacGorman and Rust, 1998). The very fast growth of the electric field amplitude and of the sizes and charges of cloud particles (drops and hail) before the first lightning flash does not find an explanation in the classical models of TC electricity (MacGorman and Rust, 1998). Many puzzles exist in the explanation of the multi-layer TC electric field structure (Rakov and Uman, 2003; Marshall and Rust, 1991, 1993; Stolzenburg et al., 1994; Marshall et al., 1995a) and of the preliminary stage of a lightning discharge (Rakov and Uman, 2003).

New questions have arisen after recent balloon experiments, which revealed intense X-ray and γ-ray emissions inside a TC (McCarthy and Parks, 1985, 1992; Eack et al., 1996a,b; Eack and Suszcynsky, 2000; Torii et al., 2002). These emissions are apparently connected with electron acceleration in a TC up to relativistic energies.

Below we consider some theoretical models, which can give qualitative and sometimes quantitative information on these questions. In Section 14.2 an isolated TC convective cell (billow) is analysed during its development taking into account a gradual growth of charges and sizes of separate cloud particles (drops and hail). In this stage of a TC formation, recirculation plays a very important role; this determines a large-scale distribution of the electrical charge in a convective cloud and can explain a periodic change and amplitude growth of a large-scale electric field. A possible mechanism for fast electric field growth during the mature state of a TC is considered in Section 14.3. This mechanism is based on an intracloud plasma beam instability, which leads to the generation of small-scale electric cells with sizes 1 to 10^2 m. Unlike the large-scale electric field, which is saturated at the level of runaway breakdown, a dynamic small-scale electric field is not saturated and can reach a conventional breakdown value. Moreover, the interaction of charged particles in a cloud is drastically changed in a cloud when instability conditions are fulfilled and charged particles of the same polarity coalesce. This results in the fast growth of a large-scale electric field, the amplitude of which can be increased by an order of magnitude during a few minutes (MacGorman and Rust, 1998).

Another very important consequence of the small-scale electric field growth, discussed in Section 14.3, is the formation of an electrical drainage system, which permits to gather electrical space charges from a cloud volume into a leader channel. A cloud metallization takes place, which is due to the development of numerous electrical microdischarges with scales $1 - 10^2$ m. The achievement of a percolation threshold is principal for such a metallization, when non-stationary fractal conducting chains appear with a length comparable to the cloud size.

Section 14.4 is devoted to an analysis of the relativistic electron acceleration in a TC due to runaway effects (Gurevich et al., 1992; Gurevich and Zybin, 2001). Together with acceleration in a constant electric field, such an acceleration is possible by a stochastic sign-alternating electric field (Trakhtengerts et al., 2002, 2003). Such an electric field appears as a result of plasma beam instability development. The acceleration by a stochastic small-scale electric field is accompanied by growth of the life time of relativistic electrons in a TC due to the effects of anomalous diffusion. In the conclusion (Section 14.5) the results obtained are summarized and some problems, which are important for TC dynamics and for the initiation of high-altitude electrical discharges, are formulated.

A list of mathematical symbols can be found at the end of this chapter.

14.2 Electric Field Generation in an Atmospheric Convective Cell

The long standing problem of electric field generation in a TC includes two basic questions: an electrical charging of separate cloud particles and a charge separation mechanism occurs. In our analysis we shall consider the charging mechanism as given and associate it with riming electrification (Takahashi, 1978; MacGorman and Rust, 1998), when large particles (drops, hail and graupel) are charged negatively and light particles (small droplets and ice crystals) are charged positively. We concentrate our attention on the charge separation mechanism, which is determined by the large-scale air flow motion. This problem is not new and was considered in several different approaches (see Levin and Tzur, 1986, and references cited therein). The main results were obtained with the help of computer simulations, but they did not elucidate two important issues, the role of recirculation and the influence of the generated electric field on the motion of charged cloud particles. These determine the temporal dynamics and large-scale distribution of space charge in a convective cloud. Real motions in a TC are rather complicated and often include multi-cell convective structures. Further, we investigate a solitary convective cell within a TC, which can be considered as an elementary generator in TC electricity.

14.2.1 The Case of a Cylindrical Convective Cell

We shall investigate the simplest 2D model of an atmospheric convective cell, suggesting a solid-like rotation of gas with cylindrical geometry (Figure 1). In this model, light positively charged particles (small droplets and ice crystals with size $r \leqslant 10\mu m$) move together with the air flow while the trajectories of negatively charged heavy particles (with $r > 10^2 \mu$m, cloud particles) differ from the trajectories of light particles. A mono-dispersed size distribution is suggested for simplicity. This problem was considered by Trakhtengerts (1992) taking into account friction, gravity forces and the self-consistent electric field, which is generated due to a non-compensated electrical space charge on the boundary of the convective cell. Two important conclusions were obtained:

1. Cloud particles are captured by the convective air flow and evolve during solid-like rotation with the same frequency around the centre that is shifted relative to the air cell centre;

2. Self-consistent effects play the principal role in determining the temporal dynamics and spatial structure of the generated electric field.

According to (Trakhtengerts, 1992) in the presence of small parameters

$$\Omega/\nu \ll 1, \quad X_c/R_c \ll 1, \tag{14.1}$$

where Ω is the rotation frequency, ν is the effective collision frequency (the friction force $F_f = M\nu u$, M is the mass of a cloud particle, u is its velocity relative to the mean air flow), R_c is the cell radius, $X_c = |g/\nu\Omega|$ is the maximal shift of the particle cell centre, the generated electric field $\mathbf{E}(E_x, E_z)$ is

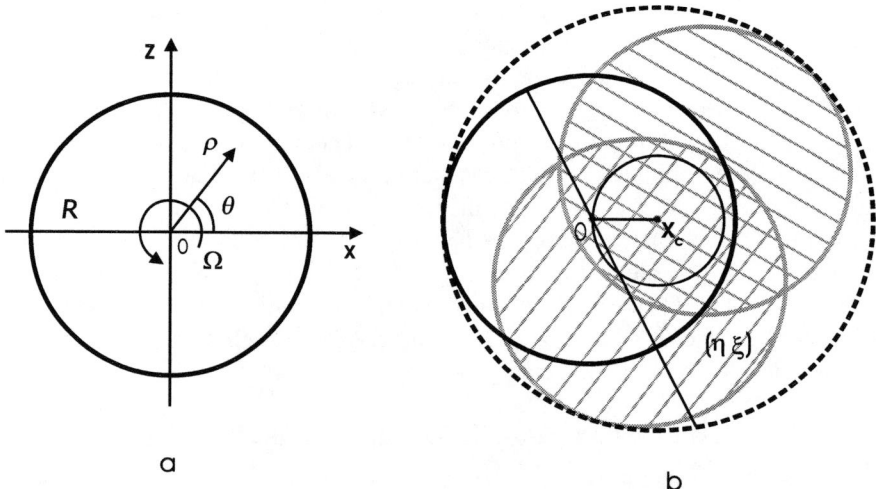

Figure 1. (a) Convective cell cross-section and coordinate system. (b) Aerosol disk "beatings" under the action of a friction force in a cylindrical convective flow and gravitational force.

determined by one parameter

$$\delta = \Omega_p^2/\Omega\nu, \tag{14.2}$$

where Ω_p^2 is particle plasma frequency (in cgs units)

$$\Omega_p = \left(\frac{4\pi Q^2 N}{M}\right)^{1/2}, \tag{14.3}$$

Q and N are the charge and concentration of cloud particles respectively. On the time scale $t \gg \nu^{-1}$ the motion of the particle rotation centre is described by the equations:

$$\xi_{\tau\tau} + 2\delta\xi_\tau + \xi\left(\frac{d\delta}{d\tau} + 1 + \delta^2\right) + \delta X_c = 0 \tag{14.4}$$

$$\eta = -\xi_\tau - \delta\xi - X_c \tag{14.5}$$

where τ is dimensionless time $\tau = \Omega t$, ξ and η are the coordinates of the particle rotation centre ($\xi \parallel \mathbf{z_0} \downarrow\uparrow \mathbf{g}$, $\eta \parallel \mathbf{x_0}$, $\mathbf{x_0}$ and $\mathbf{z_0}$ are the unit vectors), and $\xi_\tau = d\xi/d\tau$. In Eq. (14.4)-(14.5) δ is an arbitrary function of τ. It follows from Eq. (14.4)-(14.5) that there is a stationary state for (η, ξ) under $\delta = const$ and $t \to \infty$, which is:

$$\eta_{st} = -\frac{X_c}{1+\delta^2}, \quad \xi_{st} = -\frac{\delta}{1+\delta^2}X_c. \tag{14.6}$$

The electric field is determined by the difference of the fields from two homogeneously charged circles with charges of opposite sign; the centre of one (positively charged) circle is $\xi = \eta = 0$ and the centre of negatively charged circle of cloud particles is determined by Eq. (14.6). In particular, the electric potential φ for the circle with $\xi = \eta = 0$ is equal to:

$$\varphi(r) = \begin{cases} -\pi\rho r^2, & r \leq R_c \\ -\pi\rho R_c^2\left(2\ln\frac{r}{R_c} + 1\right), & r \geq R_c \end{cases}, \tag{14.7}$$

where $\rho = NQ$ is the space charge density. A similar formula determines the electric potential φ_1 (of the opposite sign), generated by the circle containing heavy particles with the shifted centre of rotation. A summary electric field is equal to $\{-\nabla[\varphi(r) + \varphi_1(r_1)]\}$, where $r_1^2 = (x+\eta)^2 + (z+\xi)^2$. Inside the cell the electric field is homogeneous and is equal to:

$$E_x = -\frac{\delta}{1+\delta^2}E_0, \quad E_z = -\frac{\delta^2}{1+\delta^2}E_0, \quad E_0 = \frac{Mg}{2Q}. \tag{14.8}$$

The sign of E_z is determined by the sign of the ratio g/Q; for the negatively charged cloud particles $g/Q > 0$, $(g < 0)$ and $E_z > 0$ (the inner vector product is positive $(\mathbf{E}, \mathbf{z_0}) > 0$). The sign of E_x depends on the sign of Ω and E_0;

for counter-clockwise rotation (Figure 1) $\Omega < 0$ and $E_x > 0$ $((\mathbf{E}, \mathbf{x}_0) > 0)$. The spatial structure of the electric field for this case is shown in Figure 2 with the conducting Earth being taken into account. The corresponding profiles of vertical and horizontal components of the electric field are shown in Figure 3. Here the height of a positive charge centre above the Earth's surface is shown for $H = 5$ km, $R_c = 2$ km, $|\eta_c| = |\xi_c| = 1$ km and $|\delta| = 1$. The value $|\delta| = 1$ can be obtained in the case when the radius of cloud particles comes to $r_0 = 1$ mm, their charge makes up $Q = -3 \cdot 10^{-11}$C, their concentration is $N \simeq 10^3 \text{m}^{-3}$, and the rotation frequency is equal to $\Omega \simeq 10^{-2} c^{-1}$ ($T \simeq 10$min).

14.2.2 Some Generalizations

In the consideration above we supposed the atmospheric electrical conductivity σ to be equal to zero. Under real conditions, electric field relaxation caused by the conductivity should be taken into account. In the simplest approximation the convective cell can be considered as an electric current generator, which accumulates space charge during a characteristic time Ω^{-1}; at the same time this space charge relaxes with a time constant $T_R = (4\pi\sigma)^{-1}$, where σ is the atmospheric ion conductivity in the TC. In such an approximation the conductivity leads to a decrease in the charge coming into Eq. (14.2)

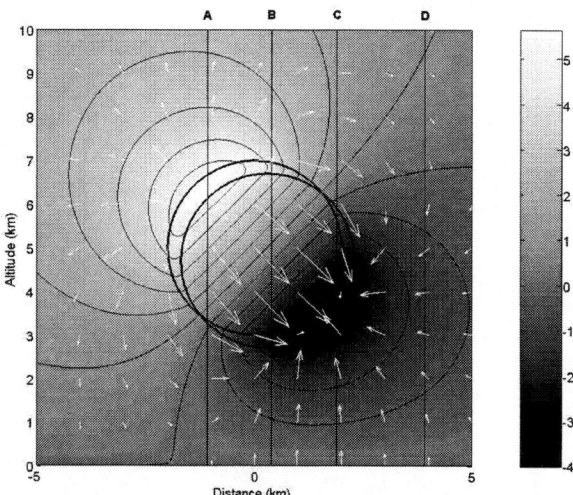

Figure 2. Spatial electric field structure in an atmospheric convective cell with cylindrical geometry and with the conducting Earth taken into account. Shades of gray correspond to different values of electric potential.

THUNDERCLOUD ELECTRODYNAMICS

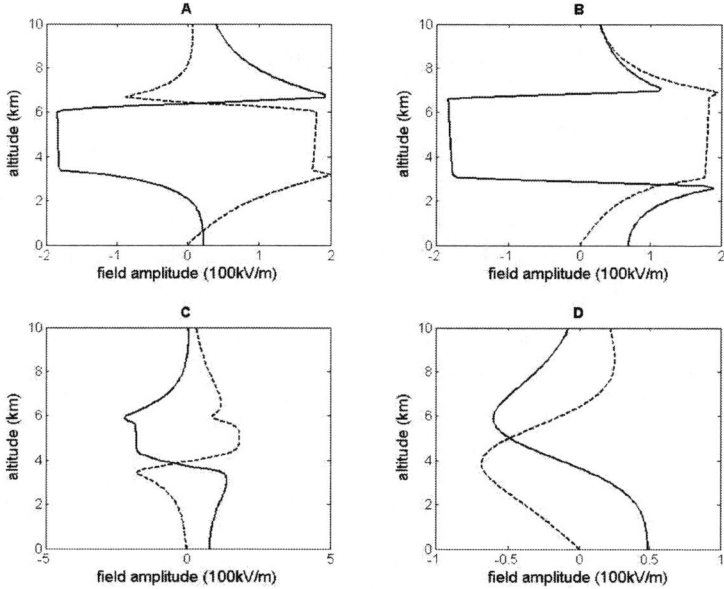

Figure 3. Profiles of vertical (solid line) and horizontal (dashed line) components of the electric field at several points labeled in Figure 2.

and, consequently, in the electric field given by the relation Eq. (14.3) by the multiplier

$$\mu = \Omega T_R (1 + \Omega T_R)^{-1}. \tag{14.9}$$

Accordingly the critical parameter δ Eq. (14.2) now takes the form

$$\delta_\sigma = \mu \delta. \tag{14.10}$$

It is not difficult to consider the elliptical trajectories of motion. For this we put the velocity components V_{x_0} and V_{z_0} of the air flow to be equal to:

$$V_{x_0} = \Omega_2 z \quad V_{z_0} = \Omega_1 x. \tag{14.11}$$

That corresponds to neutral gas motion along elliptical trajectories

$$z^2/\Omega_1 + x^2/\Omega_2 = const. \tag{14.12}$$

We obtain the same solutions Eq. (14.6) and Eq. (14.8) with the following modifications:

$$X_c^e = \frac{g}{\Omega_1 \nu}, \quad \delta_e = \frac{\Omega_p^2}{\Omega_H \nu}, \quad \mu_e = \frac{\Omega_H T_R}{1 + \Omega_H T_R}, \tag{14.13}$$

where $\Omega_H^2 = \Omega_1\Omega_2$. The solution for the electric field spatial structure is shown in Figure 4. The corresponding profiles of vertical and horizontal components of the electric field are shown in Figure 5. Here we suggested that cloud particles outside the convective cell are lost, so the full positive charge is bigger than the negative one (inside the cell $\rho_+ = \rho_-$). This structure of the space charge is interesting as a possible model for a positive cloud-to-ground lightning discharge (see Section 14.5).

14.2.3 Dynamics of the Large-Scale Electric Field in a Convective Cloud

During the mature state of a TC, which is accompanied by the growth of particle concentration N, charge Q and size r_0 (mass M) of cloud particles, the electric field increases. Inside a cloud (convective cell) this field is given by the following expressions:

$$E_x = -2\pi QN\eta, \quad E_z = -2\pi QN\xi, \quad (14.14)$$

where the coordinates (η, ξ) of the particle rotation centre satisfy the Eq. (14.4)-(14.5). It is possible to find the solution of Eq. (14.4)-(14.5) for an arbitrary dependence $\delta(\tau)$. Putting the initial conditions as

$$\tau = 0, \quad \eta_0 = \xi_0 = 0 \quad (14.15)$$

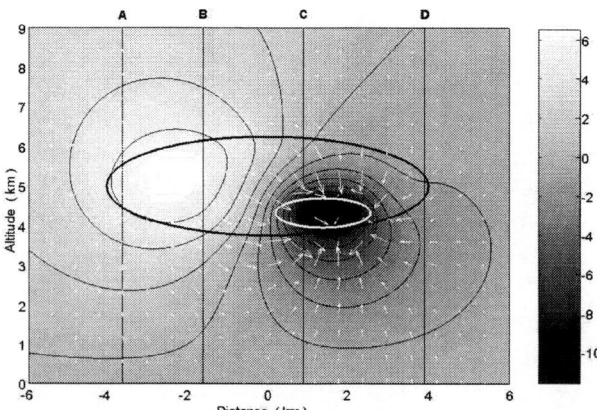

Figure 4. Relative spatial electric field structure in an elliptic convective cell with dissipation.

THUNDERCLOUD ELECTRODYNAMICS

we find:

$$x_c \equiv \eta(\tau) = -X_c \left(1 - \cos(\tau) \exp\left\{-\int_0^\tau \delta(\tau')d\tau'\right\}\right)$$
$$-X_c \int_0^\tau \delta(\tau')d\tau' \exp\left\{-\int_{\tau'}^\tau \delta(\tau'')d\tau''\right\} \cos(\tau - \tau') \quad (14.16)$$

$$z_c \equiv \xi(\tau) = -X_c \sin(\tau) \exp\left\{-\int_0^\tau \delta(\tau')d\tau'\right\}$$
$$+X_c \int_0^\tau \delta(\tau')d\tau' \exp\left\{-\int_{\tau'}^\tau \delta(\tau'')d\tau''\right\} \sin(\tau - \tau') \quad (14.17)$$

where, as earlier, $\tau = \Omega t$ and $X_c = |g/\Omega \nu|$.

In the case of a "geometrical optic" approach, when the inequality

$$2\pi \left|\frac{d\delta}{d\tau}\right| \equiv \frac{2\pi}{\Omega} \left|\frac{d\delta}{dt}\right| \ll 1 \quad (14.18)$$

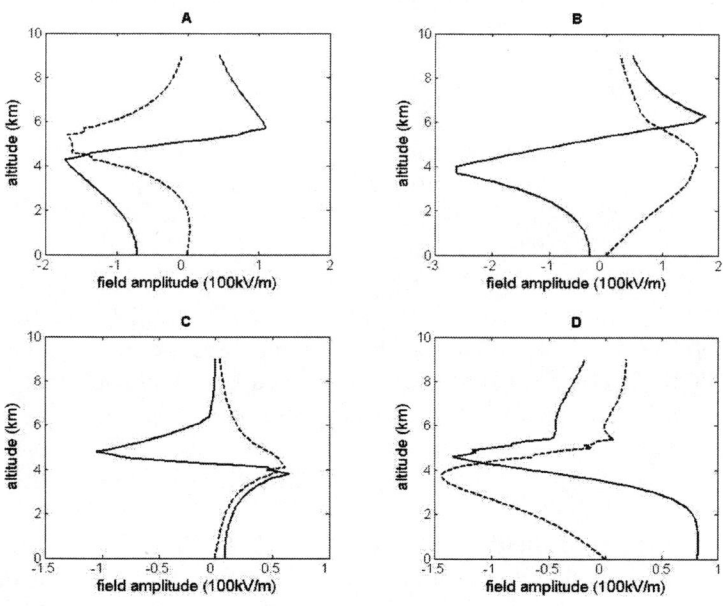

Figure 5. Profiles of vertical (solid line) and horizontal (dashed line) components of the electrical field at several points labeled in Figure 4.

is fulfilled, the formulae Eq. (14.16)-(14.17) can be simplified. In particular, for E_z Eq. (14.14) we obtain in the case Eq. (14.18):

$$E_z(\tau) = E_0 \frac{\delta}{1+\delta^2} \left[\delta \left(1 - \exp\left\{ -\int_0^\tau \delta(\tau')d\tau' \right\} \cos\tau \right) \right.$$
$$\left. - \exp\left\{ -\int_0^\tau \delta(\tau')d\tau' \right\} \sin\tau \right], \qquad (14.19)$$

where definitions are the same as in Eq. (14.8).

The results of calculations of $E_z(\tau)$ in the case of a linear dependence of δ on τ, $\delta = \alpha_E \tau$, are shown in Figure 6. Oscillations of E_z with change of sign are seen, which remind of the experimental behavior of E_z MacGorman and Rust (1998). Oscillations of the electric field appear during the decaying stage of a TC as well. This can be seen from Eq. (14.14)-(14.17), if we take, for example, the step-like dependence of δ on τ:

$$\delta = \begin{cases} \delta_0, & \tau \leqslant \tau_0 \\ 0, & \tau \geqslant \tau_0 \end{cases}. \qquad (14.20)$$

A sharp decrease of δ can be obtained due to the precipitation of heavy cloud particles. The electric field amplitude E_z is written in the case Eq. (14.20) as

$$E_z(\tau) = -\frac{8\pi QN}{1+\delta_0^2} X_c \delta_0 \sin\left(\frac{\tau_0}{2}\right) \left\{ \sin\left(\tau - \frac{\tau_0}{2}\right) - \delta_0 \cos\left(\tau - \frac{\tau_0}{2}\right) \right\}, \tau \geqslant \tau_0$$
$$(14.21)$$

where Q and N decrease with time during the decaying stage of the TC. One can see from Eq. (14.21) the occurrence of damped oscillations of E_z.

The maximum value of the electric field amplitude is reached in accordance with Eq. (14.8), when $\delta > 1$; under that condition $|E_z| = \delta |E_x| > |E_x|$. For the example considered in Subsection 14.2.1, $Q \simeq -3 \cdot 10^{-11}$C, $r_0 = 1$ mm, $N \simeq 10^3$ m^{-3} and $\delta = 1$, we obtain: $|E_z| = |E_x| = |Mg/4Q| \approx 167$kV/m, $|E| \approx 235$ kV/m. This value is close to the breakeven threshold value $E_B = 230$ kV/m (for normal atmospheric pressure, see Gurevich et al., 1992), which serves as the upper limit for E in the case of a slowly changing electric field as in our case of a convective cell. For $E \geqslant E_B$ the cloud conductivity σ should grow due to the appearance of relativistic electrons (see Section 14.4). In our model the growth of σ decreases $\delta_\sigma = \delta \cdot \mu$ Eq. (14.10) that leads to the decrease of the electric field in accordance with Eq. (14.8).

A question arises in this connection: how does a lightning discharge develop in a TC, if $E_{max} \sim E_B$ is much less than the threshold for conventional discharge? Another important problem is how to explain the observed very fast growth of the electric field during the mature stage of a TC, when E grows by an order of magnitude during several minutes, from 10 kV/m up to 10^2 kV/m. Some possible answers to these questions are discussed in the next section.

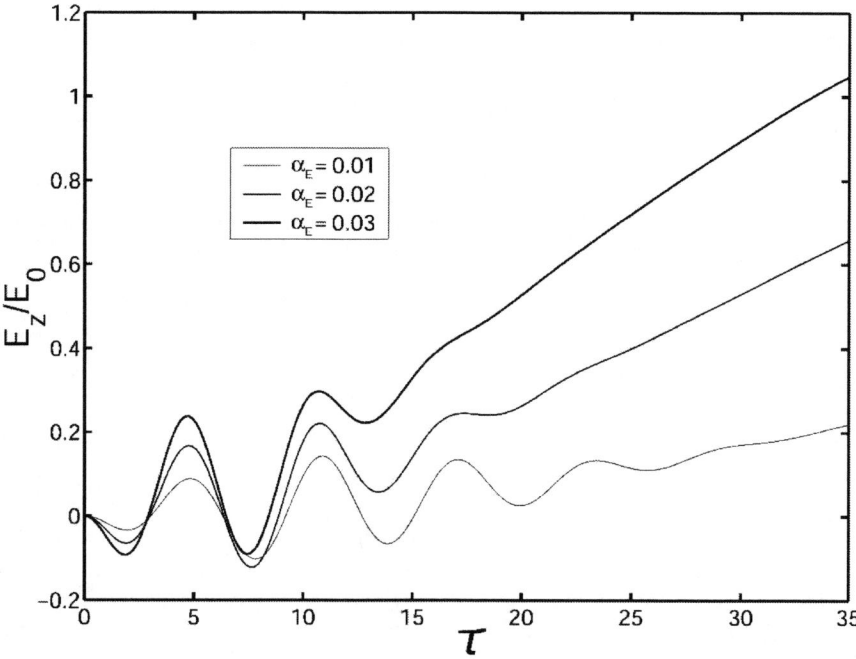

Figure 6. Variation of vertical electric field component E_z as a function of τ with varying values for α_E.

14.3 Fine Structure of Electric Fields in a Thundercloud
14.3.1 Multi-Flow Electrical Instability in a TC

Balloon *in situ* measurements of the electric field reveal rather complicated electrical structures, which include different vertical and horizontal scales from tens of meters up to several kilometers. This seems to reflect very important internal TC activity. The activity which can be produced by the plasma beam electrical instability is analysed in this section. We find that this instability can not only explain intracloud fine-scale electrical structure but can also change intracloud micro-processes, such as electrification and coagulation of cloud particles and initiate the formation of a lightning leader.

To illustrate the effect of electric cell generation we consider a simple TC model, which includes two charged components: heavy particles (drops and hail-stones) suspended in the updraft air flow and mainly in the lower half of the cloud, and the fraction of light particles (air ions, small droplets and ice crystals) that are carried to the upper part of the cloud.

The initial system of equations includes the equation of motion for charged heavy and light particles

$$\begin{aligned}\frac{\partial \boldsymbol{v_a}}{\partial t} + (\boldsymbol{v_a}\nabla)\boldsymbol{v_a} &= -(q/M)\nabla U + \boldsymbol{g} - \boldsymbol{F_{fr}}/M \\ \frac{\partial \boldsymbol{v}}{\partial t} + (\boldsymbol{v}\nabla)\boldsymbol{v} &= -(e/m)\nabla U + \boldsymbol{g} - \nu_n \boldsymbol{u} + \boldsymbol{F_D}/M\end{aligned}, \quad (14.22)$$

the continuity equations for the species, including the continuity equations for the space charge ρ and the electric current $\boldsymbol{j}$:

$$\frac{\partial \rho}{\partial t} + div\boldsymbol{j} = 0, \quad (14.23)$$

where

$$\rho = en + QN, \quad \boldsymbol{j} = en\boldsymbol{v} + QN\boldsymbol{v_a}, \quad (14.24)$$

and Poisson's equation for the electric potential U:

$$\nabla^2 U = -4\pi\rho. \quad (14.25)$$

In the system Eq. (14.22) $\boldsymbol{v_a}$ is the velocity, Q and M are the charge and mass of heavy particles, g is the acceleration due to gravity, and $\boldsymbol{F_{fr}}$ is the friction force in the updraft air flow; $\boldsymbol{v}$ is the velocity, e and m are the charge and mass of light particles, and $\boldsymbol{F_D}$ is the force creating the updraft air flow. In the relation Eq. (14.24) n and N are the densities of light and heavy particles, respectively, and $\boldsymbol{u} = \boldsymbol{v_a} - \boldsymbol{v}$.

We shall analyse the initial stage of electric cell generation, when variations of the concentration and the velocity of heavy and light particles are small, that is

$$N/N_0 \ll 1, \quad |n|/n_0 \ll 1, \quad |\Delta v|/v_0 \ll 1, \quad |\Delta v_a|/v_0 \ll 1, \quad (14.26)$$

where N_0, n_0 are the stationary concentrations of heavy and light charged particles, which follow from independent calculations (see, for example Mac-Gorman and Rust, 1998; Trakhtengerts, 1992; Straka and Anderson, 1993; Solomon and Baker, 1996), v_0 is updraft velocity, Δv_a and Δv are the velocity variations of heavy and light charged particles, respectively. In this approach from the initial system of Eq. (14.22)- (14.25) we obtain Poisson's equation for the electric potential with a self-consistent space charge distribution, which is not taken into account in traditional considerations (MacGorman and Rust, 1998; Trakhtengerts, 1992; Straka and Anderson, 1993; Solomon and Baker, 1996; Mazur and Ruhnke, 1998; Mansell et al., 2002):

$$\begin{aligned}\left[\frac{\partial}{\partial t} + (\boldsymbol{v_0}\nabla) + 4\pi\sigma\right]\nabla^2 U &= 4\pi q\left[\frac{\partial}{\partial t} + (\boldsymbol{v_0}\nabla)\right]N \\ \frac{\partial}{\partial t}\left(\frac{\partial}{\partial t} + \nu\right)N &= (N_0 q/M)\nabla^2 U\end{aligned}, \quad (14.27)$$

where σ is air conductivity (without charged heavy particles) and ν is the effective collision frequency $\nu = g/v_0$ for suspended particles. If we are interested in the spontaneous generation of electric cells with a scale much less than the cloud scale, the boundary conditions are not important. Using Eq. (14.27) the development of electric cells manifests itself as the instability that leads to the exponential growth of small variations of U. The wave potential U and the variation $\tilde{N}$ during the linear stage of instability development can be presented in the form $U, \tilde{N} \sim \exp(-i\omega t + i\boldsymbol{k} \cdot \boldsymbol{r})$, where $\boldsymbol{r}$ and t are the spatial coordinate and time, respectively. Substituting this into Eq. (14.27), we obtain the following dispersion relation between the wave vector $\boldsymbol{k}$ of the electric wave and its frequency ω (Trakhtengerts, 1989, 1994; Mareev et al., 1999):

$$1 - \frac{\Omega_p^2}{\omega(\omega + i\nu)} + \frac{4\pi i \sigma}{\omega - \boldsymbol{k} \cdot \boldsymbol{v_0}} = 0 \qquad (14.28)$$

where $\Omega_p^2 = 4\pi Q^2 N_0/M$ is the square of the "heavy particle gas" plasma frequency. The instability threshold is given by the condition

$$(\Omega_p/\nu)^2 \geq 1 \qquad (14.29)$$

and the instability growth rate is derived as $\gamma \simeq 2\pi\sigma(\Omega_p/\nu)$ for the optimal spatial period $a \sim \pi/k \simeq \pi v_0/\Omega_p$. For a particle radius ~ 0.1 cm and $N \sim 5 \cdot 10^3$ m^{-3}, the instability threshold is achieved for $q \sim 10^{-10}$ C. Putting $v_0 \sim 10$ m/s and $\Omega_p \sim 2\nu$, we find $\nu \sim 1$ s^{-1}, $a \sim 10$ m and $\gamma \sim 4\pi\sigma$ s^{-1}.

Further investigations of this instability (Grach et al., 2005) demonstrated its very important property that the instability is not suppressed by the growth of the air conductivity and also takes place for cloud particles with a spread of particle sizes. Figures 7 and 8 show the dependencies of the growth rate γ, on the conductivity σ and the wavelength $\lambda = 2\pi/k$. This means that the electric field wave amplitude continues to grow while the large-scale electric field is saturated at the level of the breakeven threshold. Another important consequence of this instability is the possibility of attraction of the same polarity charged particles. This leads to the bunching of particles of the same polarity and to the generation of electric waves. The possible role of this instability in TC dynamics is discussed in the next subsections.

14.3.2 A Mechanism of Electric Field Fast Growth during the TC Mature Stage

During the mature stage of a TC the number N, charge Q and mass M of cloud particles grow. The large-scale electric field changes accordingly as, for example, shown in Section 14.2. In our model, when the parameter $\delta > 1$, the

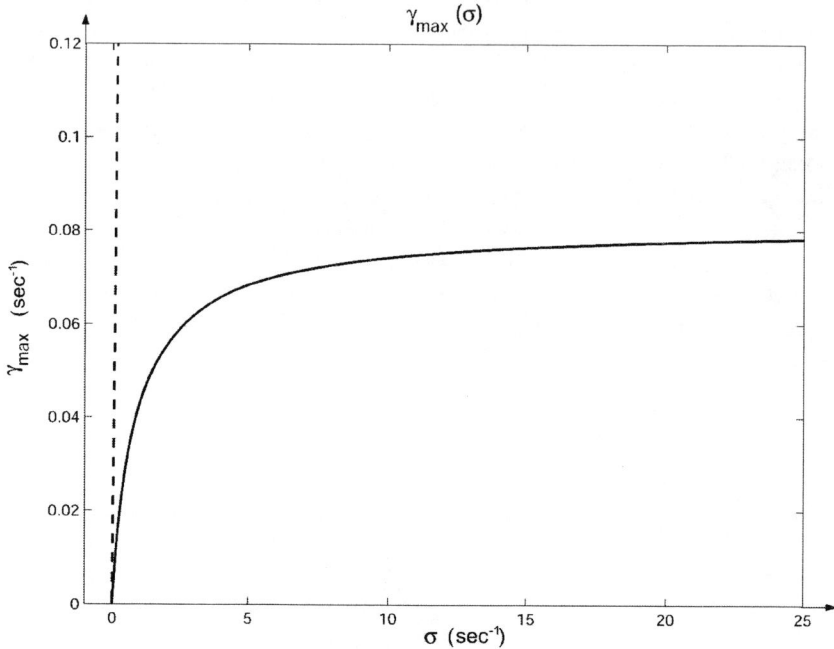

Figure 7. Dependence of the instability growth rate γ on air conductivity σ.

electric field is stabilized and determined by the expression:

$$|E| \sim \left|\frac{Mg}{2Q}\right|. \tag{14.30}$$

The plasma beam instability threshold Eq. (14.29) is reached later. As the particle size grows, its motion changes from viscous to turbulent when the Reynolds number $Re > 1$. This transition takes place for the particle size

$$70\mu < r < 150\mu \tag{14.31}$$

In particular, if we balance the gravity and friction forces (in the turbulent regime)

$$F_{fr} \equiv 0, 2\pi\rho_0 r^2 u^2 = M\nu u \simeq Mg, \tag{14.32}$$

where ρ_0 is the air density and u is the relative velocity between the cloud particles and the mean air flow, we can find u and ν and check the motion regime. Substituting $r_0 = 10^2 \mu$ we find that

$$u = 2\frac{m}{s}, \quad \nu \simeq 5s^{-1}, \quad Re \sim 10 \tag{14.33}$$

THUNDERCLOUD ELECTRODYNAMICS

The transition from the viscous to turbulent regime seems to be most favourable for reaching of the instability threshold Eq. (14.29). Putting $r_0 \simeq 10^2 \mu$ we estimate the necessary values of N and Q. From Eq. (14.29) the threshold value for the charge is equal to:

$$Q_{thr} \simeq \left(\frac{\nu^2 r^3 \rho_k}{3N}\right)^{1/2}. \qquad (14.34)$$

Substituting $r \simeq 10^2 \mu$, $\nu \simeq 5s^{-1}$, $\rho_k \simeq 0.8 \text{gr/cm}^3$ and $N \simeq 5 \cdot 10^4$ m^{-3}, we find $Q_{thr} \simeq 3.5 \cdot 10^{-12}$C. This gives the electric field on the particle surface $E_s \simeq 3000 kV/m$, which is close to the corona discharge threshold for a charged sphere (MacGorman and Rust, 1998).

Transition to the unstable state can play a principal role for cloud microphysics. As a matter of fact, the electric wave generation means attraction of the same polarity charged particles, which clusterize. This means, in particular,

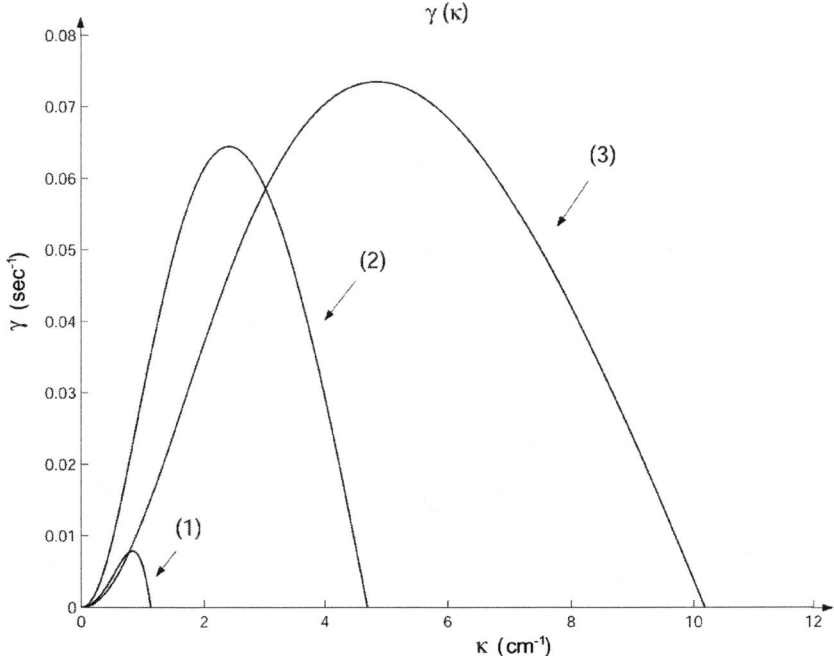

Figure 8. Dependence of the instability growth rate γ on the wave number k; the different curves correspond to different ratios $4\pi\sigma/\gamma$: (1) $4\pi\sigma/\gamma = 0.09$; (2) $4\pi\sigma/\gamma = 3.6$; (3) $4\pi\sigma/\gamma = 9$ (see Grach et al., 2005, for details).

that charged particles of the same polarity coalesce with the rapid growth of their mass and charge. The physical mechanism of such an attraction is that a self-consistent wave electric field forces charged cloud particles to gather into bunches. Moreover, when the instability threshold is reached, the electric field of an individual cloud particle changes drastically due to the surrounding ion space charge increasing the attraction of the same polarity particles (Gaponov-Grekhov et al., 2005). This idea demands further quantitative consideration but, if it takes place, we can obtain the fast growth of the electric field amplitude in accordance with Eq. (14.30). Of course, the drop mass grows as r^3. At the same time Q is growing as r^2, if the corona discharge limits the maximum value of Q. So, the electric field amplitude is changing as

$$E \sim \left(\frac{r}{r_0}\right) E_0 \sim \left(\frac{r}{r_0}\right) \cdot \left|\frac{M_0 g}{2Q}\right|_0, \qquad (14.35)$$

where the subscript "0" means the initial values. In our example, $M_0 \sim 3 \cdot 10^{-4}$gr, $Q_0 \sim 10^{-2}$cgs and $r_0 \sim 100\mu$m, which corresponds to $E_0 \sim$ 10kV/m. The growth of r up to 2 mm gives $E \simeq 100$ kV/m.

14.3.3 Fractal Dynamics of Micro-Discharges in a TC

Spatial-temporal TC dynamics reveal some peculiar features, which are now characterized by a common definition as fractal dynamics. The source of free energy in a TC is updraft convective flow, which forms multi-flow streams consisting of air molecules, light droplets, ice crystals and heavy hail stones. The interaction of these streams with one another leads to electrical charging of the cloud particles and to the generation of an electric field with different scales. A lightning flash that includes leader progression, return strokes and micro-discharges inside small-scale cells supply dissipation and a sink of free energy in a TC.

The processes in a TC are very diverse and complicated. A classical cloud to ground (CG) lightning discharge includes three stages: Preliminary breakdown, leader formation and return stroke (MacGorman and Rust, 1998; Rakov and Uman, 2003). Existing theoretical models of lightning discharges are based on its similarity with a laboratory long spark. There is a very important difference at the preliminary stage of the discharge; in the case of a laboratory spark electrical charge is accumulated on a conducting wall of a discharge space and easily flows into the spark channel. It is not clear what mechanism could provide the electric charge gathering over a considerable part, or all, of the intra-cloud volume to the leader channel. Apparently a certain important process which supplies this electric charge gathering, takes place during the preliminary breakdown stage. In its most developed phase, preliminary breakdown stage lasts approximately one tenth of a second, and consists of numerous (up to 10,000) relatively weak discharges (Rakov and Uman, 2003).

Widespread experimental efforts have demonstrated several peculiarities at the preliminary stage, proving it to be a very complex and puzzling phenomenon. Firstly, two subintervals (of approximately equal duration) may be selected in the preliminary breakdown stage. The first subinterval contains very high frequency (VHF) pulses, which appear without any visible DC field changes. A gradual DC field change accompanies VHF radiation during the second subinterval, and is closely connected with progression of the leader. Secondly, the results obtained using the VHF source location systems reveal pulse duration changes, which are connected with the Doppler effect applied to the radiation of a fast (up to 10^7 m/s) moving source. Sometimes VHF emission is observed as a coherent signal. Thirdly, the precursor activity spreads VHF sources throughout much of the cloud. Finally, the universality of the frequency spectrum of electromagnetic signals emitted by discharges was also established (Rakov and Uman, 2003; Gardner, 1990).

An important question is: what physical mechanism could result in the occurrence of such an intricate preliminary dynamic? It is particularly difficult to answer this question, since the median electric field strength in a cloud is an order of magnitude below the value for a conventional breakdown threshold.

A possible scenario of the preliminary stage development can be as follows. The plasma beam instability leads to the generation of small-scale electric cells, in some of which the electric field amplitude can reach the breakdown value. An electrical discharge in one cell can initiate micro-discharges in neighbouring cells. Many micro-discharges lead to the "metallization" conductivity of a TC, and serve as a drainage system for creating an electrical macro-charge over a cloud to the leader channel.

This TC activity was considered on the basis of a cellular automaton model (Iudin et al., 2003). Taking into account that the size of the active part of the thundercloud is about a few kilometers and the scale of an electric cell is about ten meters, the linear size of the model lattice should be about a few hundred of the spatial period. Each site of the lattice is related to a time-dependent scalar U_{ij} characterizing the electric potential. In our model the potential differences between the neighbouring sites grow occasionally in amplitude and sign due to the effects of the instability.

The potential difference is limited to some critical value U_c. As soon as this critical value is reached for any two neighbouring sites on the lattice, breakdown between the sites takes place and the lattice bond between the sites becomes a conductor. Its conductivity exponentially increases for a few model time steps and the corresponding potential difference decreases. We assume that such a fine scale spark discharge can initiate breakdowns of the neighbouring lattice bonds if the potential difference between the cells exceeds some activation level U_a, which is less than a critical level. This process is termed cell activation. The broken cells form a conductive cluster of short lifetime. The electric field for activation is considerably less than the breakdown field due to the appearance of a sharp heterogeneity of conductivity and of fast electrons. This is confirmed by experiments on the initiation of gas breakdown by a laser pulse. The ionization occurs in one time period step in the model. According to the experimental streamer velocity (Rakov and Uman, 2003; Proctor et al., 1988), the "burning front" in our model runs up to 10^5 km/s. It means that the lifetime of a conductive cluster with a spatial scale comparable to the cloud size is $\sim 10^{-3}$ s. This value is much less than an electric cell growth time. In other words, the process connected with the external driving of the system is much slower than the internal relaxation processes. It is the separation of time scales that ensures the fractal dynamics in the system. In our computer simulations this separation of time scales leads to:

$$\frac{D\tau_d}{U_a^2} \ll 1, \qquad (14.36)$$

where τ_d is the conducting cluster lifetime, D is the dispersion of the random additions that are added for nearest neighbour potentials at every step of model time. So, each bond of the lattice can be in one of four different states: the bond can be an insulator with a voltage drop that is less than the activation level U_a – a passive bond; the bond can be an insulator with a voltage drop that exceeds the activation level U_a and is less than the critical value U_c – an activated bond; the bond can be an insulator with a voltage drop that exceeds the critical value U_c – a critical bond; the bond can be a conductor – a metallized bond. The lattice is updated in parallel according to the following algorithm: (a) An activated bond becomes a metallized bond if one of its nearest neighbour sites is a metallized bond. (b) A critical bond becomes a metallized one in the succeeding model time step. (c) A metallized bond becomes a passive one if one or both of its ends has neither metallized nor activated nearest neighbour bonds. (d) The potential difference for random growth ensures transitions from activated bonds to critical bonds and between passive and activated bonds. Open boundary conditions are assumed.

Nonlinear interaction of neighbouring cells under the growth process discussed leads to the formation of dynamical chains of micro-discharges, which reveal a fractal behaviour over a wide range of TC parameters. Figure 9 shows the time evolution of the model discussed. This is the separation of time scales that turns our system into the Self-Organized Critically (SOC)-like dynamical state (see Gardner, 1990, for details). The separation of time scales is closely connected with the existence of the breakdown threshold. The fine-scale electrical field has to build up enough to exceed a certain critical value. This occurs over a much longer period of time than the short breakdown time interval. As was mentioned above, the preliminary stage of a lightning flash starts often in form of numerous discharges without any visible DC field change. Experiments (Proctor et al., 1988; Mazur and Ruhnke, 1998; Warwick, 1994; Boulch et al., 1990) give the frequency at which these microdischarges appear as about 10^5s^{-1}. The model discussed allowed us to estimate the appearance rate β of microdischarges (activated cells discharges). This rate is determined by the relation

$$\beta \simeq \gamma \frac{V}{a^3} p,$$

where γ describes the growth rate of the electric field in a cell, V is the volume of the active part of the TC, in which an instability develops, and p is the fraction of the activated cells. During the stage considered, when the activated-cluster length is comparable with the cloud size, $p \sim p_c$, where $p_c = 0.25$ corresponds to the percolation threshold in the three-dimensional case. Assuming $V \sim 10^{10}$ m^3, $\gamma \sim 0.1$ s^{-1}, and $a \sim 10$ m, we obtain $\beta \sim 2.5 \cdot 10^5$ s^{-1}, which agrees well with experiments. The important question is about the scale a of the electric cells. In the experiments of Proctor et al. (1988) the size a corresponds to the scale of an elementary discharge. Based on the spatial resolution achieved it is possible to say that $a < 60$ m. Our investigations of the Eq. (14.1) for the real parameters of a TC (charge and size of cloud particles, background conductivity, updraft flow velocity) gives $a \sim 1$ to 10^2 m, which is in qualitative agreement with the experiment. Measurements (Proctor et al.,

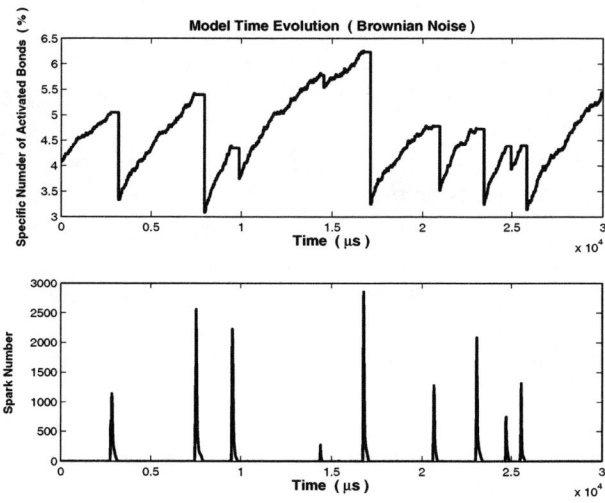

Figure 9. The time evolution of model parameters. The top panel shows a specific number of activated couples, and the bottom panel shows the temporal evolution of the number of sparks in the same time interval.

1988; Mazur and Ruhnke, 1998; Warwick, 1994; Boulch et al., 1990) show that separate pulses with a duration $\sim 1\mu s$ occur in groups, which include tens and hundreds of individual microdischarges. Interferometric measurements show that these bursts are equivalent to spatial trains that extend several kilometers with a velocity of $2 \cdot 10^7 - 10^8$ m/s. We can associate these trains with separate clusters which appear in our model. Near the percolation threshold a cluster length runs up to cloud size. If a microdischarge corresponds to the spark discharge with characteristic length a, we can estimate the tip velocity as $v_t \sim a/\tau_d \sim 10^7$ m/s, where $\tau_d \sim 1$ μs is the duration of an elementary discharge. This estimation does not contradict experimental data. Precise measurements of the fine electric structure in a TC with a spatial resolution of better than one meter are needed to obtain more sophisticated quantitative results.

14.4 Acceleration of Relativistic Electrons during a Thunderstorm

14.4.1 Runaway Breakdown in a Constant Electric Field

One of the most fascinating recent discoveries was X- and γ-ray emissions initiated by a thunderstorm (Marshall et al., 1995b; McCarthy and Parks, 1985, 1992; Eack et al., 1996a,b; Eack and Suszcynsky, 2000; Alexeenko et al., 2001; Torii et al., 2002). This phenomenon testifies the effective acceleration of electrons by the TC electrical field, up to relativistic energies. The energetic electrons connected directly with TC activity were observed in satellite experiments (Bratolyubova-Tsulukidze et al., 2001). Wilson (1925) was the first who paid attention to the possibility of electron acceleration by a TC electric field. An acceleration mechanism was suggested by Gurevich et al. (1992) and elaborated in numerous papers (see review Gurevich and Zybin, 2001, and references therein). The acceleration mechanism is based on the fact that the friction force Φ, which acts on the energetic electrons in the air, decreases with the growth of the electron energy. This so-called runaway effect takes place when the electron energy $\mathcal{E}$ is much greater than the ionization threshold in a molecule, ($\mathcal{E} \gg Z_i e I_i$, Z_i is the number of electrons in a molecule, I_i and is the ionization potential). In the range of non-relativistic energies the friction force is proportional to $\mathcal{E}^{-1}$. For $\mathcal{E} \simeq 1.4$ MeV the friction force Φ reaches its minimal value and then grows logarithmically. This dependence of Φ on $\mathcal{E}$ is shown in Figure 10. The analytical dependence of Φ in the energy range of our interest has the form (Gurevich et al., 1992; Gurevich and Zybin, 2001):

$$\Phi = \frac{\Phi_0 \alpha^2}{\alpha^2 - 1} \ln \Lambda, \qquad (14.37)$$

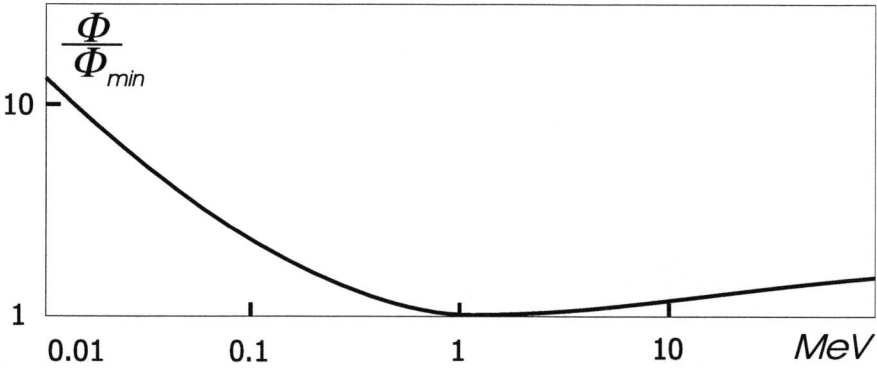

Figure 10. Dependence of the normalized friction force between energetic electrons and air on the electron energy.

where α is the dimensionless electron Lorentz factor, the energy $\mathcal{E} = mc^2(\alpha - 1)$, $\Lambda \approx \dfrac{(\alpha - 1)mc^2}{\varepsilon_i}$ and

$$\Phi_0 = 4\pi e^4 Z N_m / mc^2. \tag{14.38}$$

The minimum value of Φ_D corresponds to $\alpha_0 = 3.8$ ($\mathcal{E}_m \sim 1.4\text{MeV}$) and is equal to:

$$\Phi_{min} \simeq b\Phi_0, \quad b \simeq 11. \tag{14.39}$$

The electric field corresponding to the runaway threshold (the breakeven threshold) is

$$E_B = \Phi_{min}/e. \tag{14.40}$$

In air under normal pressure

$$E_B \simeq 216\text{kV/m} \sim 0.1 E_c, \tag{14.41}$$

where $E_c \simeq 2.3$ MV/m is the threshold value for the conventional air electrical breakdown. Coming from Eq. (14.37), Gurevich, Milikh and Roussel-Dupré suggested a very interesting idea of a runaway breakdown, which seems to be very important for TC dynamics, especially for high-altitude electrical discharges. When $E > E_B$, an avalanche of relativistic electrons develops: their number grows exponentially in time and space. Accordingly the air conductivity grows exponentially due to the ionization process. The necessary conditions for such a breakdown are that:

1. The electric field amplitude must satisfy the inequality $E > E_B$;

2. Electrons with energy $\mathcal{E} \gtrsim \mathcal{E}_m = 1.4$ MeV should be present in the initial stage;

3. The scale of the region with $E > E_B$ should exceed the characteristic scale l_a for avalanche formation.

According to Gurevich et al. (1992), l_a can be written as

$$l_a = \frac{2mc^2}{\Phi_0}\frac{E_B}{E} \simeq 50m\frac{E_B}{E}\left(\frac{N_m}{2.7 \cdot 10^{19}\text{cm}^{-3}}\right)^{-1}, \qquad (14.42)$$

where N_m is the air density per cm^{-3}. The parameter l_a^{-1} determines the spatial growth rate of the electron concentration N ($E \gtrsim E_B$):

$$N = N_0 \exp\{s/l_a\}, \qquad (14.43)$$

where s is the spatial coordinate along **E**.

Of course, runaway breakdown demands a more rigorous consideration based on the kinetic equation for the distribution function F of electrons in momentum and coordinate space. Different approaches to this problem were developed in many papers (Gurevich et al., 1992; Milich et al., 1995; Bell et al., 1995; Pasko et al., 1996; Taranenko et al., 1996; Yukhimuk et al., 1999; Gurevich and Zybin, 2001), especially in the application to high-altitude discharges (red sprites and blue jets). There is some experimental evidence that the runaway effects are important for the formation of the large-scale electric field inside a TC. In particular, Figure 11 shows the height profile of the electric field in relation to balloon measurements (Marshall et al., 1995b) and the breakeven value E_B (runaway breakdown threshold). One can see that E_B is the maximum value for $E \lesssim E_B$. It is clear that the development of a TC and of high-altitude electrical discharges depends strongly on the spatial structure and temporal dynamics of the electric field. The experiment and the basic regularities described in Section 14.2 show that the large-scale electric field in a TC grows rather slowly (over some minutes). If the plasma beam instability condition is reached before E reaches E_B, the initiation of a lightning discharge occurs through the fractal dynamics of micro-discharges, as described in Section 14.3. In this case the runaway breakdown can determine the activation level of a small-scale electric field E_a. When the value E_B is reached, the electric field is saturated on this level due to the growth of the air conductivity σ. It should be emphasised that the plasma beam instability is not saturated through the conductivity growth. In this case micro-discharge activity can be developed much more easily on the background of the large-scale electric field $E \lesssim E_B$. This point is of great interest nowadays, and it can also be important for an explanation of high-altitude discharge fine structure. It is interesting to

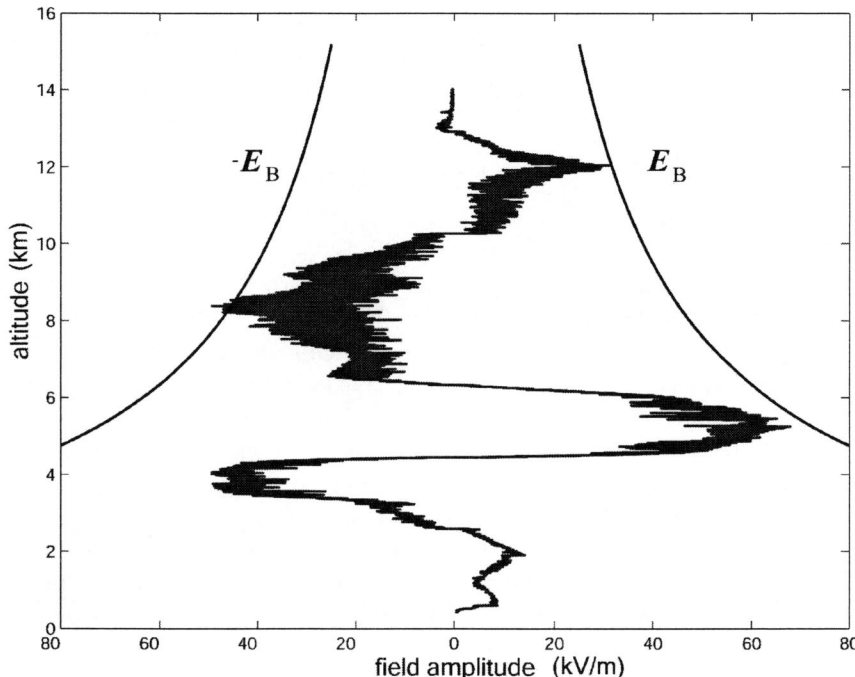

Figure 11. Electric field height profile in the balloon measurements (Marshall et al., 1995a) and the runaway breakdown threshold E_b.

discuss the properties of the large-scale electric field following from theoretical estimations, Section 14.2, in the saturation regime, when E reaches E_B. This occurs when the amplitude E_0 in Eq. (14.8) exceeds E_B:

$$E_0 \equiv \frac{Mg}{2Q} > E_B = 216 \left(\frac{N_m}{2.7 \cdot 10^{19} \text{cm}^{-3}} \right) \frac{\text{kV}}{\text{m}}. \qquad (14.44)$$

For example, in the case $r_0 \simeq 1$ mm, $\rho_d \simeq 0.8 \text{gr/cm}^3$ and $Q \simeq 10^{-10}$C, we have $E_0 \simeq 320 \frac{\text{kV}}{\text{m}} > E_B$. According to Eq. (14.8)-(14.10), the electric field amplitude in the saturated state is

$$|E| = E_0 \left(\frac{\delta_\sigma^2}{1 + \delta_\sigma^2} \right) \simeq E_B, \quad \frac{|E_z|}{|E_x|} \simeq \delta_\sigma. \qquad (14.45)$$

Putting $E_0/E_B \sim 2$, that corresponds to $h \simeq 5$ km, we find: $\delta_\sigma \simeq 1/\sqrt{3}$ and $|E_z|/|E_x| \simeq 1/\sqrt{3}$. In the case $\Omega T_R = \Omega/4\pi\sigma \ll 1$ δ_σ is given by:

$$\delta_\sigma \simeq \frac{\Omega_p^2}{4\pi\sigma\nu} \simeq \frac{Q^2 N}{M\sigma\nu}. \qquad (14.46)$$

For our example, putting $N \sim 5 \cdot 10^2 \text{m}^{-3}$ and $\nu \simeq 0.5 s^{-1}$ we find $T_R \simeq 8$ s. It would be interesting to compare the values of E_0, δ_σ, $|E_z|/|E_x|$ and T_R obtained from the model with *in situ* measurements of these parameters.

The TC large-scale electric field can be considered, after E_B is reached, as quasi stationarity in space and time. Another situation takes place in the case of high-altitude electrical discharges. They usually appear after a sufficiently strong positive cloud-to-ground (CG) lightning discharge and are connected with a sharp increase (during some ms) of an electric field in the middle atmosphere above a TC system. Nobody knows how this occurs, but it is suggested that a negative macroscopic charge is switched on at the same height above a TC top, and its parameters (total charge Q_Σ, spatial and temporal evolution) are chosen arbitrarily to explain the observed properties of high-altitude discharges.

A possible scenario of high-altitude discharge initiation can be suggested on the basis of a model of large-scale electric field formation, as considered in Section 14.2. We shall come from the model of a convective cloud with dissipation, when some part of the heavy negatively charged particles (drops and hail) have been lost from a cloud. The structure of the large-scale electric field is shown in Figure 3. The cloud carries a non-compensated positive charge, which produces non-zero electric field on the ground surface at the initial stage. At the mature stage this electric field produces numerous corona discharges at the Earth's surface, which inject free negative charge (negative ions) into the surrounding atmosphere. These ions are trapped by convection and move to the cloud top, which extends the ideas of Vonnegut (1953). The structure of the space charge in this case is shown diagrammatically in Figure 12. The maximum value of this negative charge Q_Σ^- can reach the value Q_Σ^+ of the cloud non-compensated positive charge. The volume occupied by the negative space charge will be essentially bigger than the cloud scale due to the divergence of convection trajectories. If lightning initiation is due to enhanced conductivity in the cloud core, as considered for the case of micro-discharge fractal dynamics (Subsection 14.3.3), the positive charge will be lowered to the cloud bottom and can be discharged on the ground. This means the appearance of a free negative charge Q_Σ^- above the cloud, which can initiate a high-altitude electrical discharge.

14.4.2 Acceleration by Stochastic Electric Fields

The acceleration of electrons by a constant electric field seems to be a transient process which terminates through air breakdown. At the same time, the experiments show that γ- and X-ray emission is a long process testifying to the continuous presence of energetic electrons inside the TC. Some additional possibilities for the production of energetic electrons are likely to appear if one takes into account electron acceleration by the stochastic electric field. Considering regions of the cloud where the electric field has different orientations, this formulation of the acceleration problem is quite natural.

We investigate electron acceleration by a constant-in-time but changing-in-space short-scale electric field of variable orientation, taking into account a spatially limited acceleration region. The kinetic equation for the velocity distribution function f of electrons with high energies (energy $W > 10$ keV) can be written for the arbitrarily oriented electric field $\boldsymbol{E}$ in the same form as for the constant $\boldsymbol{E}$:

$$\frac{\partial f}{\partial t} + \boldsymbol{v}\frac{\partial f}{\partial \boldsymbol{r}} - e\boldsymbol{E}(\boldsymbol{r})\frac{\partial f}{\partial \boldsymbol{p}} = \left(\frac{\partial f}{\partial t}\right)_c \quad (14.47)$$

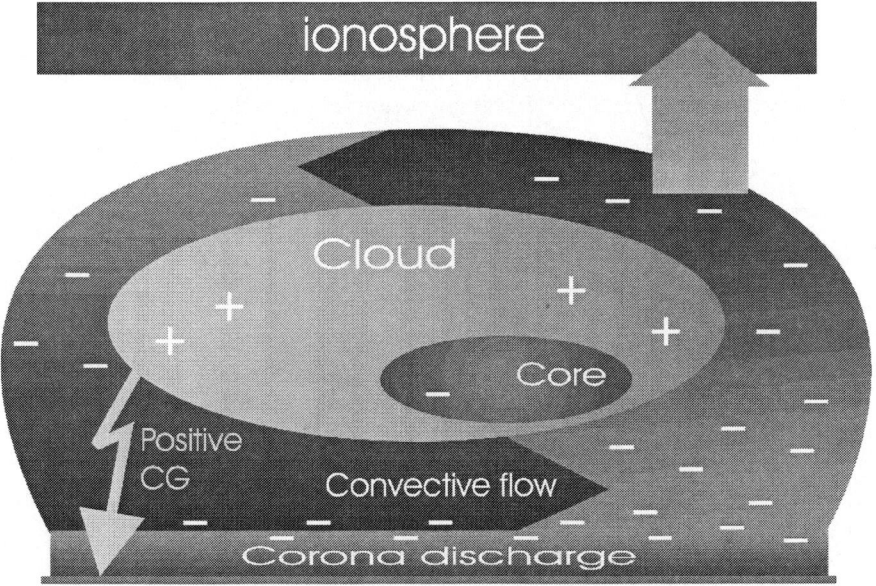

Figure 12. Qualitative structure of a possible scenario for the initiation of a high-altitude discharge.

where p is the electron momentum, r and t are the space coordinate and time, and $(\partial f/\partial t)_c$ is the collision integral for the electron-air interaction. The collision integral can be written as follows:

$$\left(\frac{\partial f}{\partial t}\right)_c = \frac{1}{p^2}\frac{\partial(p^2\Phi f)}{\partial p} + \nu_{eff} f \qquad (14.48)$$

where $\Phi(p)$ is determined by the expression (37) and ν_{eff} describes the collisional isotropization of the distribution function. The collisions in our problem are crucial because of the considerable and irreversible change of electron energy by a stochastic electric field which is possible only under the joint action of electric field acceleration and collision. It is further suggested that the change of f during one stochastic period of the electric field is small. This means that the distribution function f can be expressed in the form:

$$f = F(p, r, t) + f_\sim \qquad (14.49)$$

where F is the regular part of the distribution function averaged over an electric field stochastic ensemble, and $f_\sim$ is a small addition ($|f_\sim| \ll F$). We further apply the iteration procedure in the form Eq. (14.49) to find the contribution of a stochastic electric field to the evolution of the regular F. This contribution can be found from the kinetic equation Eq. (14.47) averaged over the electric field ensemble. The details of this averaging are described in paper (Trakhtengerts et al., 2002, 2003). Here we give the final result. The kinetic equation for the regular part F of the distribution function is written in the following form:

$$\frac{\partial F}{\partial t} = \frac{\partial}{\partial z}\left(\frac{v^2}{6(D_1+D_2)}\frac{\partial F}{\partial z}\right) + \frac{1}{p^2}\frac{\partial(p^2\Phi F)}{\partial p} + $$
$$+ \frac{1}{p^2}\frac{\partial}{\partial p}\left\{\frac{e^2 I_0}{k_0^2 v^2}\Phi\frac{(Z+2)p}{4\alpha}\frac{\partial F}{\partial p} - \frac{e^2 I_0}{k_0^2 v^2}\frac{\partial}{\partial p}\left(\frac{p^2\Phi}{v}\frac{\partial F}{\partial p}\right)\right\} + J \quad (14.50)$$

where J is the source of electrons, k_0^{-1} is the characteristic scale of electric cells, $I_0 = \langle E^2(r)\rangle$ is the averaged over the ensemble electric field intensity, the coefficient D_1 can be written as

$$D_1 \simeq e^2 I_0/k_0 v p^2, \qquad (14.51)$$

and D_2 is equal to

$$D_2 = \nu_{\text{eff}} = \frac{Z+2}{8\alpha p}\Phi. \qquad (14.52)$$

Equation (14.49) is valid when the layer occupied by electric cells is sufficiently thick, and an electron changes its direction of motion many times under

THUNDERCLOUD ELECTRODYNAMICS

the influence of collisions and the stochastic electric field. An additional condition that should be fulfilled for the validity of Eq. (14.49) is (Trakhtengerts et al., 2002, 2003)

$$I_0 \gtrsim (\Phi_{min}/e)^2, \tag{14.53}$$

where the minimum value of the friction force Φ_{min} is determined by (39). The condition (53) coincides with the condition for the appearance of runaway electrons in a constant electric field $E > E_B$ (40). The corresponding to (53) relativistic factor $\alpha_m = 3.8$. According to (50), the characteristic scale p_0 of the distribution function F in p space is of the order

$$p_0 \sim eI^{0.5}/k_0 c. \tag{14.54}$$

The non-stationary solution of Eq. (14.49) is shown in Figure 13 for the case, when the first term on the left side of Eq. (14.49) can be neglected. The dimensionless variables $\tilde{x}$ and $\tilde{y}$ in this diagram are taken in the form

$$\tilde{x} = p/p_0, \quad \tilde{y} = \Phi_{min} t/p_0^3. \tag{14.55}$$

According to (Trakhtengerts et al., 2002) the distribution function in Figure 13 acquires the universal dependence on $\tilde{x}$ under $\tilde{x} \gg 1$, which corresponds to the stationary solution of Eq. (14.49) without a source, and its amplitude grows linearly with time. The asymptotic behaviour of F is $\tilde{x}^{-1} \cdot \exp(-\tilde{x})$, when $\tilde{x} \to \infty$. The lifetime of energetic electrons in a TC is determined by the relation

$$\tau_e \simeq 6 z_m^2 (D_1 + D_2)/c^2, \tag{14.56}$$

where z_m is the vertical scale of a cloud. The ratio D_1/D_2, characterizing the slowing-down of the electron diffusion in a cloud, is determined from (51) and (52) ($\alpha \gg 1$)

$$D_1/D_2 = 8e^2 I_0/k_0 c^2 m(Z+2)\Phi. \tag{14.57}$$

Let us consider the quantitative example for the height $h = 5$ km. For air, $Z = 14.5$ and $N_m = 1.32 \cdot 10^{19}$ cm^{-3}, and we have from Eq. (14.39) and (39) the threshold electric field intensity $\sqrt{I_{0thr}} \geq 104$ kV/m. Using the relation (54) and the equality $(p_0/mc) \geq \alpha_m = 3.8$ we find the scale of electric cells

$$k_0^{-1} \lesssim \alpha_m (mc^2/\Phi_{min}) \sim 20 \, m. \tag{14.58}$$

In this case the ratio $D_1/D_2 \sim \alpha_m \cdot (mc^2/\Phi_{min}) > 1$. The lifetime in the case $D_1/D_2 \gg 1$ grows with I_0 as

$$\tau_e \sim 6 z_m^2 e^2 I_0/k_0 c^3 p^2. \tag{14.59}$$

So, if we fix the energy of the electrons, the flux density in this energy channel grows proportionally to I_0. Simultaneously, the width of the distribution function grows according to (54). Substitution of p_0 into (59) gives the lifetime as

$$\tau_e \sim 6k_0 z_m^2/c. \qquad (14.60)$$

In the case $z_m \sim 3\text{km}$ and $k_0^{-1} \sim 20\text{m}$, we have $\tau_e \sim 10^{-2}\text{s}$. This particular example shows that a TC electric field, which reaches values $(1-2.5) \cdot 10^2$ kV/m and demonstrates a rather complicated and multi-layer structure, can essentially influence the intensity of an energetic electron component. More detailed consideration of the case, when the source J is due to cosmic rays, gives the increase of the electron flux density with energies $\gtrsim$MeV by $(1+D_1/D_2)$ times, that is equal to 5 for $I_0^{0.5} \sim 200$ kV/m. New transient sources of energetic electrons appear during TC activity, such as electrical micro-and macro-discharges, but a consideration of these is reserved for the future.

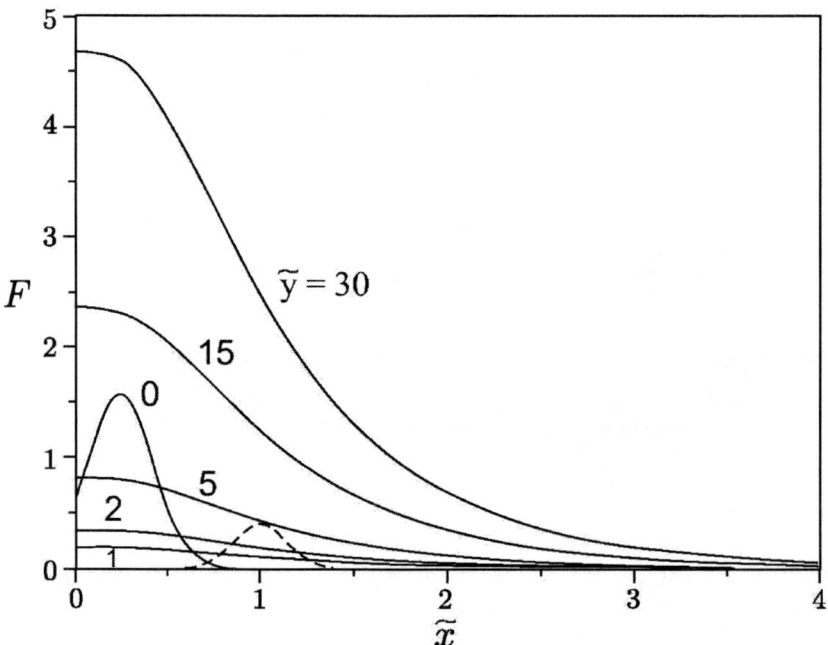

Figure 13. The velocity distribution function of electrons for the relativistic case; the different curves correspond to different values of the dimensionless variable $\tilde{y}\phi_{min}t/p_0^3$.

14.5 Conclusions

We have considered some simple analytical models for a qualitative description of TC electrodynamics. In the frame of this consideration we have already obtained some important results. In particular, a mechanism of large-scale electric field oscillations during the initial and decaying stages of a TC. It was shown that the recirculation of charged particles in the process of cloud convection can serve as a natural cause for these oscillations. Important consequences follow from the recirculation model for the distribution of space electrical charge in a TC and the formation of a self-consistent model for the initiation of high-altitude discharges.

Intracloud electrodynamic instabilities can change the microphysics of a cloud substantially. The plasma beam instability considered is a good example of such instabilities. It leads to the generation of multi-cell electrical structure, with scales $\lambda \sim 10 - 100$ m, and can drastically change the coalescence of drops and the growth of ice crystals in a TC.

Runaway breakdown is a key process for TC electrodynamics. Apparently, it is responsible for the acceleration of relativistic electrons during TC activity and may play a very important role in the origin of high-altitude discharges (sprites). The hot question is: how does runaway breakdown manifest itself in a micro-discharge activity? This activity is another key problem in TC dynamics. It seems to be responsible for the formation of a drainage system for charge redistribution in a TC and for lightning leader formation.

Additional experimental and theoretical investigations are needed to quantitatively develop the problems discussed above. The correlated measurements at spaced points seem to be useful to select the nature of large-scale electric field variations during the initial and decaying stages of a thunderstorm, temperature changes or large-scale rotation of an electrical space charge, e.g., considered in Section 14.3. More sophisticated radar measurements of convection velocity maps are desirable inside TC cells and mesoscale convective systems with good spatial and temporal resolution. As for small-scale electric field and microdischarges it is necessary to reach a spatial resolution up to $l_{min} \sim 1$ m. A similar resolution is desirable for multipoint ground-based VHF measurements, creating the radar picture of a TC. The very important information could be obtained from in situ measurements of relativistic electrons and cloud conductivity.

In theoretical studies we would like to pay attention to two questions: The first is the dynamics of large-scale electric fields in a convective cell. Here computer modelling is important by using more realistic cloud models of cloud convection and realistic cloud particle size distributions. The second question is a quantitative investigation of the collective coalescence of same-polarity particles under multi-flow instability conditions.

Acknowledgments

This work was partly supported by the Russian Foundation for Basic Research (project No. 04-02-17405) and by the Russian Academy of Science in the framework of the Program "Physics of the atmosphere: electric processes, radio-physical methods".

List of Symbols

α	relativistic factor
β	appearance rate of microdischarges
γ	instability growth rate
δ	critical parameter
ν	effective collision frequency
ν_{eff}	describes collisional isotropization of the distribution function
σ	conductivity
τ	dimensionless time
τ_e	life time of energetic electrons in a thundercloud
τ_d	duration of an elementary intracloud discharge
φ	electric potential
ξ, η	coordinates of the particle rotation center
a	lattice spatial period in cellular automaton model
c	velocity of light
CG	cloud to ground
D_1, D_2	diffusion coefficients
$\mathcal{E}$	electrons energy
$\mathcal{E}_m$	runaway threshold energy
E	electric field
E	electric field amplitude
E_0	electric field characteristic amplitude
E_B	breakeven value
E_c	conventional air electrical breakdown value
E_x, E_z	components of electric field **E**
f	distribution function
$f_\sim$	distribution function irregular part
F	distribution function regular part
F_{fr}	friction force for aerosol particles
Φ	friction force for runaway electrons
l_a	spatial increment of the electron concentration N
I_i	ionization potential
I_0	intensity of a stochastic electric field
J	source of electrons
M	cloud particle mass
N	concentration of cloud particles
N_e	electron concentration
Q	charge of cloud particles
s	coordinate along **E**
t	time
TC	thunderstorm cloud

T_{osc}	typical period of thundercloud electric field oscillations
T_R	space charge relaxation time
k_0^{-1}	characteristic scale of electric cells
r	size of charged particles
Ω	convective cell rotation frequency
Ω_1, Ω_2	rotation frequencies for elliptical trajectories of air flow
Ω_p	heavy particle plasma frequency
u	cloud particle velocity relative to the mean air flow
R_c	convective cell radius
g	acceleration due to gravity
U_a, U_c	activation and critical values in cellular automaton model
X_c	maximum shift of convective cell center
x, y	coordinates
$\tilde{x}, \tilde{y}$	dimensionless variables
z_m	vertical scale of a cloud
Z	mean molecular charge
Z_i	number of electrons in a molecule

Bibliography

Alexeenko, V.V., Khaerdinov, N.S., Lidvansky, A.S., and Petkov, V.B. (2001). Gamma ray emissions observed in a thunderstorm anvil. *Proceedings of ICRC Copernicus Gesellschaft*, 1.

Bell, T. F., Pasko, V. P., and Inan, U. S. (1995). Runaway electrons as a source of red sprites in the mesosphere. *Geophys. Res. Lett.*, 22:2127.

Boulch, M., Hamelin, J., and Weidman, C. (1990). UHF-VHF radiation from lightning. In Gardner, R. L., editor, *Lightning electromagnetics*. Hemisphere, New York.

Bratolyubova-Tsulukidze, L.S., Grachev, E.A., Grigoryan, O.R., and Nechaev, O.Yu. (2001). Near-equatorial electrons as measured onboard the Mir space station. *Cosmic Research*, 39(6):564–573.

Eack, K. B., Beasley, W. H., Rust, W. D., Marshall, T. C., and Stolzenburg, M. (1996a). Initial results from simultaneous observation of X rays and electric fields in a thunderstorm. *J. Geophys. Res.*, 101:29637–29640.

Eack, K. B. and Suszcynsky, D. M. (2000). X-ray pulses above a mesoscale convective system. *Geophys. Res. Lett.*, 27(2):185–188.

Eack, K. B. and W. H. Beasley, Rust, W. D., Marshall, T. C., and Stolzenburg, M. (1996b). X-ray pulses observed above a mesoscale convective system. *Geophys. Res. Lett.*, 23:2915–2918.

Gaponov-Grekhov, A.V., Iudin, D.I., and Trakhtengerts, V.Y. (2005). Attraction mechanism of like-charged aerosol particles in a moving conductive medium. *J. Exp. Theor. Phys.*, 101(1):177–185.

Gardner, R. L., editor (1990). *Lightning electromagnetics*. Hemisphere, New York.

Grach, V. S., Demekhov, A. G., and Trakhtengerts, V. Y. (2005). Kinetic instability of charged-particle flow in a thunderstorm cloud. *Radiophysics and Quantum Electronics*, 48(6):435–446.

Gurevich, A. V., Milikh, G. M., and Roussel-Dupré, R. A. (1992). Runaway breakdown mechanism of air breakdown and preconditioning during a thundercloud. *Phys. Lett. A*, 165:463.

Gurevich, A. V. and Zybin, K. P. (2001). Runaway breakdown and electric discharges in thunderstorms. *Phys. Usp.*, 44(11):1119.

Iudin, D. I., Trakhtengerts, V. Yu., and Hayakawa, M. (2003). Fractal dynamics of electric discharges in a thundercloud. *Phys. Rev.*, E68(1):Art. no. 016601.

Levin, Z. and Tzur, I. (1986). *The Earth's Electrical Environment*, pages 131–145. National Acad. Press, Washington D.C.

MacGorman, D. R. and Rust, W. D. (1998). *The electrical nature of storms*. Oxford Univ. Press.

Mansell, E. R., MacGorman, D. R., Ziegler, C. L., and Strake, J. M. (2002). Simulated three-dimensional branched lightning in a numerical thunderstorm model. *J. Geophys. Res.*, 107(D9):doi:10.1029/2000JD00244.

Mareev, E. A., Sorokin, A. E., and Trakhtengerts, V. Yu. (1999). Effects of collective charging in a multiflow aerosol plasma. *Plasma Phys. Rep.*, 25:289–300.

Marshall, T. C., McCarthy, M. P., and Rust, W. D. (1995a). Electric field magnitudes and lightning initiations in thunderstorms. *J. Geophys. Res.*, 100(D4):7097.

Marshall, T. C., Rison, W., Rust, W. D., Stolzenburg, M., Willett, J. C., and Winn, W. P. (1995b). Rocket and balloon observations of electric field in two thunderstorm. *J. Geophys. Res.*, 100:20815–20828.

Marshall, T. C. and Rust, W. D. (1991). Electric field sounding through thunderstorms. *J. Geophys. Res.*, 96:22297–22306.

Marshall, T. C. and Rust, W. D. (1993). Two types of vertical electrical structures in stratiform precipitation regions of mesoscale convective systems. *Bull. Am. Met. Soc.*, 74:2159–2170.

Mazur, V. and Ruhnke, L. H. (1998). Model of electric charges in thunderstorms and associated lightning. *J. Geophys. Res.*, 103(D18):23299–23308.

McCarthy, M. P. and Parks, G. K. (1985). Further observations of X-rays inside thunderclouds. *Geophys. Res. Lett.*, 12:393–396.

McCarthy, M. P. and Parks, G. K. (1992). On the modulation of X-ray fluxes in thunderclouds. *J. Geophys. Res.*, 97:5857–5864.

Milich, G. M., Papadopoulos, K., and Chang, C. L. (1995). On the physics of high-altitude lightning. *Geophys. Res. Lett.*, 22:85.

Pasko, V. P., Inan, U. S., , and Bell, T. F. (1996). Sprites as luminous columns of ionization produced by quasi-electrostatic thundercloud fields. *Geophys. Res. Lett.*, 23:649.

Proctor, D. E., Uytenbogaargt, R., and Meredith, B. M. (1988). VHF radio pictures of lightning flashes to ground. *J. Geophys. Res.*, 93(D 10):12683–12727.

Rakov, V. A. and Uman, M. A. (2003). *Lightning*. Cambridge Univ. Press.

Solomon, R. and Baker, M. B. (1996). A one-dimensional lightning parametrization. *J. Geophys. Res.*, 101:14983–14990.

Stolzenburg, M., Marshall, T. C., Rust, W. D., and Smull, B. F. (1994). Horizontal distribution of electrical and meteorological conditions across the stratiform region of a mesoscale convective system. *Mon. Wea. Rev.*, 122:1777–1797.

Straka, J. M. and Anderson, J. R. (1993). Numerical simulations of microburst-producing storms - some results from storms observed during COHMEX. *J. Atmos. Sci.*, 50:1329–1348.

Takahashi, T. (1978). Riming electrification as a charge generation mechanism in thunderstorms. *J. Atmos. Sci.*, 35:1536.

Taranenko, Y. N., , and Roussel-Dupré, R. A. (1996). High altitude discharges and gamma ray flashes: A manifestation of runaway air breakdown. *Geophys. Res. Lett.*, 23:571.

Torii, T., Takeishi, M., and Hosono, T. (2002). Observation of gamma-ray dose increase associated with winter thunderstorm and lightning activity. *J. Geophys. Res.*, 107(D17):doi:10.1029/2001JD00938.

Trakhtengerts, V. Y. (1989). About origin of electric cells in a thundercloud. *Dokl. Akad. Nauk SSSR*, 308:584.

Trakhtengerts, V. Y. (1992). Electric field generation in atmospheric convective cells. *J. Atmos. Terr. Phys.*, 54(3/4):217.

Trakhtengerts, V. Y. (1994). The generation of electric fields by aerosol-particle flow in the middle atmosphere. *J. Atmos. Terr. Phys.*, 56:337–342.

Trakhtengerts, V. Yu., Iudin, D. I., Kulchitsky, A .V., and Hayakawa, M. (2002). Kinetics of runaway electrons in a stochastic electric field. *Phys. Plasmas*, 9(6):2762–2766.

Trakhtengerts, V. Yu., Iudin, D. I., Kulchitsky, A .V., and Hayakawa, M. (2003). Electron acceleration by a stochastic electric field in the atmospheric layer. *Phys. Plasmas*, 10(8):3290.

Vonnegut, B. (1953). Possible mechanism for the formation of thunderstorm electricity. *Bull. Am. Met. Soc.*, 34:378.

Warwick, J. W. (1994). A new model of lightning. In Kikuchi, H., editor, *Dusty and Dirty Plasmas, Noise and Chaos in Space and in the Laboratory*, pages 284–293. Plenum Press, New York.

Wilson, C. T. R. (1925). The electric field of a thundercloud and some of its effects. *Proc. Cambridge Philos. Soc.*, 37(32D):534–538.

Yukhimuk, V., Roussel-Dupré, R. A., and Simbalisty, E. M. D. (1999). On the temporal evolution of red sprites: Runaway theory versus data. *Geophys. Res. Lett.*, 26(6):679.

POSTER ABSTRACTS

T. Farges
Commissariat à l'Energie Atomique, Bruyèresle Châtel, France.

15.1 Introduction

The poster session is divided in three main topics: (1) observations from the ground, (2) observations from space and (3) theoretical modelling.

15.2 Observations from the Ground

15.2.1 Automated, Remote-Controlled Optical Observation Systems in TLE Research

T. H. Allin[1], T. Neubert[2], S. Laursen[2] and I. L. Rasmussen[2]
[1]*Measurement and Instrumentation Systems, Oersted DTU, Bldg. 327, DK-2800 Lyngby,*
[2]*Danish National Space Center, Juliane Maries Vej 30, DK-2100 Copenhagen Ø*

The observation phase of Transient Luminous Event (TLE) research has, for the optical part, been dominated by exhaustively long periods of waiting for events, or careful review of tape recordings of the night sky. The 2003 version of the Spritewatch observation system (Allin et al., 2004) represent the first automatic event detection system used in TLE research. Recording both digital video files and single event snapshot images, the system provided both high-quality documentation of TLE events, as well as continuous recordings for back-tracing on interesting times. Within the envelope of a standard personal computer, such an advanced, low cost observation system can be created in short time if no guidance is needed – e.g., to monitor the same region of interest over time. For the 2004 European Sprite campaign, a remotely guided 4-camera system was built, providing for the first time simultaneous recordings of broad-band, wide FOV "patrol" images and three-color, narrow FOV images of TLEs. The system is currently installed and operating from Pic du Midi de Bigorre, 42.9N, 0.01E, 2877 m above MSL.
Refs: T. H. Allin, J. L. Joergensen, T. Neubert and S. Laursen, The Sprite-

watch – A semi-automatic, remote-controlled observation system for transient luminous events, *IEEE Trans. on Measurement and Instrumentation*, submitted, 2004.

15.2.2 Observation of Schumann Resonance Transients at Nagycenk, Hungary

József Bór[1]

[1]*Geodetic and Geophysical Research Institute of the Hungarian Academy of Sciences, Hungary H-9400 Sopron, Csatkai u. 6-8.*

jbor@ggki.hu

Schumann Resonance (SR) transients are temporary excitations of the Earth-ionosphere waveguide by powerful electromagnetic pulse sources such as high energy lightning flashes in the Extremely Low Frequency (ELF) band ($\sim$3 Hz-3 kHz). These transient excitations can be observed as coherent amplitude excursions of the electromagnetic field components. Geographical location, polarity and charge moment change of a vertical discharge type source can be estimated by processing the recorded time series. Investigation of these properties of a lightning flash has been of great interest as energetic lightning strokes are responsible for the occurrence of transient luminous events (TLEs) in the u pper atmosphere (sprite, ELVES, etc.) SR transients have been monitored at the Széchenyi István Geophysical Observatory (NCK) near Nagycenk, Hungary, since 1998. The poster presents the recording system for SR transients and covers the steps of data processing applied for a typical SR transient.

15.2.3 Post Filtering of Unwanted Powerline and Lightning Effects in VLF

A. Kero[1], J. Manninen[1] and T. Turunen[1]

[1]*Sodankylä geophysical observatory, Tähteläntie 62, FIN-99600 Sodankylä, Finland*

antti.kero@sgo.fi

The dynamic range used in VLF-recordings is normally dominated by the disturbance peaks associated to the global lightning discharges. In addition, in the vicinity of powerlines, the 50 Hz harmonics and the powerline control signals produce disturbances with distinct frequecies to the data. In most cases the magnitude of these unwanted components is decades above the signatures associated to magnetospheric effects in VLF.

Software based post filtering techniques for the used orthogonal VLF-antenna measurement are presented in this poster. Used methods are based on filtering of the single frequencies for the powerline harmonics, and the linear polarisation filtering for the lightning signals.

15.2.4 Infrasonic Signatures of Thunder

Thomas Farges[1], Alexis Le Pichon[1], Elisabeth Blanc[1], Torsten Neubert[2], Thomas Allin[3] and Stéphane Pedeboy[4]

[1] CEA/DASE/LDG, BP 12, 91680 Bruyères le Châtel, France, [2] Danish National Space Center, Juliane Maries Vej 30, DK-2100 Copenhagen Ø, [3] Measurement and Instrumentation Systems, Oersted DTU, Bldg. 327, DK-2800 Lyngby, [4] Météorage

thomas.farges@cea.fr

In the framework of the implementation of the Comprehensive Test Ban Treaty, infrasound monitoring of the waves produced by strong explosions will be realized with 60 stations unfolded all around the world. To differentiate natural events from a nuclear test event, we have to study the signature of different kind of natural events frequently observed.

We study here the signature of the infrasonic waves generated by lightning, e.g. the waveform, the frequency content, the variability of the waveform with distance and lightning current, and the lightning detectability. We research also the infrasound waves which could be produced by sprites.

This study is carry out using the data of the Flers infrasound station (France, 48.75N, 0.5W), composed of four microbarographs, obtained during the summer 2003. The detected signals are associated with lightning characteristics (localization, dating, lightning current) provided by Météorage. We realize a statistical study for the summer data and a specific study for a large thunderstorm occurring near the Flers station.

In conclusion, we present a method to differentiate infrasonic waves generated by lightning from those likely produced by sprites.

15.2.5 On the Absorption of ELF Signals in the Earth-Ionosphere Waveguide

Steven Golden[1]

[1] Institute for Meteorology and Geophysics, University of Frankfurt, Germany

The weakly conductive lower atmosphere, bounded by the higher conductive ground and ionosphere, acts as a waveguide for electromagnetic waves in the ELF and VLF frequency ranges. Transmission properties are especially good in the lower ELF range between approximately 8 Hz and 1 kHz, where wave propagation takes place in the lowest, quasi-transverse, waveguide-mode. The attenuation in this frequency band is low enough, that strong electromagnetic pulses caused by lightning (sferics), can travel along the waveguide for several thousand kilometers. The frequency dependent phase velocity and absorption rate of these guided waves depend on the varying conductivity profile of the atmosphere, especially on the contribution by the lower ionosphere, and should therefore be useful parameters for characterizing this system. In this study, a simple method for determining the absorption rate from observations on ELF sferics was tested, which uses amplitude measurements of the same sferic at different sites (Golden, 2001).

If the only reliable informations on a sferic are its amplitude at a single station and the location of its generating lightning, the determination of absorption rates or source currents requires rather detailed additional assumptions on the atmosphere and the lightning. Using additional amplitude measurements at a second site greatly simplifies these requirements: assuming only single mode wave propagation and a laterally homogeneous atmosphere between the sites, the dependence of amplitude on distance should be only affected by an exponential decay due to absorption and by a decay due to geometric spreading. Knowing the locations of lightning and stations, the contribution due to geometric spreading can be easily determined and subtracted. The ratio of the so corrected amplitudes can be set into relation to the difference of the source distance, as seen from the different stations, to determine the exponential decay factor, which represents the absorption rate.

The method was tested on a dataset collected during the Sprites'98 campaign: during July and August of 1998 three ELF-measurement stations were operated across western North America. Each station continuously recorded the horizontal magnetic field components, as measured by induction coil magnetometers, with a sampling rate of 2048 Hz. This dataset could have been used for the determination of the necessary lightning locations, as was demonstrated by Füllekrug and Constable (2000). However, for this study this information was taken from the National Lightning Detection Network (NLDN), which was also used for an automatic selection of the used lightnings (only strong CG's, larger than 500 LLP = 89.7 kA). After a first test, using only 16 manually selected, especially strong lightnings, the method was repeated for six automatically selected datasets, three of them using only day data, the other three using only night data.

A comparison of the results shows a significant difference between the determined day or night absorption rates: during daytime values from 3.5 to 3.9 dB/Mm were observed, while during nighttime values range from 2.5 to 2.8 dB/Mm. These results are in good agreement with values estimated earlier by Hughes and Gallenberger (1974), using a different method. The presented method thus is proved to work at least to first order accuracy. A possible application could be the use within a global ELF based lightning detection network, to continuously monitor absorption rate variations. Since these are related to large scale variations of the lower ionosphere, such measurements could form a useful contribution to more localized information, as is already gained from other measurements.

Refs: Füllekrug, M. and Constable, S., Global Triangulation of Intense Lightning discharges, *Geophys. Res. Lett.*, 27(3), 333-336, 2000.

Golden, S., Messung der Absorption atmosphärischer ELF-Impulse, *Diploma* thesis (in German), Johann Wolfgang Goethe-Universität, Frankfurt am Main, 2001.

Hughes, H. G. and Gallenberger, R. J., Propagation of extremely low-frequency (ELF) atmospherics over a mixed day-night path. *J. Atmos. Terr. Phys.*, 36, 1643-1661, 1974.

15.2.6 A Global Lightning Location Algorithm Based on Electromagnetic Signatures in the Schumann Resonance Band

Eran Greenberg[1] and Colin Price[2]

[1]*Department of Electrical Engineering, Tel-Aviv University, Israel*, [2]*Department of Geophysics and Planetary Sciences, Tel-Aviv University, Israel*

A new, improved algorithm has been developed to geolocate intense lightning flashes around the globe. The method uses ELF (Extremely Low Frequency) radiation propagated in the Earth-Ionosphere cavity to locate unusually intense lightning strokes. Two parameters are needed to locate the lightning discharge from a single station: bearing and source-observer distance (SOD). The bearing was obtained using the Poynting vector ($E \times H$), while the SOD was obtained based on the modeling of the electric and magnetic ELF spectra, and the comparison with experimental data. To check the accuracy of our algorithm we used primarily infrared cloud top temperature images of deep convective storms from geostationary satellites (METEOSAT, GOES, GMS). Analysis of ELF data from our field station of 147 events gave an average source-observer distance error of 660 km (7.05%) and azimuth error of 1.9°. Of the 147 events 72% had positive polarity, while the majority (59%) occurred over the oceans.

15.2.7 Neutral and Charged Particles at Low Latitudes. Is their Connection with Thunderstorms Possible ?

Oleg Grigoryan[1], Maria Usanova[2] and Vasiliy Petrov[2]

[1]*Skobeltsyn Institute of Nuclear Physics, M. V. Lomonosov Moscow State University, Leninskie Gory, Moscow, 119992 Russia* [2]*Faculty of Physics, M. V. Lomonosov Moscow State University, Leninskie Gory, Moscow, 119992 Russia*

Various electrodynamics processes above thunderstorms give rise to a variety of phenomena, such as energetic electron fluxes, gamma rays, quasi-electrostatic fields, electromagnetic pulses, VLF/LF signals, optical emissions ('red sprites' and 'blue jets'), and so on. Some of them are discussed here in detail. The connection between electron fluxes and lightning flashes at $L > 1.8$-2.0 is verified theoretically and experimentally (Inan et al., 1988; Inan and Carpenter, 1987). The electron fluxes (Ee 100keV) appeared to coincide with the locations of thunderstorms near the equator were observed on board MIR orbital station (Grigoryan et al., 1997; Bratolyubova-Tsulukidze et al., 2001). However, the mechanism of electron production at high altitudes is not yet evident.

The possibility of satellite neutron observation is of great importance. It was observed during a high-powered exploding wire discharge experiment that 2.10^{10} neutrons were produced. Atmospheric lightning discharges are also high powered and the lightning power properties are similar to the exploding wire discharge plasma. Therefore it was suggested (Stephankis et al., 1972; Libby and Luken, 1973; Libby and Luken, 1975) that $\sim 10^{15}$ neutrons can be produced by lightning discharge. Also on the basis of the experiments (Shyam and Kaushik, 1999) it was concluded, that neutron bursts ($10^8 - 10^9$ magnitudes) are associated with lightning. One of the possible neutron production mechanisms could be protons and deuterons accelerated in lightning discharge and neutron production in nuclear reactions with other elements of the environment. The similar result was obtained using a neutron detection system (the detectors square of 1 m^2, En<0.5 eV) at Lomonosov Moscow State University. The observed neutron bursts were associated with lightning (Bratolyubova-Tsulukidze et al., 2003). The approximate distance of lightning discharges from our detector was $\sim$1 km. Therefore the total number of neutrons produced by one typical lightning discharge may be estimated as 2.5 10^{10}.

15.2.8 Sprites Observed over France on 23 July 2003 in Relation to their Parent Thunderstorm System

L. Knutsson[1], S. Soula[1], O. van der Velde[1], T. Neubert[2] and T. Allin[3]
[1]*Laboratoire d'Aerologie, UMR 5560 UPS/CNRS, OMP, 14 av. E. Belin 31400 Toulouse, France.* [2]*Danish Space Research Institute, Juliane Maries Vej 30, Copenhagen O, 2100, Denmark,* [3]*Danish Technical University, Bygn.327, Kgs. Lingby 2800, Denmark*
vdvo@aero.obs-mip.fr

For our research of meteorological aspects associated with sprite events, 23 July 2003 was particularly interesting because of the availability of both cloud-to-ground and intracloud lightning activity data. A large thunderstorm with two convective areas developed over southern France during the late afternoon. About two hours after sunset, the first sprite was detected. In about three hours, 13 sprites were observed, 7 over the northern part and 6 over the southern part of the mesoscale convective complex. Twelve of the thirteen sprites were associated with positive cloud-to-ground flashes, and one could be associated with an intracloud flash. Sprites tended to occur over the stratiform region of the storm system in the area with the coldest (highest) cloud tops. The associated positive flashes were also within or close to this portion of the storm. Sprite-associated +CGs had higher peak currents on average, but the relationship was rather weak. Sprites occurred in the late stages of the two storm areas, with decreasing -CG and IC lightning activity, while +CG activity remained stable or increasing. In these stages, there was a high ratio of +CG flash rate to total flash rate. Overall, the intracloud activity was low during the sprite-producing periods, but sudden bursts of VHF sources produced by intracloud lightning have frequently been observed at the moment of the sprite. The areal coverage of the radar echo was calculated. The result supports the idea that sprite events tend to appear almost exclusively over large thunderstorm systems.

15.2.9 Sprite Observations from Langmuir Laboratory, New Mexico

Robert A. Marshall[1] and Umran S. Inan[1]
[1] *Space, Telecommunications, and Radioscience Laboratory, Stanford University*

The goals of this project are:
1. Image sprites using a high-speed CCD camera (1000 fps +) mounted to a Dobsonian telescope, yielding fine temporal and spatial resolution images of sprites. These images can be used to more closely analyze sprite initiation and streamer propagation, including the transition from column streamers to branches. Combine these images with Wide field-of-view images, photometric data, and charge moment calculations from broadband ELF measurements of sferics.
2. Use photometric data of sprite halos to correlate their occurrence with so-called Early/Fast VLF ionospheric disturbances.
3. Use wide field-of-view images from Langmuir and Jemez Mountains to triangulate sprite locations and deduce accurate altitudes.

Results so far show over 60 sprite images, with two appearing in the high-speed / telescopic field of view. Sprites and sprite halos are visible in photometer data. VLF data has yet to be analyzed.

15.2.10 VLF Signatures Associated with Sprites

Ágnes Mika[1]
[1] *University of Crete, Department of Physics*

Observations on the nights of July 21-24, 2003 of the ionospheric effects of thunderstorms located in Central France are reported. A camera system in the Pyrenees captured 49 sprites. A narrowband VLF receiver located at Crete observed subionospheric VLF signals from six ground-based transmitters. The amplitude of one of the VLF signals exhibited rapid onset perturbations occurring in nearly one-to-one relationship with the optical sprites. The results of the data analysis are grouped as follows:
1. Study the relationship between sprites and associated VLF signal perturbations

2. Quantify the onset durations of sprite related VLF events
3. Consider the recovery phases and study their variation with time
4. Investigate the non-ducted LEP events observed during these storms and in relation to the sprites.

15.2.11 Stratospheric Electric Field, Magnetic Field and Conductivity Measurements Above Thunderstorms: Implications for Sprite Models

Jeremy N. Thomas[1], Robert H. Holzworth[1], Michael P. McCarthy,[1] Osmar Pinto Jr.[2] and Mitsuteru Sato[3]

[1]*Dept. of Earth and Space Sciences, University of Washington, Seattle, WA,* [2]*Instituto Nacional de Pesquisas Espaciais, INPE, São Paulo, Brazil,* [3]*Graduate School of Science, Tohoku University, Sendai, Japan*

Electric and magnetic fields and conductivity were measured in the stratosphere (32-34 km altitude) above thunderstorms as part of the Sprite Balloon Campaign 2002-2003 in southeastern Brazil. During the two balloon flights, the payloads measured hundreds of electric field transients, some as large as 130 V/m, correlated with lightning events. Also, conductivity variations over thunderstorms of at least a factor of two different from the fair weather values were measured. Since optical verification of sprites was not achieved during these flights and recent studies (Hu et al. 2002) have found that the probability of sprite production is proportional to the charge moment of the corresponding positive cloud-to-ground (+CG) stroke, we rely on charge moment estimates from remote ELF observations in Onagawa, Japan and Syowa, Antarctica to indicate possible sprite events. Here we present a detailed study of several nearby (<60 km) positive +CG strokes indicated by the ELF charge moment estimate to be probable sprite generators. By using a quasi-static electric field model (Pasko et al. 1997), we show how these large field changes correlated with +CG strokes, along with varying conductivity over thunderstorms, may provide the necessary conditions for sprite development.

15.2.12 Triggering of Positive Lightning and High-Altitude Atmospheric Discharges

Leonid V. Sorokin[1]

[1]*Peoples' Friendship University of Russia, Moscow*

Triggering of electrical atmospheric discharges can be caused by a number of physical mechanisms, among them the non-linear effects concerning the earthquakes.

The cases of earthquake triggering by seismic waves from other earthquakes at the big angular distances in 30-year period have been computed. All evaluations for definition of possible phases of seismic waves and evaluations of their travel times were conducted with the use of model AK135, built on the basis of model IASPEI91.

The space-time coupling of earthquakes and electrical discharges in atmosphere have been found at big angular distances. The angular distribution of positive cloud to ground discharges (+CG) has been analysed on the Malaysian LDN data, and space-time coupling with exact seismic waves from the earthquakes has been described.

On the base of actual data records, the cases of electromagnetic pulses generation at the big angular distances by seismic waves from the earthquakes have been discovered. The detected signals have sub-millisecond duration and high intensity. The amplitude and time characteristics of sub-millisecond electromagnetic pulses have also been studied. Electromagnetic pulses related with seismic waves can provoke positive lightning and transient luminous events. One of the provoking factors, promoting development +CG and transient luminous events, is passage of seismic waves through a place of event. The known video observations (Mitchell at al. 1997; Vaughan at al. 1997) have been studied together with the NLDN data. Space-time coupling between exact earthquake seismic waves and eight high-altitude electric discharges in atmosphere (6 Red Sprites, Vertical Pulse, Elve) was found on the big angular distances.

15.3 Observations from Space

15.3.1 Transient Luminous Events Explored by the ROCSAT-2/ISUAL Instrument: Observation with the Array Photometer

Toru Adachi,[1] Hiroshi Fukunishi,[1] Yukihiro Takahashi,[1] Rue-Ron Hsu,[1] Han-Tzong Su,[1] Alfred Bing-Chih Chen,[1] Stephen B. Mende,[1] Harald U. Frey,[1] and Lou-Chuang Lee[4]

[1]*Department of Geophysics, Tohoku University, Japan,* [2]*Physics Department, National Cheng Kung University, Taiwan,* [3]*Space Sciences Laboratory, University of California at Berkeley, USA,* [4]*National Space Program Office, Taiwan*

In this study, we report preliminary results of TLE observations using the Array Photometer (AP) onboard the ROCSAT-2 satellite. The ROCSAT-2 satellite with a scientific payload named ISUAL (Imager of Sprites/Upper Atmospheric Lightning) is the first satellite which observes transient luminous events from space. The ISUAL is composed of an imager, a spectrophotometer, and an array photometer (AP). The AP provides us spectral information by measuring two wave length ranges of 360-470 nm and 520-750 nm selected by blue and red filters, respectively. Each photometer has 16 channels aligned in vertical and spatial resolution corresponds to ~11 km in the cases of sprites occurring at the limb point 3315 km away from the satellite. The time resolution of 50 or 500 μs enables us to detect fast temporal variation of sprites which have average durations of several to tens of ms. The AP succeeded in detecting clear upward/downward vertical motions of sprites in both the blue and red channels with quite high signal-to-noise ratios compared with ground-based observations. By calculating the Blue/Red (B/R) emission intensity ratio, it is clarified that sprites are bluer in the earlier stage and at lower altitudes.

15.3.2 Fractal Analysis Method Applicability to Terrestrial Gamma-Ray Flashes

I. V. Arkhangelskaja[1]

[1]*Astrophysics Institute, Moscow Engineering Physics Institute (State University), Kashirskoe shosse 31, Moscow, 115409, Russia*

irene.belousova@usa.net

The TGF (Terrestrial Gamma Flashes) were first detected by BATSE experiment (based on 8 NaI(Tl) detector modules consists of 2 detector each: Large Area Detector $\oslash$20" and 0.5" thick and Spectroscopy detector $\oslash$5" and 3" in thick) onboard CGRO satellite. Observed TGF had duration $0.01 \div 20$ ms. TGF are produced in a process of bremsstrahlung of the runaway electrons in the middle atmosphere which are accelerated upward by the thundercloud electric field and collide with the material constituting the atmosphere. These electrons are very energetic ($E_e \sim 1$ MeV) and the γ-radiation pattern is directed along the beam. Some short events with duration $1 \div 16$ ms are registered by AVS-F apparatus. The AVS-F (amplitude-time Sun spectrometry) apparatus is intended to study characteristics of fluxes of hard X-rays, γ-rays and neutrons from the Sun and solar flares and to detect and record events like γ-ray bursts. The experiment is carry out at the CORONAS-F special-purpose automatic station (NORAD catalog number 26873, International Designator 2001-032A) that had been launched from Russian kosmodrom Plesetsk at 11:00 UT of 31 July 2001 into a circular orbit oriented towards the Sun with inclination 82.5° and altitude $\sim$500 km. The AVS-F apparatus uses signals produced by the SONG-D detector (based on the CsI(Tl) crystal $\oslash$ 20 cm and height of 10 cm, energy deposition ranges of $0.1 \div 17.0$ MeV and and $4.0 \div 94.0$ MeV by up to date calibration data), XSS-1 detector (the semiconductor detector with 3-30 keV energy deposition range) and the anticoincidence signal generated by the plastic scintillation counter of SONG-D. According the AVS-F experiment description, observed short events may be TGF or can caused by following apparatus reasons: fluorescence of CsI(Tl) after charged particle pass through it; fluctuations of photoelectron number or ions of photomultiplier filler after charged particle pass through it; transient processes in electronic system caused by great energy deposition in detector. We use the fractal analysis of temporal profiles for separation TGF because all background fluctuation (including ones leads to apparatus short events) in AVS-F apparatus are Poisson in energy region 0.1÷17 MeV outside the radiation belt and SAA and counts rates time intervals had exponential distribution. Fractal dimension of such temporal profiles is equal to 1.5 if it is possible to consider the mean count rate as constant within studying time interval. We have analyzed 100 temporal profiles with duration 4096 ms in energy region of 0.1÷17 MeV registered by AVS-F apparatus in equatorial regions The mean fractal dimension for 97 analyzed time intervals is D=1.50±0.03 and in 3 cases different values were obtained: D=1.63±0.03, D=1.37±0.03, D=1.62±0.03. These short events are similar to TGF on temporal profiles shape and duration and one of them were registered during strong tropical cyclone Beni. The location of this short event is near cyclone center. Now we study weather condition during other two events.

15.3.3 Searching for Lightning-Induced Terrestrial Gamma Ray Bursts on CORONASF Satellite

R. Bučík[1,2], K. Kudela[1], S. N. Kuznetsov[3], I. N. Myagkova[3] and B. Yu. Yushkov,[3]
[1]*Institute of Experimental Physics, Košice, Slovakia,* [2]*Space Research Centre, Warsaw, Poland,* [3]*Institute of Nuclear Physics, Moscow, Russia*

There are few low-altitude satellite or balloonborne measurements (e.g. SAMPEX, LACE) indicating both atmospheric loss cone energetic electrons (Blake et al., 2001) and daughter bremsstrahlung hard X/gamma ray enhancements (Feldman et al., 1996) suggested to be associated with thunderstorms. Seasonal variations in precipitating trapped electrons have been reported those emissions are most enhanced during the northern summer months. We analyze data received by the SOlar Neutron and Gamma rays (SONG) device aboard low altitude (∼500 km) polar orbiting CORONASF satellite, which provides high time resolution measurements (1 s in burst mode) of X/gamma rays in range of 0.037 MeV. The omnidirectional sensor consists of large area CsI scintillator (20×10 cm) surrounded by a 2 cm thick anticoincidence shield for charged particles. X/gamma-ray fluxes measured on satellite altitude are surveyed to discuss the competitive sources of lightning induced gamma ray terrestrial flashes. The preliminary geographic maps of average intensity during October December 002 in the four channels (30-60 keV, 60-150 keV, 150-500 keV and 500-1500 keV) demonstrate intense gamma ray emissions distributed along all longitudes under the outer radiation belt (L∼36). Since these magnetic storm driven enhancements mask the lightning induced electron precipitation events we concentrate on inner zone electron belt (L<2) as well as on the slot region (L∼23). The longitudes outside the South Atlantic Anomaly region which is the region of stable trapped particle population at the satellite altitude are the most convenient to follow. During the geomagnetic quiet time in the studied three months period the CORONASF data reveals either conjugate bounce loss cone and drift loss cone enhanced fluxes observed at

L$\sim$2.2 perhaps associated with underlying lightning documented by the Lightning Imaging Sensor (LIS) on board TRMM satellite.
Refs: Blake, J. B., Inan, U. S., Walt, M., Bell, T. F., Bortnik, J., Chenette, D. L. and Christian, H. J., Lightning-induced energetic electron flux enhancements in the drift loss cone, *J. Geophys. Res.*, 106, 29733-29744, 2001.
Feldman, W. C., Symbalisty, E. M. D., and Roussel-Dupré, R. A., Hard X ray survey of energetic electrons from low Earth orbit, *J. Geophys. Res.*, 101, 5195-5209, 1996.
Acknowledgments: This work was supported by the Slovak Scientific Grant Agency VEGA under contract 2/4064/04. R.B. was supported by the European Commission through contract no. HPRNCT200100314.

15.3.4 Seismo-electromagnetic Emissions

A. Buzzi[1], L. Conti[1], A. M. Galper[2], S. V. Koldashov[2], V. Malvezzi[1], A. M. Murashov[2], P. Picozza[3], R. Scrimaglio[4], V. Sgrigna[1] and L. Stagni[5] (The ESPERIA Collaboration).

[1] *University of Roma Tre, Department of Physics, 84 Via della Vasca Navale, I-00146 Rome, Italy,* [2] *MEPhI, Institute of Cosmic Physics, 31 Kashirskoe Shosse, 115409 Moscow, Russian Federation,* [3] *University of "Tor Vergata", Department of Physics & Sez. INFN, 1 Via della Ricerca Scientifica, I-00133 Rome, Italy,* [4] *University of L'Aquila, Department of Physics, Via Vetoio, I-67010 Coppito-L'Aquila, Italy,* [5] *University of Roma Tre, Department of Ingegneria Meccanica ed Automatica, 79 Via della Vasca Navale, I-00146 Rome, Italy*

In the last decades a possible influence of electromagnetic fields of seismic origin in the ionosphere-magnetosphere transition region has been reported in literature. In recent years, a few space experiments also revealed anomalous bursts of charged particles precipitating from the lower boundary of the inner radiation belt. They were thought to be caused by low-frequency seismo-electromagnetic emissions. A temporal correlation between earthquakes with M$\geq$5.0 and anomalous particle bursts collected by the PET-SAMPEX satellite mission is critically investigated and presented in this chapter. A short-term seismic precursor of $\sim 4 \div 5$ hours is observed in the histogram of the time difference between the time occurrence of earthquakes and that of particle burst events. The best correlation is obtained only when considering high-energy electrons (E$\geq$4 MeV) near the loss cone. At present are available only data from space experiments not dedicated to observation of seismo-electromagnetic emission and related phenomena. Results suggest the importance of coordinated and simultaneous ground-based and space investigations specifically dedicated to the subject. The new ESPERIA satellite mission devoted to investigate the seismic influence on the ionosphere-magnetosphere transition region is presented. Proposals for other experiments on board of ESPERIA will allow to study: 1) Relevant phenomena caused by external sources (sun and cosmic rays) or generated inside the magnetospheric cavity, 2) Luminous emissions (Sprites, Blue jets, Elves, Trimpis,...) during thunderstorm activity, 3) Atmospheric & ionospheric structure and dynamics.

15.3.5 First Results of Transient Luminous Event Observations by ISUAL

Harald U. Frey[1], Stephen B. Mende[1], Rue-Ron Hsu[2], Han-Tzong Su[2], Alfred B. Chen[2], Lou-Chuang Lee[2,3], Hiroshi Fukunishi[4] and Yukihiro Takahashi[4]

[1]*Space Sciences Laboratory, University of California at Berkeley, USA,* [2]*Physics Department, National Cheng Kung University, Taiwan,* [3]*National Applied Research Laboratories, Taiwan,* [4]*Department of Geophysics, Tohoku University, Japan*

The Imager for Upper Atmospheric Lightning (ISUAL) contains a Spectrophotometer with six individual photometers covering the spectral range from the far ultraviolet to the near infrared. The photometers point towards the limb and integrate the light in a field of view of 20×5 degrees. Sudden changes in the amplitude of the output signal are used to trigger the other ISUAL components (Imager and Array Photometer) to collect data. The photometers cover well known spectral ranges of TLE as for instance the N2-1P band at 623-750 nm or the lightning signature at 777.4 nm. In addition there are two photometers for the far-UV (150-280 nm) and near-UV (250-390 nm) that are aimed at spectral signatures of TLE that are only observable from space due to the absorption by atmospheric O_2 towards ground-based instruments. After commissioning of the instrument, the routine observations started on July 1, 2004. We present some of the first detected sprites and elves and discuss their specific spectral signatures and appearance in the images.

15.3.6 Detection of Terrestrial Gamma-ray Flashes with the RHESSI Spacecraft

Liliana I. Lopez[1], Robert P. Lin[1], David M. Smith[2] and Christopher P. Barrington-Leigh[3]

[1]*Space Sciences Laboratory, UC Berkeley,* [2]*UC Santa Cruz,* [3]*University of British Columbia*

We report the detection of Terrestrial Gamma-Ray Flashes (TGFs) with the Reuven Ramaty High Energy Solar Spectroscopic Imager (RHESSI), the second instrument ever to observe this phenomenon. The first instrument to detect these events was the Burst and Transient Source Experiment (BATSE), launched on board the Compton Gamma-Ray Observatory (CGRO) in 1991. We have analyzed 5 months and present here the preliminary results of two TGF burst events. The data show photon energies extend to greater than 15 MeV. Time duration for the events analyzed so far is less than 1 millisecond. RHESSI records every interaction in 1 microsecond time bins and into one of 9 germanium detectors. These events may be caused by electrical discharges in the upper atmosphere related to large lightning storms although there have been alternate models proposed. A relationship with high-altitude optical phenomena, namely sprites and blue jets, has been considered.

15.3.7 ENVISAT Capabilities of Observing High Altitude Optical Phenomena

Christian Muller[1]
[1] *Institut d'Aéronomie Spatiale de Belgique*
Christian.Muller@oma.be

The ENVISAT satellite covers the whole Earth in a heliosynchronous polar orbit and was launched on March 1st 2002 and its ten Earth observation instrument began to deliver commissioned data in late 2002. The variety of atmospheric sensors, especially the limb sounders SCIAMACHY, MIPAS and GOMOS is superior to any previous payload for their study. Also, as polar troposphere and lower stratosphere studies demand the knowledge of the emissions from the upper atmosphere, they must be eliminated by a combination of their detection in the signal itself and their theoretical modelling. Outside the auroral regions, high altitude emissions exist too and are worth studying. This is the case for SPRITE related emissions. The final result will be the understanding of the gaseous, energetic and meteoric structure of the mesosphere and thermosphere.

SPRITE studies can be addressed by the eclipse mode of SCIAMACHY, where nadir observations between 220 nm and the middle infrared are performed for dark current study. MIPAS, as an infrared sounder is also a nighttime active instrument. GOMOS as a pure limb sounder has broad footprint and would probably miss SPRITEs and did not report these phenomena until now.

SCIAMACHY nadir night side observations were conducted between December 2002 and September 2004, when the necessity for continuous dark current studies stopped. The data has unfortunately not yet been distributed despite that it is requested by several AO proposals, this is due to the priority given to the validation data base as the ENVISAT atmospheric instruments are not yet fully validated, also environmental objectives in the upper troposphere and lower stratosphere are given priority for distribution as they lead to operational applications. The data are however archived and sample cases will continue to be requested so that a resumption of this mode could be requested on the basis of the existing data.

15.4 Theoretical Modelling
15.4.1 Do sprites Impact Climate? An atmospheric Coupling Approach.

E. Arnone[1] and P. Berg[2]

[1]*Department of Physics and Astronomy, University of Leicester, Leicester, U.K.*, [2]*Danish Meteorological Institute, Copenhagen, Denmark*

Understanding of climate change requires a knowledge of the atmospheric response to perturbations. Sprites, believed to be NO_x sources, may cause local perturbations of no large entity. However, because of the highly non linear response of the middle atmosphere to chemical perturbations, feedback processes may cause indirectly important changes in the forcing to the troposphere. If this results in an impact on climate depends on the possibility of the forcing to propagate to the troposphere.

For this reason, we have started a joint effort to model coupling mechanisms of the lower and upper atmosphere. We use a global circulation model (GCM) to study the lower atmosphere and a stratosphere-mesosphere model for the upper regions. Together, they provide an extended view from the troposphere (climate processes) to the middle atmosphere (directly affected by solar forcing and sprites).

Here, we present some of the evaluation results from the SMM model (middle atmosphere) and from the attempt to simulate downward propagation processes as coupling of atmospheric layers.

15.4.2 The Sodankylä Ion Chemistry model: Application of Coupled Ion-neutral Chemistry Modelling

Carl-Fredrik Enell[1], Pekka Verronen[1], Annika Seppälä[1], Thomas Ulich[1], Antti Kero[1], Tero Raita[1] and Esa Turunen[1]

[1]*Sodankylä geophysical observatory, Tähteläntie 62, FIN-99600 Sodankylä, Finland*

This poster presents the Sodankylä Ion Chemistry (SIC) model, which is presently being modified in order to model the chemical effects of transient luminous events (TLEs) and artificial ionospheric heating.

The effects of ionisation and electron temperature enhancements during short (10 ms and 100 ms) bursts are shown. For the highest peak ionisation rates applied in the SIC runs, 10^7 $cm^{-3}s^{-1}$ and 10^9 $cm^{-3}s^{-1}$, the concentrations of NO after the bursts are found to be enhanced by 0.3-30 times relative to the background run in the lower altitude range (around 70 km). There the background NO concentration is normally very low. At the NO maximum, around 80 km, the relative enhancement is lower, from a few up to 5 times in the extreme model case.

Although further work is needed to incorporate the known major source of NO_x, reactions with excited N_2 and O_2 molecules, the results indicate that

TLEs can be an important local source of chemically and radiatively active species, such as NO_x. The global significance, however, cannot be estimated without taking into account the effects of transport and the occurrence rate of TLEs.

15.4.3 Simulation of Streamer Propagation Using a PIC-MCC Code: Application to Sprite Ignition

Olivier Chanrion[1] and Torsten Neubert[1]
[1]*Danish National Space Center, Juliane Maries Vej 30, DK-2100 Copenhagen Ø*
chanrion@spacecenter.dk

In order to study electron kinetics in sprite ignition we experiment with the use of Particle In Cell methods with Monte Carlo Collision (PIC-MCC). The aim of our model is to simulate the propagation of a streamer-like discharge, as in Pasko, Raizer, Rycroft and Roussel-Dupre works. A discharge model based on a kinetic description of electrons (Boltzmann-Poisson model) is presented. Then, we present the computational methods used in 'PIC-MCC'. The calculation of a streamer propagation and of a "sprite-like" test case are shown.

15.4.4 Characteristics of Transient Luminous Event Streamers in Weak Electric Fields

Ningyu Liu[1] and Victor P. Pasko[1]
[1]*Dept. of Electrical Engineering, the Pennsylvania State University, University Park, PA*

It is well established by now that transient luminous events (TLEs) observed at different altitudes above thunderstorms commonly consist of large numbers of needle-shaped filaments of ionization, called streamers (e.g., Gerken and Inan, 2003; Su et al., 2003; Pasko, 2003; and references cited therein). The strong electric fields $E > E_k$ are needed for the initiation of streamers. The initiated streamers, however, are capable of propagating in fields substantially lower than E_k [e.g., Allen and Ghaffar, 1995], and one of the important questions of the current TLE research, which directly relates to the evaluation of the total volumes of atmosphere affected by the TLE phenomena and possible role of TLEs in establishing a direct path of electrical contact between the tropospheric and mesospheric/lower ionospheric regions, is related to the determination of the minimum electric fields ($E < E_k$) required for the propagation of streamers in air at different pressures. In this talk, we will present our latest simulation results on propagation of streamers in weak ($< E_k$) fields. The results indicate that the peak electric fields and electron densities of streamers in weak fields can be as low as 50% and 10% of those in strong fields, respectively. The velocities of streamers can drop down to as low as $\sim$100 km/s, which agree with those reported in existing streamer literature (e.g., Bazelyan and Raizer, 1998, p. 154) and with the observed speeds of vertical development of blue jets (e.g., Wescott et al., 1995; Pasko et al., 2002). At ground pressure, the decay of plasma in the streamer channel has been noticed in previous studies (e.g., Morrow and Lowke, 1997; Aleksandrov and Bazelyan, 1999). These observations have been attributed to the effects of three-body attachment and recombination processes, which are believed to be the major loss mechanisms of electrons in the streamer channel. In this talk we will discuss the effects of these two processes on streamer dynamics at low air pressures. We will also discuss the model determined minimum fields required for the propagation of streamers of different polarities at TLE altitudes.

15.4.5 Three-Dimensional Subionospheric VLF Electromagnetic Field Scattering by a Highly Conducting Cylinder and Its Application to the Trimpi Effect Problem

E. V. Moskaleva[1], O. V. Soloviev[1] and V. I. Ivanov[1]
[1]*Institute of Radio Physics, St.Petersburg State University, Ulyanovskaya 1/1, Petrodvorets, 198504, St.Petersburg, Russia*

By means of three-dimensional full-wave analytical-numerical technique we consider the diffraction of VLF point source field by a localized perturbation in the lower ionosphere, which is taken as a truncated highly conducting cylinder. There are no restrictions on the shape and sizes of its horizontal cross-section. We are dealing with short distance propagation path (1500 km at most), this is why in our model the Earth's curvature is ignored. The surface impedance concept is additionally used to model the Earth-ionosphere waveguide using given electron density and collision frequency day or night height profiles for unperturbed ionosphere. Our approach is based on the preliminary asymptotic integration of the rigorous two-dimensional integral equation in order to facilitate its further numerical handling, and an original computational algorithm is proposed to obtain a solution of the approximate equation. With the efforts of reducing CPU time, the developed procedure enables us to study both small and comparatively large irregularities.

Our calculations indicate that scattering may be significant not only in the forward, but also in the backward and out of the way directions. A comparison of the behaviours of the field at night and day indicates that the perturbation is more clearly recognized at night than at day. The analysis of the numerical results shows that field has the essential dependence from direction of wave propagation concerning the azimuth of magnetic field. These effects and the influence of the property lower surface are most noticeable for situation of the location of irregularity corresponds to minimum of the amplitude of the attenuation function. The same results were obtained using International Reference Ionosphere model.

15.4.6 Changes of the Lower Ionospheric Electron Concentration due to Solar Cosmic Rays

A. Ondraskova[1]

[1]*Department of Astronomy, Physics of the Earth, and meteorology, Faculty of Mathematics, Physics and Informatics, Bratislava, Slovakia*
ondraskova@fmph.uniba.sk

Solar proton event (SPE) is an event caused by solar energetic particles or the co-called solar cosmic rays (SCR) with proton energies from an interval of 1-100 MeV. The geomagnetic field bends trajectories of SCR in such a way that they enter the atmosphere in a circle around the geomagnetic pole down to about 65 deg. Energetic particles interact with molecules of the air and cause additional dissociation and ionization. The maximum ionization rate depends on particles energy and on latitude and occurs at upper stratospheric or mesopheric heights. Additional NO_x, OH_y and ions are formed and enter chemical and ion-molecular reactions. Induced changes of the ionopheric D-layer are modeled using a 1-D model photochemical time-dependent model for neutral species (Krivolutsky et al., Adv.Space Res. 27, 1975-1981, 2001) and a 1-D lower ionosphere model with chemistry (Ondraskova, Studia geoph. et geod. 37, 189-208, 1993). Changes of the electron and ion densities, and the most important ionospheric parameters are calculated after SPE with the onset on July 14, 2000 and the results are compared with results obtained previously for the October 19, 1989 SPE. It is found that: the increase of electron density N(e) after the both SPEs is greater than three orders of magnitude at 50-60 km; the increase of N(e) was greater for the July 2000 SPE than for the October 1989 SPE and there is an agreement with absorption measurements; changes of other parameters also show that the magnitude of the ionospheric response depends on season.

Index

Activated cluster, 359
Activation level, 358, 362
Adiabatic water content, 62, 64
Air mass, 40
Altitude above ground level (AGL), 29
Array Photometer (AP), 123, 139
Associated Electronics Package (AEP), 127
Atmosphere, 1
  absorption, 153
  atmosphere-ionosphere coupling, 162
  pressure, 239–240
Attachment, 177, 182
Blue jet, 38, 255
Blue starter, 38, 256, 285
Bow Echo and Mesoscale Convective Vortex Experiment (BAMEX), 23
Branching streamer, 273
Breakdown mechanism, 257
Breakeven threshold, 361
Capacitive discharge, 238, 245
Cathode glow, 241
Cellular automaton, 357
Charge Coupled Device (CCD), 101–102, 127
  camera, 110
  charge transfer, 103
  frame transfer, 103
  full frame transfer, 104
  full well capacity, 103
  integration, 103
  interline transfer, 104
  noise, 104
  pixel, 103
  quantum efficiency, 105
  readout, 103
  sensitivity, 103
Charge moment, 8, 218, 270
Charge moment change, 29, 192, 214
Charge transfer, 222
Circuit
  AC, 5
  DC, 4
Circulation, 87
Cloud, 3
  Cloud Ionosphere Discharge (CID), 174

cloud-to-ground (CG) discharge, 4
  core, 364
  droplet, 60
  electrification, 72
  life-cycle, 72
  modeling, 314
  particles, 342
  sprite-producing, 57
  stratiform cloud system, 40
  structure, 61
  top, 364
  water content, 62
  winter monsoon cloud, 41
Coalescence, 342, 356
Collision, 58, 60
Colorimeter PR-650, 240
Colour filter, 115
Condensation nuclei (CCN), 21
Conductive cluster, 358, 360
Conductivity, 228, 243, 357
  TLE produced, 174
Conjugate sprite, 43
Constant Altitude Plan Position Indicator (CAPPI), 225
Continuing current, 27, 218
Continuity equation, 352
Convection, 21, 65
  Convective Available Potential Energy (CAPE), 22
  convective cell, 343
  Convective Inhibition (CIN), 22
  convective scale, 65
Conventional breakdown, 193, 357, 361
Convolution, 203
Coriolis force, 20, 88
Corona discharge, 355–356, 364
Corona streamer, 255, 281
Cosmic rays, 2
Critical
  bond, 358
  parameter, 347
  value, 357
Cumulonimbus, 23
Current density, 242

Current moment, 10, 192, 215
Dark band, 242
Debye length, 6
Defense Meteorological Satellite Program (DMSP), 30
Detection efficiency (DE), 29
Dewpoint temperature, 20
Digital Processing Unit (DPU), 128
Digital Signal Processor (DSP), 127
Dipole
  inverted, 62
  positive, 61–62
  tilted, 69
Discharge physics, 12
Distribution function, 362, 367
Doppler on Wheels (DoW), 25
Doppler radar, 25
Drainage system, 357, 369
D-region, 170
Earth-ionosphere waveguide, 196
Einstein coefficient, 244
Electrical conductivity, 167
Electric current, 346, 352
Electric field, 351, 353
  acceleration, 366
  intensity, 367
  oscillation, 350
  profile, 362
  relaxation, 346
  spatial structure, 348
  stochastic, 343, 365
Electric potential, 345, 357
Electrode, 237–238
Electrodynamic coupling, 227
Electromagnetic pulse (EMP), 5, 179, 214
Electromagnetic sensors, 196
Electromagnetic waves, 212
Electron acceleration, 365
Electrostatic field change, 196, 198
Elve, 8, 32, 36
Energetic electrons, 360, 367
Energy balance, 85
Environmental lapse rate, 21
Event detection, 108, 117
Exposure, 101, 110
Extremely Low Frequency (ELF), 196, 212
  radiation, 10
  transient, 211, 224
Faraday dark space, 238
Far Ultra-Violet (FUV), 140
Field of View (FOV), 126, 141
Flash, 27–28, 33
  cloud-to-ground (CG), 130
  positive, 62
Fractal
  behavior, 358
  dynamics, 356

model, 272
Frame grabber, 111
Free energy, 356
Frequency spectrum, 357
Friction force, 344, 360
Front
  cold, 90–91
  polar, 85
  warm, 89–91
Giant jet, 30, 39
Gigantic jet, 253, 256, 285
Global electric circuit, 4, 95, 330
Global Positioning System (GPS), 115, 199
Global thunderstorm activity, 213
Glow discharge, 237
  tube, 238
Graupel, 58
Growth rate, 353
Hadley circulation, 88–89
Hail, 62, 65
Halo, 38–39
High-altitude discharge, 362, 369
Ice, 60
  crystal, 58
  nuclei, 21
  particle, 62
Image
  intensification, 105
  noise, 105
  profiling, 117
Impulse response, 202–203
In-cloud (IC), 27
Infrasound, 26, 32
Initiation, 288
Instability threshold, 353
International Space Station (ISS), 152–153
Intertropical convergence zone (ITCZ), 26, 87
Inverse problem, 202
Inverted dipole, 62
Ionization, 227
  attachment, 177, 182
  recombination, 177, 182
  relaxation, 181, 244
  temperature, 181
  threshold, 360
Ionosphere, 2, 202, 214
  D-region, 170
  perturbation, 227
  subionospheric propagation, 169
  Wait, 182
Irradiance, 102
Isotherm, 212, 221
Kinetic equation, 362, 365–366
Large-scale electric field, 363–364
Latent heat, 20
Leader, 263, 351, 356
Lens

INDEX

aperture, 102
   F-number, 102
   focal length, 102
   spherical, 102
   spot size, 110
Level of free convection (LFC), 22
Lifted condensation level (LCL), 21
Lifted Index (LI), 22
Lightning, 5, 8, 264
   cloud-to-air (CA), 27
   cloud-to-cloud (CC), 27
   cloud-to-ground (CG), 27, 195, 212
   discharge, 26, 335, 356
   electromagnetic pulse, 179
   flash, 356, 359
   in-cloud (IC), 27
   initiation, 341
   Lightning Detection and Ranging System (LDAR), 30
   Lightning Detection Network (LDN), 28
   Lightning Imaging Sensor (LIS), 159
   Lightning Mapping Array (LMA), 30
   positive, 348, 364
   satellite observations, 92
   spider, 65
   sprite-producing, 66, 191
   stroke, 27
   total, 27, 30
   upward, 39
   winter, 222
Linux, 112
   kernel, 112
   threads, 117
Locational accuracy (LA), 28
Long spark, 356, 360
Low light imaging, 101
Low-light television (LLTV), 31
Low Pass Filter (LPF), 139
Luminous efficiency, 242
Lyman-Birge Hopfield band (LBH), 137
Macro-scale, 20
Magnetic direction finding (MDF), 28
Mesocyclones, 25
Mesoscale, 20, 68
Mesoscale convective complex (MCC), 24, 64, 91
Mesoscale convective system (MCS), 11, 23, 64, 91
Metallized bond, 358
Micro Channel Plate (MCP), 129, 135
Microphysics, 61
Microscale, 20
Midcourse Space Experiment (MSX), 137
Middle Ultra-Violet (MUV), 123
Mixed phase, 60
Modelling, 12
Molecular spectrum, 240
Monsoon, 87–88
Multicellular thunderstorms, 40

National Lightning Detection Network (NLDN), 28
Negative flash, 73
Negative glow, 238, 243
Network time protocol, 114
Nitrogen
   first negative emission, 238
   first positive emission, 238, 240
Non-inductive charging, 58
Nonlinear
   diffusion, 332
   dynamics, 315
   methods, 314
NO_x production, 42
Optical spectrum, 240
Optical Transient Detection (OTD), 30
Pan/tilt, 108
Peak current, 29, 222
Percolation threshold, 359–360
Photoionization, 267, 274
Photo multiplier tube (PMT), 139
Photon, 245
Pic du Midi de Bigorre France, 107
Planetary boundary layer (PBL), 20
Plasma, 243, 247
Plasma beam instability, 342, 362
Point discharge, 4
Poisson equation, 352
Polarity, 224, 227
Positive
   charge, 66
   column, 238, 244
   dipole, 61
   flash, 62
   lightning, 348, 364
Poynting flux, 216
Precipitation, 62, 65
Preliminary breakdown, 356–357
Pressure, 238, 241
Pressure gradient forces, 20
Q-burst, 32
QE thunderstorm fields, 177
QTEM waveguide mode, 200
Quasi electrostatic (QE), 177, 227
Quasi-stationary current, 330, 332
Radar
   bright band, 64
   cross section, 71
   reflectivity, 70
Radiance, 102, 240, 244
Radio frequency (RF), 27
Rapid Onset Rapid Decay (RORD), 180
Recombination, 177, 182
Reflectivity, 25
Relativistic electrons, 263, 350, 369
Relaxation, 244
Remote Sensing Instrument (RSI), 125
Return stroke, 356

Reynolds number, 354
Riming, 58
Rossby wave, 20
Runaway breakdown, 193, 263, 360
Runaway electron, 263, 289
Schumann Resonance (SR), 196, 213, 215
Screening, 6
Severe Thunderstorm Electrification and Precipitation Study (STEPS), 22, 33
Sferic, 169, 191
Solar cycle, 1
Solar maximum, 2
Solar minimum, 2
Solar system, 2
Space Shuttle, 42
Space weather, 3
Spectrophotometer (SP), 124, 131
Spider lightning, 35, 65
Sprite, 8, 31, 253, 362
    carrot-type, 223
    column, 223
    columnar, 34
    conjugate, 43
    current, 195
    generation, 317
    halo, 38, 255
    initiation, 322
    initiation point, 32
    observation from space, 151–152
    parent +CGs (SP+CGs), 29
    red, 238
    spectrum, 9
    sprite-producing lightning, 191
Spritewatch
    program, 117
    system, 107
Squall line, 24, 40
Stochastic
    electric field, 343, 365
    ensemble, 366
    period, 366
Stratiform
    cloud systems, 40
    precipitation, 65
    region, 63
Streamer
    branching, 273, 291
    corona, 255, 281
Striations, 241–242
STS-107, 32
Subionospheric propagation, 169
Sublimation, 60
Sun, 2
Superbolt, 30

Supercell, 23, 40
Supercooled water, 61
Suspended particle, 351, 353
Telescopic observation, 244
Television and infrared observation satellite (TIROS I), 26
Temperature
    relaxation, 174, 181
Tendril, 34
Terminator
    day, 171
    night, 171
Terrestrial gamma ray flash (TGF), 264
Terrestrial gamma-rays, 42
Thunderstorm, 3, 62
    decaying stage, 342, 369
    electrification, 58, 315
    global thunderstorm activity, 92–93, 213
    mature state, 348, 353
    self-organized, 227
Time of arrival (TOA), 28
Time synchronization, 114
Tornado vortex signature, 25
Towering cumulus, 23
Transient luminous event (TLE), 19, 124, 253
Trimpi
    classic, 172, 179
    early, 174, 177
Tropical
    cyclone, 41
    intertropical convergence zone (ITCZ), 26
    Tropical Rainfall Measurement Mission (TRMM), 30, 156
Tropopause, 22
Turbulent regime of motion, 354–355
Updraft flow velocity, 359
Upper positive charge, 69
Very Low Frequency (VLF), 167, 196, 227
    amplitude, 172
    perturbation, 172
    phase, 172
    radiation, 5
    sprite, 174
    transmitter, 169
Viscous regime of motion, 354–355
Wait ionosphere, 182
Wave impedance, 217–218
Weather satellite, 26
Whistler-induced electron precipitation (WEP), 172
Wide-angle scattering, 174
Winter monsoon cloud, 41
Winter storm, 68–69
X and γ emission, 360
X-ray, 293, 342
Yucca Ridge Field Station (YRFS), 31